T0259408

Multilingual Speech Processing

Multilingual Speech Processing

Schultz & Kirchhoff

AMSTERDAM • BOSTON • HEIDELBERG • LONDON
NEW YORK • OXFORD • PARIS • SAN DIEGO
SAN FRANCISCO • SINGAPORE • SYDNEY • TOKYO
Academic Press is an imprint of Elsevier

Academic Press is an imprint of Elsevier
30 Corporate Drive, Suite 400, Burlington, MA 01803, USA
525 B Street, Suite 1900, San Diego, California 92101-4495, USA
84 Theobald's Road, London WC1X 8RR, UK

This book is printed on acid-free paper. ∞

Library of Congress Cataloging-in-Publication Data

Application submitted.

British Library Cataloguing-in-Publication Data

A catalogue record for this book is available from the British Library.

ISBN 13: 978-0-12-088501-5
ISBN 10: 0-12-088501-8

For information on all Elsevier Academic Press publications visit our Web site at www.books.elsevier.com

Contents

List of Figures

List of Tables

Contributor Biographies

Dr. Tanja Schultz received her Ph.D. and Masters in Computer Science from University Karlsruhe, Germany in 2000 and 1995, respectively, and earned a German Masters in Mathematics, Sports, and Education Science from the University of Heidelberg, Germany in 1990. She joined Carnegie Mellon University in 2000 and is a faculty member of the Language Technologies Institute as an Assistant Research Professor. Her research activities center around human-machine and human-human interaction. With a particular area of expertise in multilingual approaches, she directs research on portability of speech and language processing systems to many different languages. In 2001 Tanja Schultz was awarded with the FZI price for her outstanding Ph.D. thesis on language independent and language adaptive speech recognition. In 2002 she received the Allen Newell Medal for Research Excellence from Carnegie Mellon for her contribution to Speech-to-Speech Translation and the ISCA best paper award for her publication on language independent acoustic modeling. She is an author of more than 80 articles published in books, journals, and proceedings, and a member of the IEEE Computer Society, the European Language Resource Association, and the Society of Computer Science (GI) in Germany. She served as Associate Editor for IEEE Transactions and is currently on the Editorial Board of the Speech Communication journal.

Dr. Katrin Kirchhoff studied Linguistics and Computer Science at the Universities of Bielefeld, Germany, and Edinburgh, United Kingdom, and was a visiting researcher at the International Computer Science Institute, Berkeley, California. After obtaining her Ph.D. in Computer Science from the University of Bielefeld in 1999, she joined the University of

Washington, where she is currently a Research Assistant Professor in Electrical Engineering. Her research interests are in automatic speech recognition, language identification, statistical natural language processing, human-computer interfaces, and machine translation. Her work emphasizes novel approaches to acoustic-phonetic and language modeling and their application to multilingual contexts. She currently serves on the Editorial Board of the Speech Communication journal.

Dr. Christopher Cieri is the Executive Director of the Linguistic Data Consortium, where he has overseen dozens of data collection and annotation projects that have generated multilingual speech and text corpora. His Ph.D. is in Linguistics from the University of Pennsylvania. His research interests revolve around corpus based language description especially in phonetics, phonology, and morphology as they interact with nonlinguistic phenomena as in language contact and studies of linguistic variation.

Dr. Mark Liberman is Trustee Professor of Phonetics in Linguistics at the University of Pennsylvania, where he is also Director of the Linguistic Data Consortium, Co-Director of the Institute for Research in Cognitive Science, and Faculty Master of Ware College House. His Ph.D. is from the Massachusetts Institute of Technology in 1975, and he worked from 1975 to 1990 at AT&T Bell Laboratories, where he was head of the Linguistics Research Department.

Dr. Khalid Choukri obtained an Electrical Engineering degree (1983) from Ecole Nationale de l'aviation civile (ENAC), and Masters Degree (1984) and doctoral degrees (1987) in Computer Sciences and Signal Processing at the Ecole Nationale Supérieure des Télécommunications (ENST) in Paris. He was a research scientist at the Signal Department of ENST, involved in Man-Machine Interaction. He has also consulted for several French companies, such as Thomson, on various speech system projects and was involved in SAM, ARS, etc. In 1989, he joined CAP GEMINI INNOVATION, R&D center of CAP SOGETI to work as the team leader on speech processing, oral dialogs and neural networks. He managed several ESPRIT projects, such as SPRINT, and was involved in many others, such as SUNDIAL. He then moved to ACSYS in September 1992 to take on the position of Speech Technologies Manager. Since 1995, he has been the Executive Director of the European Language

Resources Association (ELRA) and the Managing Director of the Evaluations and Language Resources Distribution Agency (ELDA) for which the priority activities include the collection and distribution of Language Resources. In terms of experience with EC-funded projects, ELDA/ELRA has played a significant role in several European projects, such as C-Oral-Rom, ENABLER, NET-DC, OrienTel, CLEF, CHIL, and TC-STAR.

Dr. Victoria Arranz holds an M.Sc. in Machine Translation and a Ph.D. in Computational Linguistics (1998) from the Centre for Computational Linguistics, University of Manchester Institute of Science and Technology (UMIST), United Kingdom, where she participated in several international projects dealing with restricted domains and corpus study and processing. She has worked as a Research Scientist both at the Grup d'Investigació en Lingüística Computacional (gilcUB) and at the Centre de Llenguatge i Computació (CLIC), Universitat de Barcelona, Spain, working on the production of language resources and coordinating the development of terminological LRs for the medical domain. Then she joined the Universitat Politècnica de Catalunya, Barcelona, where she has been a Visiting Researcher, a Senior Researcher, and also a Lecturer of Computational Linguistics within the Natural Language Processing and Cognitive Science Ph.D. program. She has also participated in a number of national and international projects regarding Terminological LRs (SCRIPTUM), LRs for Speech-to-Speech Translation (LC-STAR), Dialogue Systems (BASURDE), Speech-to-Speech Translation (ALIADO, FAME, TC-STAR), and other NLP techniques. Currently, she is the Head of the Language Resources Identification Unit at ELDA, Paris, France, in charge of the BLARK and UNIVERSAL CATALOGUE projects, whose aims relate to the compiling of the existing LRs and the production of LRs in terms of language and technology needs.

Dr. Lori Lamel joined LIMSI as a permanent CNRS Researcher in October 1991 (http://www.limsi.fr/Individu/lamel/). She received her Ph.D. degree in Electrical Engineering and Computer Science from the Massachusetts Institute of Technology in May 1988. She obtained her 'Habilitation a diriger des Recherches' [Document title: Traitment de la parole (Spoken Language Processing)] in January 2004. Her research activities include multilingual studies in large vocabulary continuous speech recognition; acoustic-phonetics, lexical, and phonological modeling; spoken dialog systems; speaker and language identification; and the design, analysis, and realization of large speech corpora. She has been a prime contributor to the LIMSI participations in DARPA benchmark evaluations, being involved in acoustic model development and responsible for the pronunciation lexicon and has been involved in many European projects on speech recognition and spoken language dialog systems. She has over 150 publications and is a member of the Editorial Board of the Speech Communication journal, the Permanent Council of ICSLP and the coordination board of the L'Association Francophone de la Communication Parle.

Dr. Martine Adda-Decker has been a permanent CNRS Researcher at LIMSI since 1990 (http://www.limsi.fr/Individu/madda). She received an M.Sc. degree in Mathematics and Fundamental Applications in 1983 and her doctorate in Computer Science in 1988 from the University of Paris XI. Her main research interests are in multilingual, large vocabulary continuous speech recognition, acoustic and lexical modeling, and language identification. She has been a principal developer of the German ASR system. She is also interested in spontaneous speech phenomena, pronunciation variants, and ASR errors related to spontaneous speaking styles. More recently she has focused her research on using automatic speech recognizers as a tool to study phonetics and phonology in a multilingual context. In particular ASR systems can contribute to describe less studied languages, dialects, and regional varieties on the acoustic, phonetic, phonological, and lexical levels. She has been involved in many national CNRS and European projects.

Dr. Sanjeev P. Khudanpur is with the Department of Electrical & Computer Engineering and the Center for Language and Speech Processing at the Johns Hopkins University. He obtained a B. Tech. from the Indian Institute of Technology, Bombay, in 1988, and a Ph.D. from the University of Maryland, College Park, in 1997, both in Electrical Engineering. His research is concerned with the application of information theoretic and statistical methods to problems in human language technology, including automatic speech recognition, machine translation and information retrieval. He is particularly interested in maximum entropy and related techniques for model estimation from sparse data.

Dr. Alan W. Black is an Associate Research Professor on the faculty of the Language Technologies Institute at Carnegie Mellon University. He is a principal author of the Festival Speech Synthesis System, a standard free software system used by many research and commercial institutions throughout the world. Since joining CMU in 1999, with Kevin Lenzo, he has furthered the ease and robustness of building synthetic voices through the FestVox project using new techniques in unit selection, text analysis, and prosodic modeling. He graduated with a Ph.D. from the University of Edinburgh in 1993, and then worked in industry in Japan at ATR. He returned to academia as a Research Fellow at CSTR in Edinburgh and moved to CMU in 1999. In 2000, with Kevin Lenzo, he started the for-profit company Cepstral, LLC in which he continues to serve as Chief Scientist. He has a wide background in computational linguistics and has published in computational morphology, language modeling for speech recognition, computational semantics, and most recently in speech synthesis, dialog systems, prosody modeling and speech-to-speech translation. He is a strong proponent of building practical implementations of computational theories of speech and language.

Dr. Jiří Navrátil received M.Sc. and Ph.D. (summa cum laude) degrees from the Ilmenau Technical University, Germany in 1994 and 1999, respectively. From 1996 and 1998 he was Assistant Professor at the Institute of Communication and Measurement Technology at the ITU performing research on speech recognition and language identification. For his work in the field of language identification, Dr. Navrátil received the 1999 Johann-Philipp-Reis Prize awarded by the VDE (ITG), Deutsche Telekom, and the cities of Friedrichsdorf and Gelnhausen, Germany. In 1999, he joined IBM to work in the Human Language Technologies Department at the Thomas J. Watson Research Center, Yorktown Heights, New York. He has authored over 40 publications on language and speaker recognition, received several invention achievement awards and has a technical group award from IBM. His current interests include voice-based authentication, particularly conversational biometrics, language recognition, and user-interface technologies.

Dr. Etienne Barnard is a research scientist and coleader of the Human Language Technologies research group at the Meraka Institute in Pretoria, South Africa, and Professor in Electronic and Computer Engineering at the University of Pretoria. He obtained a Ph.D. in Electronic and Computer Engineering from Carnegie Mellon University in 1989, and is active in the development of statistical approaches to speech recognition and intonation modeling for the indigenous South African languages.

Dr. Marelie Davel is a research scientist and coleader of the Human Language Technologies research group at the Meraka Institute in Pretoria, South Africa. She obtained a Ph.D. in Computer Engineering at the University of Pretoria in 2005. Her research interests include pronunciation modeling, bootstrapping of resources for language technologies, and new approaches to instance-based learning.

Dr. Silke Goronzy received a diploma in Electrical Engineering from the Technical University of Braunschweig, Germany, in 1994. She joined the Man-Machine Interface group of the Sony Research Lab in Stuttgart, Germany, to work in the area of automatic speech recognition on speaker adaptation, confidence measures, and adaptation to non-native speakers. In 2002 she received a Ph.D. in Electrical Engineering from the University of Braunschweig. At Sony she also performed research in the area of multimodal dialog systems, and in 2002, she lead a team working on personalization and automatic emotion recognition. In 2004, she joined 3SOFT GmbH in Erlangen, Germany, where she is leading the Speech Dialog Systems team that is developing HMI solutions for embedded applications. She also gives lectures at the University of Ulm, Germany.

Dr. Laura Mayfield Tomokiyo holds a Ph.D. in Language Technologies and an M.S. in Computational Linguistics from Carnegie Mellon University. Her undergraduate degree is from the Massachusetts Institute of Technology in Computer Science and Electrical Engineering. She has held positions at Toshiba and the Electrotechnical Laboratories (ETL) in Japan. Currently, she is Director of Language Development at Cepstral, LLC, where she is responsible for expansion to new languages and enhancement of existing languages in text-to-speech synthesis. Her research interests include multilinguality in speech technology, application of speech technology to language learning, and the documentation and preservation of underrepresented languages.

Dr. Seichii Yamamoto graduated from Osaka University in 1972 and received his Masters and Ph.D. degrees from Osaka University in 1974 and 1983, respectively. He joined Kokusai Denshin Denwa Co. Ltd. in

April 1974, and ATR Interpreting Telecommunications Research Laboratories in May 1997. He was appointed president of ATR-ITL in 1997. He is currently a Professor of Doshisha University and invited researcher (ATR Fellow) at ATR Spoken Language Communication Research Laboratories. His research interests include digital signal processing, speech recognition, speech synthesis, natural language processing, and spoken language translation. He has received Technology Development Awards from the Acoustical Society of Japan in 1995 and 1997. He is also an IEEE Fellow and a Fellow of IEICE Japan.

Dr. Alex Waibel is a Professor of Computer Science at Carnegie Mellon University, Pittsburgh and at the University of Karlsruhe, Germany. He directs the Interactive Systems Laboratories at both Universities with research emphasizing speech recognition, handwriting recognition, language processing, speech translation, machine learning, and multimodal and multimedia interfaces. At Carnegie Mellon University, he also serves as Associate Director of the Language Technology Institute and as Director of the Language Technology Ph.D. program. He was one of the founding members of the CMU's Human Computer Interaction Institute (HCII) and continues on its core faculty. Dr. Waibel was one of the founders of C-STAR, the international consortium for speech translation research and served as its chairman from 1998–2000. His team has developed the JANUS speech translation system, the JANUS speech recognition toolkit, and a number of multimodal systems including the meeting room, the Genoa Meeting recognizer and meeting browser. Dr. Waibel received a B.S. in Electrical Engineering from the Massachusetts Institute of Technology in 1979, and his M.S. and Ph.D. degrees in Computer Science from Carnegie Mellon University in 1980 and 1986. His work on the Time Delay Neural Networks was awarded the IEEE best paper award in 1990, and his work on speech translation systems the "Alcatel SEL Research Prize for Technical Communication" in 1994.

Dr. Stephan Vogel studied physics at Philips University in Marburg, Germany, and at Imperial College in London, England. He then studied History and Philosophy of Science at the University of Cambridge, England. After returning to Germany he worked from 1992 to 1994 as a Research Assistant in the Department of Linguistic Data Processing, University of Cologne, Germany. From 1994 to 1995 he worked as software developer at ICON Systems, developing educational software. In 1995 he joined the research team of Professor Ney at the Technical University of Aachen, Germany, where he started work on statistical machine translation. Since May 2001, he has worked at Carnegie Mellon University in Pittsburgh, Pennsylvania, where he leads a team of students working on statistical machine translation, translation of spontaneous speech, automatic lexicon generation, named entity detection and translation, and machine translation evaluation.

Dr. Helen Meng is a Professor in the Department of Systems Engineering & Engineering Management of The Chinese University of Hong Kong (CUHK). She received her S.B., S.M., and Ph.D. degrees in Electrical Engineering, all from the Massachusetts Institute of Technology. Upon joining CUHK in 1998, Helen established the Human-Computer Communications Laboratory, for which she serves as Director. She also facilitated the establishment of the Microsoft-CUHK Joint Laboratory for Human-Centric Computing and Interface Technologies in 2005 and currently serves as co-director. Her research interests include multilingual speech and language processing, multimodal human-computer interactions with spoken dialogs, multibiometric user authentication as well as multimedia data mining. Dr. Meng is a member of the IEEE Speech Technical Committee and the Editorial Boards of several journals, including *Computer Speech and Language, Speech Communication*, and the *International Journal of Computational Linguistics*, and *Chinese Language Processing*. In addition to speech and language research, Dr. Meng is also interested in the development of Information and Communication Technologies for our society. She is an appointed member of the Digital 21 Strategy Advisory Committee, which is the main advisory body to the Hong Kong SAR Government on information technology matters.

Dr. Devon Li is a chief engineer in the Human-Computer Communications Laboratory (HCCL), The Chinese University of Hong Kong (CUHK). He received his B.Eng. and M.Phil. degrees from the Department of Systems Engineering and Engineering Management, CUHK. His Masters thesis research focused on monolingual and English-Chinese cross-lingual spoken document retrieval systems. This work was selected to represent CUHK in the Challenge Cup Competition in 2001, a biennial competition where over two hundred universities across China compete in terms of their R&D projects. The project was awarded Second Level Prize in this competition. Devon extended his work to spoken query retrieval during his internship at Microsoft Research Asia, Beijing. In 2002, Devon began to work on the "Author Once, Present Anywhere (AOPA)" project in HCCL, which aimed to develop a software platform that enables multi-device access to Web content. The user interface is adaptable to a diversity of form factors including the desktop computer, mobile handhelds, and screenless voice browsers. Devon has developed the CU Voice Browser (a bilingual voice browser) and also worked on the migration of CUHK's Cantonese speech recognition (CU RSBB) and speech synthesis (CU VOCAL) engines toward SAPI-compliance. Devon is also among the first developers to be proficient with the emerging SALT (Speech Application Language Tags) standard for multimodal Web interface development. He has integrated the SAPI-compliant CUHK speech engines with the SALT framework.

Foreword

Speech recognition and speech synthesis technologies have enjoyed a period of rapid progress in recent years, with an increasing number of functional systems being developed for a diverse array of applications. At the same time, this technology is only being developed for fewer than twenty of the world's estimated four to eight thousand languages. This disparity suggests that in the near future, demand for speech recognition and speech synthesis technologies, and the automated dialog, dictation, and summarization systems that they support, will come to include an increasingly large number of new languages. Due to current globalization trends, multilingual speech recognition and speech synthesis technologies, which will enable things like speech-to-speech translation, and the ability to incorporate the voice of a speaker from one language into a synthesizer for a different language (known as polyglot synthesis), will become increasingly important.

Because current speech recognition and speech synthesis technologies rely heavily on statistical methods, when faced with the challenge of developing systems for new languages, it is often necessary to begin by constructing new speech corpora (databases). This in turn requires many hours of recording and large amounts of funding, and consequently, one of the most important problems facing minority language researchers will be how to significantly advance research on languages for which there are either limited data resources or scant funds. Moreover, because all languages have unique characteristics, technologies developed based on models of one language cannot simply be ported "as-is" to other languages; this process requires substantial modifications. These are a few of the major handicaps facing current efforts to develop speech technology and extend it into new areas and to new languages. If speech recognition and synthesis systems could be easily and effectively ported between different languages, a much greater number of people might share the benefits that this technology has to offer.

This book spans the state-of-the-art technologies of multilingual speech processing. Specifically, it focuses on current research efforts in Europe and America; it describes new speech corpora under development and discusses multilingual speech recognition techniques from acoustic and language modeling to multilingual dictionary construction issues. Regarding the issue of language modeling, it even touches on new morphological and lexical segmentation techniques. On the topic of multilingual text-to-speech conversion, it discusses possible methods for generating voices in new languages. Language identification techniques and approaches to recognition problems involving non-native speakers are also discussed. The last portion of the book is devoted to issues in speech-to-speech translation and automated multilingual dialog systems.

Presently enrolled at my laboratory here at Tokyo Tech are not only Japanese students but also students from eleven other countries, including the United States, England, France, Spain, Iceland, Finland, Poland, Switzerland, Thailand, Indonesia, and Brazil. Despite this large number of representative languages, the ones for which we are easily able to obtain comparatively large- scale spoken corpora are limited to widely researched languages like English, Japanese, and French. Developing new speech recognition and synthesis systems for any other languages is considerably expensive and time consuming, mainly because researchers must begin by first constructing new speech corpora. In our lab, for the purpose of developing an automated dialog system for Icelandic–a language spoken by approximately 300,000 people worldwide–we are currently conducting research in an effort to automatically translate a series of written English corpora into Icelandic. This is possible both because the English data is abundantly available and because the two languages' similarity in terms of grammatical structure makes the idea more feasible than with many other language pairings.

Constructing corpora is not, however, solely a problem for minority languages. In order to expand the application of current speech recognition systems, instead of just recognizing carefully read text from written manuscripts, the ability to recognize spontaneous speech with a high degree of precision will become essential. Currently available speech recognition technology is capable of achieving high levels of accuracy with speech read aloud carefully from a text; however, when presented with spontaneous speech, recognition rates drop off dramatically. This is because read speech and spontaneous speech are dramatically different, both from a linguistic standpoint and in terms of their respective acoustic models. This difference is actually greater than the difference between two closely related languages. In order to improve recognition accuracy for spontaneous speech, large scale spontaneous speech corpora are necessary, such as are currently available only for a limited number of widely spoken and recorded languages, like English or Japanese. But even the substantial size of spoken corpora in these languages is several orders of magnitude below what they have to offer in terms of written text. The reason for this is, again, the colossal amount of money required to construct a new corpus, which requires many hours of recording, followed by faithful transcription and detailed annotation. From this point on the most pressing difficulties in both spontaneous speech recognition and multilingual speech technology will go hand in hand.

Seen from these various perspectives, the publication of this new book appears not only very timely, but also promises to occupy a unique place among current speech-related textbooks and reference manuals. In addition to the respect and gratitude I have for the two young researchers who have made its publication a reality, in the years to come I expect this book to become an important milestone in the course of further speech-related research.

Sadaoki Furui
Tokyo Institute of Technology
August 26, 2005

Chapter 1

Introduction

In the past decade, the performance of automatic speech processing systems (such as automatic speech recognizers, speech translation systems, and speech synthesizers) has improved dramatically, resulting in an increasingly widespread use of speech technology in real-world scenarios. Speech processing systems are no longer confined to research labs but are being deployed in a wide variety of applications, including commercial information retrieval systems (e.g., automatic flight status or stock quote information systems), personal dictation systems, and household devices (e.g., voice dialing in cell phones, translation assistance), as well as military and security applications.

One of the most important trends in present-day speech technology is the need to support multiple input and output languages, especially if the applications of interest are intended for international markets and linguistically diverse user communities. As a result, new models and algorithms are required that support, for instance, the simultaneous recognition of mixed-language input, the generation of output in the appropriate language, or the accurate translation from one language to another. The challenge of rapidly adapting existing speech processing systems to new languages is currently one of the major bottlenecks in the development of multilingual speech technology.

This book presents the state of the art of multilingual speech processing, that is, the techniques that are required to support spoken input and output in a large variety of languages. Speech processing has a number of different subfields, including speech coding and enhancement, signal processing, speech recognition, speech synthesis, keyword spotting, language identification, and speaker identification. Possible applications of techniques from these fields include automated dialog systems, speech-to-speech translation, speech summarization, and audio browsing. This book excludes those fields that are largely language independent, such as speech coding, enhancement, signal processing, and automatic speaker identification. Moreover, we have excluded discussions of techniques that go significantly beyond the speech level, which includes most areas of natural language processing. Within this range, however, the book provides a comprehensive overview of multilingual issues in acoustic modeling, language modeling, and dictionary construction for speech recognition, speech synthesis, and automatic language identification. Speech-to-speech translation and automated dialog systems are discussed in detail as example applications, as are linguistic data collection, non-native speech, dialectal variation, and interface localization. Throughout the book, two main topics are addressed: (1) the challenges for current speech processing models posed by different languages, and (2) the feasibility of sharing data and system components across languages and dialects of related types. In addition to describing modeling approaches, the book provides an overview of significant ongoing research programs as well as trends, prognoses, and open research issues.

This book is intended for readers who are looking for a comprehensive introduction to research problems and solutions in the field of multilingual speech processing, from both a theoretical and a practical perspective. It is appropriate for readers with a background in speech processing, such as researchers and developers from industry and academia, instructors, graduate-level students, or consultants on language and speech technologies. A basic introduction to speech processing is not provided here; publications directed at the beginner include e.g. Huang et al. (2001).

This book is organized in eleven chapters. **Chapter 2** describes language characteristics in greater detail, including language distributions, speaker populations, and the typological classification of languages. The similarities and dissimilarities between languages are analyzed in terms of spoken and written forms and their relation, and consequences for multilingual speech processing are discussed.

Chapter 3 gives an overview of available language resources, from both U.S. and European perspectives. Multilingual data repositories as well as large ongoing and planned collection efforts are introduced, along with a description of the major challenges of collection efforts, such as transcription issues due to inconsistent writing standards, subject recruitment, recording equipment, legal aspects, and costs in terms of time and money. The overview of multilingual resources comprises multilingual audio and text data, pronunciation dictionaries, and parallel bilingual/multilingual corpora.

Chapter 4 introduces the concept of multilingual acoustic modeling in automatic speech recognition. It describes sound inventories that are suitable as basic units for multilingual acoustic models, investigates techniques and algorithms to develop these models, and gives examples of applications of multilingual acoustic models in speech recognition, with special emphasis on the use of these models for rapid porting to new languages.

Chapter 5 gives an overview of dictionary modeling and generation issues in the context of multilingual speech processing. Suitable units for dictionary modeling are discussed in the light of similarities and dissimilarities between languages. Dictionary generation techniques are illustrated, along with the pros and cons of automatic or semiautomatic procedures.

Chapter 6 addresses major questions of language modeling in the context of multilinguality, such as the portability of existing language modeling techniques to less well investigated languages, issues of morphological complexity and word segmentation, and the feasibility of cross-linguistic language modeling.

Chapter 7 introduces the concepts of multilingual text-to-speech generation and synthesis, and describes the needs and challenges for the voice-building process in new languages. Strategies for building multilingual voices are also explained. A description of large evaluation efforts across languages complements this chapter.

Chapter 8 gives a detailed introduction of past and present approaches to automatic language identification and discusses their strengths and weaknesses with respect to their practical applicability. It also introduces standard databases and evaluation procedures used in this field, and outlines current trends and future challenges for language identification.

Chapter 9 focuses on problems posed by non-native speech input as well as accent and dialect variation with respect to acoustic modeling, dictionaries, and language modeling in speech recognition systems. Pragmatic strategies, such as handling code-switching, and the design of voice-based user interfaces for speakers with different cultural backgrounds are discussed.

The book is complemented by two chapters describing concrete applications of multilingual speech processing. **Chapter 10** addresses the theory and practice of speech-to-speech translation, including different translation approaches and strategies for coupling speech recognition and translation. The second part of the chapter describes a state-of-the-art end-to-end speech translation system and its evaluation. **Chapter 11** reviews problems and challenges in developing multilingual automated dialog systems, such as adapting system components to different languages. In addition, it presents empirical observations of system-user interactions and describes an example implementation of a multilingual spoken dialog system based on the VoiceXML platform.

Chapter 2

Language Characteristics

Katrin Kirchhoff

2.1 Languages and Dialects

How many languages exist in the world? The perhaps surprising answer is that the exact number is not known; any published figures represent estimates rather than precise counts. There are several reasons for this—the first being the fact that even today, new languages are still being discovered. Second, there needs to be agreement on what should be counted; for instance, it is not clear whether to include only living languages or extinct languages as well. This distinction is in itself problematic: when exactly does a language count as "extinct"? Textbook definitions (Fromkin et al., 2003, p. 524) label a language as extinct when it has no living native speakers—that is, speakers who acquire the language during childhood. However, some languages (such as the Kiong language in Nigeria) only have a small number of surviving speakers and face the prospect of becoming extinct when those speakers die, whereas other languages, such as Latin or Ancient Greek, do not have native speakers but are still very much alive in educational or liturgical settings. A third reason for widely differing language statistics is the difficulty of drawing a distinction between genuinely different languages and dialects of the same language. A **dialect** is typically defined as a regional variant of a language that involves modifications at the lexical and grammatical levels, as opposed to an **accent**, which is a regional variant affecting only the pronunciation. British Received Pronunciation (RP), for instance, is an accent of English, whereas Scottish English would be considered a dialect since it often exhibits grammatical differences, such as "Are ye no going?" for "Aren't you going?" Dialects of the same language are assumed to be mutually intelligible, primarily spoken varieties without a literary tradition. Different **languages**, by contrast, are not mutually intelligible but need to be explicitly learned by speakers of other languages. In addition, they have a distinct literary tradition.

These conventional definitions are, however, greatly simplified. First, many languages do not use a writing system and thus do not have a literary tradition but are nevertheless living languages, often with a rich oral history. Second, the language versus dialect distinction is a continuum rather than a clear-cut binary division, and the classification of a particular variety as either a language or a dialect is often arbitrary and motivated more by sociopolitical than by linguistic considerations. China, for example, has a large number of geographical varieties that differ strongly in pronunciation, vocabulary, and grammar while being unified by a common writing standard (see Section 2.4). Most linguists would argue that these are different languages since they are often mutually unintelligible. A speaker from Beijing speaking Eastern Mandarin, for instance, might not understand a speaker of Min, a variety spoken in Fujian Province in Mainland China, Taiwan, the Philippines, and other parts of Southeast Asia. In order to promote the concept of national unity, however, the spoken varieties of China are not officially labeled different languages but dialects, the proper Chinese term being *făngyán* (regional speech) (DeFrancis, 1984). An example of the use of the term "language" to reinforce the opposite concept, namely national independence, is the case of Moldovan and Romanian. Moldovan is both historically and linguistically a dialect of Romanian, with which it is mutually intelligible. The concept of a separate Moldovan language did not exist until it was introduced as a result of the Soviet annexation of Moldavia in 1945. In 1994, however, the parliament of the newly independent Moldovan Republic voted to preserve the term "Moldovan" instead of Romanian for the national language, and in 2003

Table 2.1 Distribution of the world's languages by geographical origin, percentage of the world's languages, and percentage of native speakers. Data from Gordon (2005).

Area	Number of languages	% (of all languages)	% speakers
Africa	2,092	30.3	11.8
Americas	1,002	14.5	0.8
Asia	2,269	32.8	61.0
Europe	239	3.5	26.3
Pacific	1,310	18.9	0.1
Total	6,912	100.0	100.0

a Moldovan-Romanian dictionary was published, although linguists in both countries agree that Romanian and Moldovan are essentially the same language. (The dictionary became a bestseller due to its curiosity value.)

Generally, the number of languages in the world is estimated to be between 4,000 and 8,000. The most recent edition of the *Ethnologue*, a database describing all known living languages (Gordon, 2005), lists 6,912 known living languages. The primary criteria used to established this classification were mutual intelligibility and ethnolinguistic identity. Mutual intelligibility enables two varieties to be grouped together as either different dialects or different languages; the degree of shared ethnolinguistic background then establishes the dividing line. Table 2.1 shows the distribution of different languages by percentage of first-language speakers and region of origin; note that this is not equivalent to the current geographical distribution. For instance, Spanish is a language of European origin but is now predominant in Latin America. Although the data shows that the languages of Asia and Europe are the most widely distributed languages, the number of speakers varies widely for different languages within those groups. Table 2.2 shows the number of speakers for the twenty most widely spoken languages; note that this too only includes first-language speakers, not second- or third-language speakers. The rank of individual languages would be very different if non-native speakers were taken into account; for example, English is probably the language with the largest combined number of native and non-native speakers worldwide.

Table 2.2 The twenty most widely spoken languages in the world according to the number of first-language speakers. Data from Gordon (2005).

#	Language	Speakers (in millions)	#	Language	Speakers (in millions)
1	Mandarin Chinese	867.2	11	Wu Chinese	77.2
2	Spanish	322.3	12	Javanese	75.5
3	English	309.4	13	Telugu	69.7
4	Arabic	206.0	14	Marathi	68.0
5	Hindi	180.8	15	Vietnamese	67.4
6	Portuguese	177.5	16	Korean	67.0
7	Bengali	171.1	17	Tamil	66.0
8	Russian	145.0	18	French	64.8
9	Japanese	122.4	19	Italian	61.5
10	German	95.4	20	Urdu	60.5

According to Gordon (2005), 5% of the world's languages are spoken by 94% of the world's population, whereas the remaining 95% are distributed over only 6% of the population.

2.2 Linguistic Description and Classification

Languages are usually classified according to two criteria: historical relatedness (language family) and linguistic characteristics (typology). These two criteria may be but are not necessarily correlated. Italian and Spanish, for example, share a large portion of their vocabulary and many grammatical features since they are both Romance languages derived from a common ancestor (Latin). On the other hand, languages that are usually grouped under the same genetic family or family subbranch may differ considerably in their grammatical characteristics. English and German, for example, though both classified as North Germanic languages, have a very different word order, with that of English being primarily SVO (subject-verb-object) and that of German permitting a less restricted word order.

In developing speech technology, knowledge of which languages are typologically related can be helpful, since the modeling problems presented by different languages of the same morphological type can often be addressed using the same techniques.

2.2.1 Language Families

The establishment of family trees charting the genetic relatedness of languages has long been a concern of historical linguistics, and there has been much debate about the categorization of particular languages within this scheme. The intention here is to provide a condensed view of the major language families, primarily for the sake of completeness. More details on the genetic classification of languages can be found in Katzner (2002) and Comrie (1990). The overview below essentially follows the classification established in Katzner (2002), which recognizes 21 major language families. The five largest and most well known of these are the Indo-European, Afro-Asiatic, Niger-Congo, Sino-Tibetan, and Austronesian families. In addition to these, there are many smaller groups, such as the Dravidian, Australian, and American Indian languages, as well as many "independent" languages—that is, languages with no known relationship to other languages, such as Basque, spoken in northern Spain and southern France, or Ainu, spoken on the Hokkaido island of Japan.

The **Indo-European** family is the world's largest family (in terms of number of speakers) and comprises most languages spoken in Europe plus many languages spoken in India and the Middle East. It is subdivided into the Germanic, Italic, Romance, Celtic, Hellenic, Slavic, Baltic, and Indo-Iranian branches, as shown in Figure 2.1. The Germanic branch includes North Germanic languages, such as Danish, Swedish, and Icelandic, and West Germanic languages, such as English, German, and Dutch. The Romance branch includes, among others, Italian, Spanish, Romanian, and Portuguese, and the Slavic branch covers the Eastern European languages, including Czech, Russian, Polish, and Serbo-Croatian. Other European languages fall into the Celtic (Gaelic, Irish), Hellenic (Greek, Albanian), or Baltic (Lithuanian, Latvian) branches. Finally, the Indo-Iranian branch includes Farsi (Persian), Kurdish, and Pashto; the classical Sanskrit language of India; and the modern Indian languages derived from it (e.g., Bengali, Hindi, Punjabi, Urdu).

The **Afro-Asiatic** family comprises a number of languages spoken in Africa and the Middle East, in particular Semitic languages, such as Arabic,

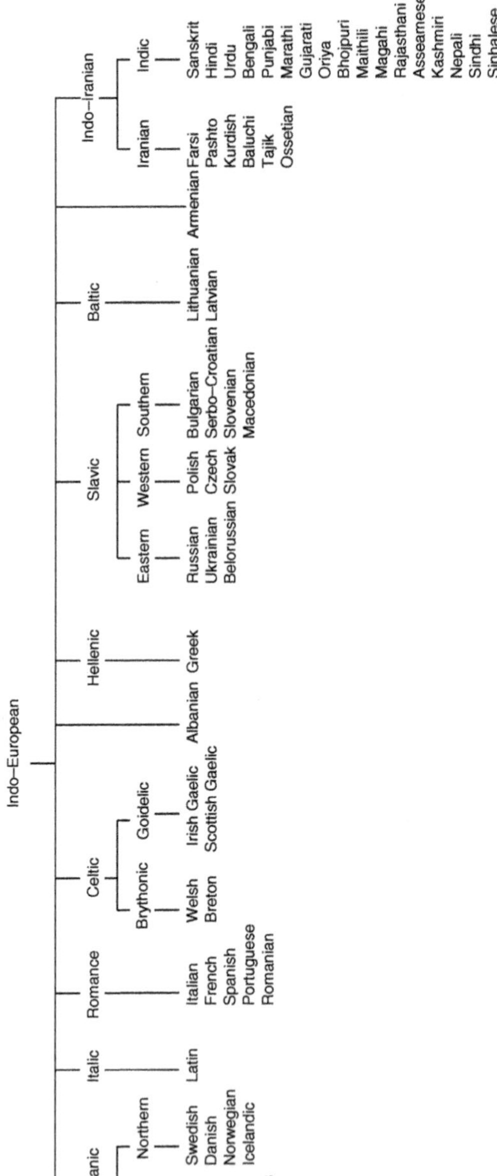

Figure 2.1: The Indo-European language family (extinct languages are not shown).

Hebrew, and Amharic, but also the Berber languages spoken in Morocco and Algeria, the Cushitic languages of Somalia and Ethiopia, and ancient Egyptian and its descendant Coptic.

The **Niger-Congo** family covers most of the languages spoken in Africa (numbering over a thousand). This includes the Benue-Congo subbranch, with the large class of Bantu and Nigerian languages including Swahili, Kikuyu, Ruanda, Yoruba, and Zulu; the Mande branch, with Malinke and Mende; the Atlantic branch, with Fulani, Wolof, Serer, and Temne; and four other branches: Volatic, Kwa, Kru, and Adamawa-Ubangi.

The **Sino-Tibetan** group consists of only two branches, Chinese and Tibeto-Burman, which includes Burmese, Tibetan, and many smaller languages spoken in China and adjacent countries, such as Yi, Tujia, and Bai (China); Karen, Chin, and Arakanese (Burma); and Newari (Nepal).

Austronesian languages are those spoken mainly in Indonesia, New Guinea, and Polynesia. There are four subbranches: Western Austronesian, Micronesian, Oceanic, and Polynesian. Prominent examples include Malay and Indonesian (spoken in Malaysia and Indonesia, respectively), Javanese, Timorese, Tahitian, and Maori—the language of the original inhabitants of New Zealand.

In addition to the "naturally developed" languages, there are a number of languages that either were created artificially or arose as a means of communication between speakers of different native languages. An example of the former is Esperanto, a language created at the end of the 19th century from Germanic, Romance, and Slavic elements. The latter category includes **pidgins** and **creoles**, which are language varieties combining elements of different existing languages. Examples of these are Haitian Creole, which is a mixture of French and West African elements, and Tok Pisin, a mixture of English and Austronesian languages spoken in Papua New Guinea.

2.2.2 Language Typology

Language **typology**—that is, the classification of languages into different categories based on their structural characteristics—is the subject of an entire subdiscipline of linguistics that has developed its own theoretical models and approaches. In this chapter, we concentrate only on those linguistic characteristics that are relevant from a speech technology point of view rather than following any particular linguistic model. For instance, a common classification criterion refers to the different types of word-formation processes that occur in the world's languages. Since these are important for understanding problems related to language modeling and dictionary development, we present this classification in detail below. Other criteria commonly used by linguists, such as the role of animacy for classifying nouns, are of little importance for the purpose of this book and are therefore not discussed here. Standard reference works on linguistic typology include Comrie (1989), Croft (1990), and Whaley (1997).

Phonetics, Phonology and Prosody

The sound structure of a language can be described in terms of its phonetic, phonological, and prosodic properties. Whereas the goal of **phonetics** is the the descriptive analysis of sound acoustics, sound production, and sound perception, **phonology** studies the functional, contrastive role of sounds in an entire system. Prosody spans both phonetics and phonology and refers to phenomena such as pitch, stress, prominence, intonation, and phrasing that typically span several sound segments. Sounds as concrete physical events are referred to as **phones**, whereas sounds as entities in a linguistic system (as the smallest units that distinguish a minimal word pair—e.g., *pin* vs. *bin*) are termed **phonemes**. The relationship between the phonetic and phonemic level is characterized by the existence of **allophones**. These are sounds that are clearly distinct phones but do not have the word-distinguishing status of phonemes. For instance, the [x] and [ç] phones in German are merely positional variants of one another and are therefore not considered phonemes.[1] The distinction between the abstract linguistic category and its concrete realization is one that is mirrored at several levels of linguistic description and can more generally be described as the relationship between a **type** (abstract) and a **token** (concrete).

The description of phones in terms of articulation is usually based on the categories of the International Phonetic Alphabet (IPA) (IPA, 1999). The IPA is an internationally used notational system for transcribing speech sounds. It was first proposed in 1888 and has since been revised several times; its most commonly used version is shown in Figure 2.2.

[1]Typically, phones are displayed as [phone], while phonemes occur as /phoneme/.

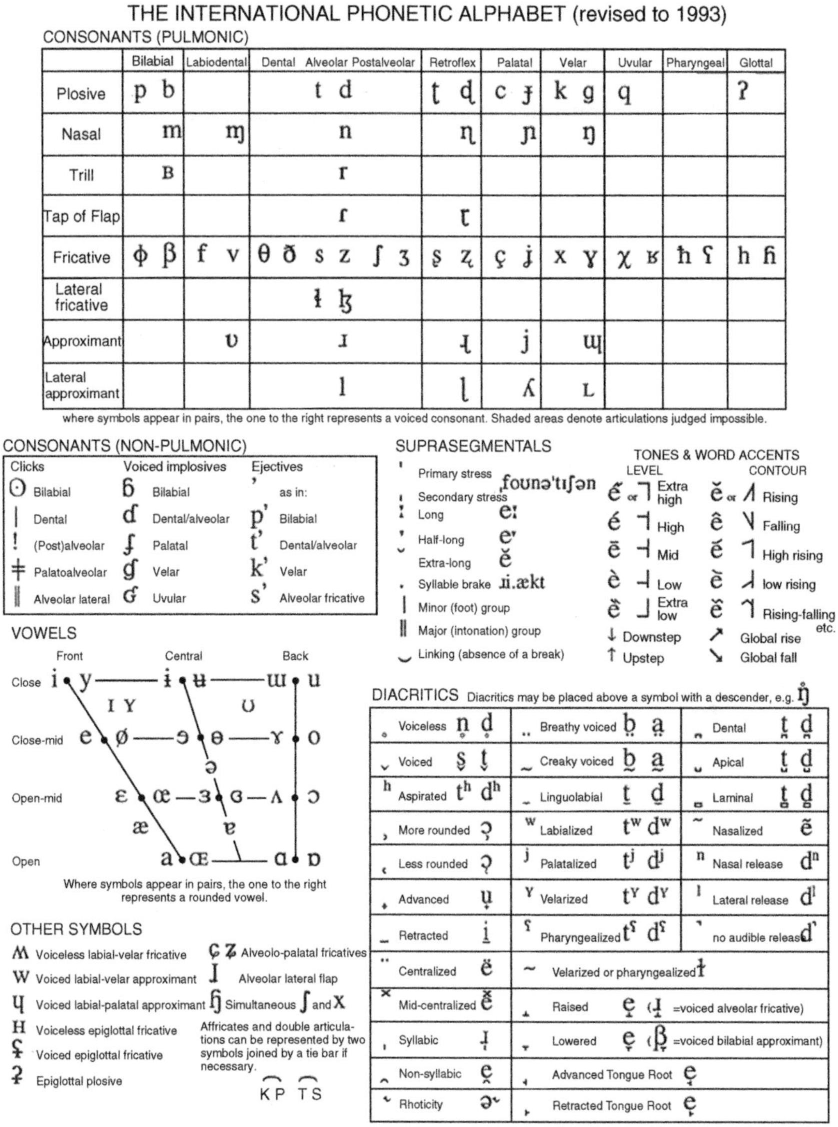

THE INTERNATIONAL PHONETIC ALPHABET (revised to 1993)

CONSONANTS (PULMONIC)

	Bilabial	Labiodental	Dental Alveolar Postalveolar	Retroflex	Palatal	Velar	Uvular	Pharyngeal	Glottal
Plosive	p b		t d	ʈ ɖ	c ɟ	k g	q		ʔ
Nasal	m	ɱ	n	ɳ	ɲ	ŋ			
Trill	ʙ		r						
Tap of Flap			ɾ	ɽ					
Fricative	ɸ β	f v	θ ð s z ʃ ʒ	ʂ ʐ	ç ʝ	x ɣ	χ ʁ	ħ ʕ	h ɦ
Lateral fricative			ɬ ɮ						
Approximant		ʋ	ɹ	ɻ	j	ɰ			
Lateral approximant			l	ɭ	ʎ	ʟ			

where symbols appear in pairs, the one to the right represents a voiced consonant. Shaded areas denote articulations judged impossible.

CONSONANTS (NON-PULMONIC)

Clicks	Voiced implosives	Ejectives
ʘ Bilabial	ɓ Bilabial	ʼ as in:
ǀ Dental	ɗ Dental/alveolar	pʼ Bilabial
ǃ (Post)alveolar	ʄ Palatal	tʼ Dental/alveolar
ǂ Palatoalveolar	ɠ Velar	kʼ Velar
ǁ Alveolar lateral	ʛ Uvular	sʼ Alveolar fricative

VOWELS

Front Central Back

Close i • y ——— ɨ • ʉ ——— ɯ • u

ɪ ʏ ʊ

Close-mid e • ø ——— ɘ • ɵ ——— ɤ • o

ə

Open-mid ɛ • œ — ɜ • ɞ — ʌ • ɔ

æ ɐ

Open a • ɶ ——— ɑ • ɒ

Where symbols appear in pairs, the one to the right represents a rounded vowel.

OTHER SYMBOLS

ʍ Voiceless labial-velar fricative

w Voiced labial-velar approximant

ɥ Voiced labial-palatal approximant

ʜ Voiceless epiglottal fricative

ʢ Voiced epiglottal fricative

ʡ Epiglottal plosive

ɕ ʑ Alveolo-palatal fricatives

ɺ Alveolar lateral flap

ɧ Simultaneous ʃ and x

Affricates and double articulations can be represented by two symbols joined by a tie bar if necessary.

k͡p t͡s

SUPRASEGMENTALS

ˈ Primary stress ˌfoʊnəˈtɪʃən

ˌ Secondary stress

ː Long eː

ˑ Half-long eˑ

˘ Extra-long ĕ

. Syllable brake ɹi.ækt

| Minor (foot) group

‖ Major (intonation) group

‿ Linking (absence of a break)

TONES & WORD ACCENTS

LEVEL		CONTOUR	
e̋ or ˥	Extra high	ě or ˩˥	Rising
é ˦	High	ê ˥˩	Falling
ē ˧	Mid	e᷄ ˧˥	High rising
è ˨	Low	e᷅ ˩˧	low rising
ȅ ˩	Extra low	e᷈ ˧˦˧	Rising-falling etc.
↓	Downstep	↗	Global rise
↑	Upstep	↘	Global fall

DIACRITICS Diacritics may be placed above a symbol with a descender, e.g. ŋ̊

◌̥ Voiceless n̥ d̥	◌̤ Breathy voiced b̤ a̤	◌̪ Dental t̪ d̪
◌̬ Voiced s̬ t̬	◌̰ Creaky voiced b̰ a̰	◌̺ Apical t̺ d̺
ʰ Aspirated tʰ dʰ	◌̼ Linguolabial t̼ d̼	◌̻ Laminal t̻ d̻
◌̹ More rounded ɔ̹	ʷ Labialized tʷ dʷ	◌̃ Nasalized ẽ
◌̜ Less rounded ɔ̜	ʲ Palatalized tʲ dʲ	ⁿ Nasal release dⁿ
◌̟ Advanced u̟	ˠ Velarized tˠ dˠ	ˡ Lateral release dˡ
◌̠ Retracted i̠	ˤ Pharyngealized tˤ dˤ	◌̚ no audible release d̚
◌̈ Centralized ë	~ Velarized or pharyngealized ɫ	
◌̽ Mid-centralized e̽	◌̝ Raised e̝ (ɹ̝ =voiced alveolar fricative)	
◌̩ Syllabic ɹ̩	◌̞ Lowered e̞ (β̞ =voiced bilabial approximant)	
◌̯ Non-syllabic e̯	◌̘ Advanced Tongue Root e̘	
˞ Rhoticity ə˞	◌̙ Retracted Tongue Root e̙	

Figure 2.2: The International Phonetic Alphabet (from the International Phonetic Association).

Under this scheme, consonants are classified according to three different criteria: (1) their place of articulation, or the point of contact of articulators within the vocal tract (bilabial, labiodental, dental, alveolar, postalveolar, retroflex, palatal, velar, uvular, pharyngeal, or glottal); (2) their manner of articulation, or the way in which the airstream is shaped or interrupted during sound production (plosive, nasal, trill, tap/flap, fricative, lateral fricative, approximant, or lateral approximant); and (3) the phonation type, or voiced or voiceless production. Vowels are classified according to tongue height, tongue advancement, lip rounding, and nasality. The IPA includes a dedicated symbol for every basic speech sound, plus a range of diacritics to annotate more fine-grained articulatory phenomena, such as aspiration, nasalization, and palatalization. Such modifications often result from **coarticulation**, which is the influence of neighboring speech sounds on the current speech sound. For speech research, several ASCII-based versions of phonetic alphabets have been developed: Arpabet (Shoup, 1980) for English, SAMPA for European languages (Wells, 1997), and Worldbet (Hieronymus, 1994), which comprises all of the world's languages and includes the IPA as well as several extensions. The IPA is used as the standard for phonetic notational symbols for the remainder of this book.

Languages are often classified by the types of phonemes and phoneme combinations they permit. First, the inventory of phonemes may vary in complexity from simple systems having 8 consonants and 5 vowels (Hawaiian) to more complex systems with 17 vowels, 3 diphthongs, and 22 consonants (German). In addition to their phoneme inventories, languages can be distinguished by patterns of phoneme combinations, **phonotactics**. Phonotactic constraints determine permissible syllable structures. Languages such as English, German, and Dutch allow a fairly large number of consonants both at the beginning (the onset) and at the end (coda) of a syllable, leading to a large number of possible syllable structures. By contrast, the Maori language spoken in New Zealand only allows syllables consisting of a vowel, two vowels, or a consonant plus a vowel. Other languages place constraints on the types of consonants and vowels that can be combined—for example, Spanish does not permit syllables beginning with /s/ + consonant.

The prosodic level of description includes phenomena that extend over more than one phonetic segment—in particular, variations in pitch, duration, and rhythm. Pitch is the perceptual correlate of fundamental frequency (the rate of opening and closing of the vocal chords) and is primarily used in two major ways: in **tone languages**, the pitch contour of a word is lexically or grammatically distinctive. Different tone heights or contours (such as high, low, rising, falling) give a different meaning to the word, as in the well-known example from Mandarin Chinese shown below:

Word	**Tone**	**Gloss**
mā	high level	*mother*
má	high rising	*hemp*
mǎ	falling-rising	*horse*
mà	falling	*to scold*

In addition to lexical differences, tone can also signal grammatical features. This is a widespread feature in many African tone languages. In the Edo language spoken in Nigeria, tone indicates tense (such as past versus present) and aspect (completed versus continuous action).

Intonation languages, by contrast, use pitch contours not in order to distinguish word meanings but to signal sentence type (question versus statement), to indicate phrase and sentence boundaries, and for contrastive emphasis. Examples of intonation languages include most of the Indo-European languages. Intonation languages often have particular word stress patterns, such as one syllable per word receiving greater prominence than all other syllables in the word. The stress pattern can be either fixed (word-final as in French or word-initial as in Hungarian) or quite intricate and variable, as in English.

Knowledge of the complexity of the sound structure of a language can be exploited for designing acoustic models, pronunciation dictionaries, and data-sharing methods for multilingual acoustic modeling (see Chapter 4). For languages with a complex syllable structure, a phone-based approach to acoustic modeling will typically be adopted, along with data-driven parameter-tying procedures to reduce the number of free parameters resulting from a large set of phonetic contexts. For languages with a limited number of simple syllable types, by contrast, syllable or demisyllable acoustic models are often used. Chinese, for instance, has about 1,700 syllables, and it is common to use complete syllable models or models for syllable-initial and syllable-final parts (Li et al., 2000). In a similar way, the use of pitch information in the language (tone or intonation) determines whether and in what way fundamental frequency measurements are integrated into a speech processing system.

Morphology

The field of **morphology** describes the process of word formation in a language, which is the process of combining the smallest meaningful parts of the language (**morphemes**) to form larger words. Similar to the situation in phonology, we can distinguish between morphemes and **morphs**—the concrete realizations of morphemes—for example, $-s$, $-z$, and $-es$ are all concrete instances of the PLURAL morpheme.

The definition of the term "word" itself is not without difficulties. A word is commonly defined as a sequence of characters delimited by whitespace, which, as a consequence, can occur in isolation or be shifted to different positions within a sentence. Words combine to become larger syntactic constructs, such as phrases and sentences, and are themselves units of meaning that can, but must not necessarily, be broken down into sets of morphemes. However, as explained below and in Section 2.4, some languages do not place whitespaces around their words but simply use sequences of characters without any explicit grouping. In other languages, words may have a complex structure equivalent to that of phrases in a language such as English, and it would be impractical to treat every sequence delimited by whitespace as a separate word.

A basic distinction can be drawn between **free morphemes**, which can occur on their own, and **bound morphemes**, which have to attach to other morphemes. For example, *clean* can occur on its own, but the *-ed* suffix must attach to a previous stem. Depending on the type of morph involved and the category of the resulting morph combination, three main types of word formation can be distinguished: **compounding**, **derivation**, and bf inflection. Compounding is the process of combining two or more free morphs to form a new word, as in *tea* + *pot*. This type of word formation occurs in virtually all languages, but some, such as German, are particularly notorious for the enormous number and length of the compounds they can form. Derivation and inflection involve combinations of bound and free morphemes, or other types of alterations to the basic word form. Derivation creates words of a different syntactic word class (part of speech), such as the adjective *inflatable* from the verb *inflate*. Inflection preserves the basic word class but changes the grammatical features of the word, such as *oscillate* + *s*. Both these examples involve attaching bound to free morphemes, which is termed **affixation**. Affixes fall into four major categories: prefixes, which precede the morpheme they attach to (the stem); suffixes, which occur after the stem; infixes, which can be inserted in the middle of a stem; and circumfixes, which consist of two parts enclosing the stem. There are a number of other types of word formation, including vowel changes, like English *sing, sang, sung*; duplication of the entire word or part of the word; and clipping (eliminating part of the word).

Languages can be classified according to their primary word-formation mechanism. The class of **isolating** languages simply form sequences of invariable free morphemes. Such languages are often said to "have no morphology." The example below is taken from Vietnamese (Comrie, 1989, p. 43):

Khi	*tôi*	*đến*	*nhà*	*bạn*	*tôi,*	*chúng*	*tôi*	*bắt đầu*	*làm*	*bài.*
When	*I*	*come*	*house*	*friend*	*I*	PLURAL	*I*	*begin*	*do*	*lesson.*

We can see that grammatical features such as past tense, number, and person are not expressed by special morphological affixes or changes to the word form but by simple juxtaposition of invariant morphemes. In isolating languages, there is often no clear-cut definition of the terms "word" or "word boundary." In the Vietnamese example, whitespaces occur after each morpheme, and each morpheme could also be accepted as an individual word. However, the sequence *bắt đầu* can arguably be considered a semantic unit and thus a word; this is similar to many cases in English in which a compound noun can be written with or without whitespaces.

Agglutinative languages combine several morphemes per word, and each morpheme can be identified by a linear segmentation of the word into its component parts. A prototypical example of an agglutinative language is Turkish, which can combine a stem with a large number of affixes. For example, the Turkish word *evleriden* analyzes as

ev	*-ler*	*-i*	*-den*
house	PLURAL	POSSESSIVE	ABLATIVE

'from their house'

So-called **fusional languages** also contain several morphemes per word; the difference to agglutinative languages is that the combination of morphemes within one word may lead to drastic alterations of the word form. For instance, the combination of English *go* with the (abstract) morpheme PAST TENSE leads to the form *went*, in which the individual morphs can no longer be identified by a simple linear segmentation process.

A special case of nonlinear morphology is presented by Semitic languages such as Arabic and Hebrew, in which word stems have a so-called **root-and-pattern** structure. Here, the basic word meaning is determined by a root morpheme consisting of three or four consonants that need not be adjacent but can be interspersed with other consonants or vowels. The consonant-vowel pattern expresses secondary grammatical categories such as number, voice (active or passive), and tense. Examples of Arabic word derivations associated with the same root are shown in Table 2.3.

Most languages do not fall neatly into one of the three categories described above but combine elements from several types. Thus, most agglutinating languages will show some type of fusion as well, and a fusional language may have structures that resemble those of isolating languages. However, the degree of morphological complexity of a language—that is, the relative preponderance of agglutinating or fusional versus isolating word formation—is of much significance for speech processing. Both agglutination and fusion contribute to the large number of possible word forms in morphologically complex languages, which in turn presents problems for dictionary construction and language modeling. Since the standard statistical models used in present-day speech

Table 2.3 Illustration of derived forms in Arabic for the roots ktb (write) and drs (study).

Root: **ktb**		Root: **drs**	
Arabic	*Gloss*	Arabic	*Gloss*
kataba	*he wrote*	**dar**asa	*he studied*
kitaab	*book*	**dars**	*lesson*
kutub	*books*	**duruus**	*lesson*
'a**ktub**u	*I write*	'a**dr**usu	*I study*
ma**ktab**	*desk or office*	ma**dr**asa	*school*
ma**ktaba**	*library*	mu**dar**ris	*teacher*
kaatib	*writer*	**dar**rasa	*he taught*

and language processing systems mostly regard each word as a separate event, a large number of different word forms increases the number of parameters in the model; this may cause unreliable probability estimates when the available training data is sparse. Isolating languages have a much smaller number of word forms but often require more sophisticated algorithms for word segmentation. These issues are discussed in more detail in Chapters 5 and 6, respectively.

Word Order

In addition to word structure, properties of phrase and sentence structure (**syntax**) are often used to describe or classify languages. Sentence structures are most often categorized by the relative ordering of subject (S), verb (V), and object (O). The six resulting possible word orders—SOV, SVO, VSO, VOS, OVS, and OSV—have all been attested in the world's languages, but the first two types have a much higher frequency than the others. Most languages are not limited to just one of these types but allow several different word orders, though they may exhibit a strong preference for one type over another. Even in languages like German or Russian, which are often said to have "free" word order, one type acts as the default type in that it occurs much more frequently. There are languages where such preferences are hardly detectable; these languages can thus be said not to have any basic word order at all. The prototypical example cited by linguists is Dyirbal, an Australian Aborigine language (with, however, only a handful of surviving speakers). Further distinctions can be drawn among languages according to their predominant phrasal structure, that is whether modifiers such as adjectives or relative clauses typically precede or follow the head words they modify.

The degree of "free versus fixed" word order often goes hand in hand with the degree of morphological complexity. Languages with relatively unconstrained word order often have rich morphology; overt grammatical inflections such as case markers help to determine the relationships among the constituents in the sentence when constituents are nonadjacent. Languages that do not make heavy use of inflections often need to follow a fixed word order to avoid ambiguities in determining the role of each constituent.

The main issue of importance for speech processing is the question of whether higher variability in the sentence and phrase structure of a language leads to less robust language models, thereby increasing the demand for training data. This problem is discussed in Chapter 6.

2.3 Language in Context

Apart from their structural aspects, languages also differ in the way they are used in actual communication. For the purpose of developing speech applications (especially dialog systems or other forms of speech-based human-computer interfaces), an understanding of these characteristics is at least as important as the accurate modeling of purely acoustic or grammatical characteristics.

The issue of regional variation (dialects) has already been addressed. In addition to regional variation, languages exhibit idiolectal and sociolectal variation. An **idiolect** are the consistent speech patterns (in pronunciation, lexical choice, and grammar) that are specific to a given speaker. A **sociolect** is a set of variations that are characteristic of a group of speakers defined not by regional cohesion but by social parameters, such as economic status, age, and profession. Idiolectal patterns may include speaker-specific recurrent phrases (e.g., a tendency to start sentences with "Well, to be honest..."), characteristic intonation patterns, or divergent pronunciations. Idiolectal features can be used for tailoring a speech application to a specific user, such as for training a speech-based automated office assistant. In addition, idiolectal features have recently been shown to be helpful in automatic speaker identification (Doddington, 2001; García-Romero et al., 2003). Similarly, sociolectal features can be taken into account when developing an application for an entire user group.

Many linguistic societies exhibit what is termed **diglossia**—the coexistence of two linguistic varieties, one of which is considered more formal or prestigious whereas the other is a low-prestige variety, such as a local spoken dialect. The former is typically used in formal situations, such as official events, nationwide media broadcasts, and classrooms, whereas the latter is used in everyday informal communication. Examples of diglossia include Standard ("High") German versus the regional dialects of Germany (such as Bavarian, Swabian, Low German), and Modern Standard Arabic versus the regional Arabic dialects (e.g., Egyptian Arabic, Gulf Arabic). A situation of diglossia does not necessarily mean that the two varieties are always used in complementary situations. Speakers will often switch between the two, sometimes within the same utterance or even within a single word, using for example, an inflectional affix from one variety and a word stem from another. This is termed **code-switching**. Code-switching may also occur between two different languages, especially among multilingual speakers, as the example in Figure 2.3 shows. This is an obvious problem for speech applications such as automatic speech recognition or language identification since it requires detecting a language change, correctly segmenting the utterance into different-language portions, identifying the languages, and, possibly, recognizing the words spoken. Chapters 8 discusses this problem in detail.

Another frequent feature of multilingual societies or regions is the use of a **lingua franca**. A lingua franca is either an existing natural language or a newly created language that serves as the primary means of communication between speakers who do not speak each others' native language and are not native speakers of the lingua franca either. One example is Tok Pisin, the pidgin language of Papua New Guinea, but even English is frequently used as a lingua franca, for example in aviation, at international organizations or conferences, and in business transactions.

An important aspect of language use is the fact that contextual features can trigger the use of particular grammatical features. In every language, speakers typically adjust their speech depending on the situational context—including, for example, who they are talking to, what they are talking about, who else is listening. In some languages, this simply results in the choice of different titles, or words indicative of a certain speech style (formal versus slang words). Some languages, however, have a very complicated system of speech levels related to social status, each with its own set of inflections, pronouns, and other word forms called **honorifics**. In Japanese, for instance, there are four common (and some additional rarer)

Moroccan Arabic/French:	*... ḥant*	*walla*	*le luxe*	*bezaaff'*	*les*	*hôtels.*
English translation:	*... because*	*has become*	*luxury*	*much*	*in*	*the hotels.*

"... because the hotels have become very luxurious."

Figure 2.3: Example of within-utterance code-switching between French and Moroccan Arabic, from Naït M'Barek and Sankoff (1988).

Table 2.4 Korean speech levels: forms of the infinitive verb *gada* (go).

Speech Level	Form of *gada*
Low	*ga*
Middle	*gayo*
High	*gamnida*
Highest	*gashimnida*

honorific affixes that are attached to a person's name, depending on the social status of the person that the speaker is referring to; for example, a person with the name Suzuki will politely be refered to as Suzuki-san, Suzuki-sensei, or Suzuki-sama, rather than just Suzuki. Similarly, there are honorific prefixes that are not used for people's names but for nouns and verbs, especially those denoting culturally significant objects or activities. Thus, *tea* (*cha*) and *rice* (*han*) will usually be referred to as *o-cha* and *go-han*. Korean has at least four different levels of politeness, each with its own set of pronouns, conjugation patterns, affixes, and sometimes entirely different words. An overview is given in Table 2.4. Additional forms can be added to these, depending not only on the status of the speakers but also on the status of the person the speakers are talking about.

Knowledge of the proper use of honorifics in a language is particularly important for speech-based human-computer interfaces—such as speech generation in automated dialog systems, and speech-to-speech translation. When developing or adapting a dialog system to a language with a rich system of honorifics, care must be taken to choose a level of formality appropriate to both the contents of the output and the intended audience. The field of cultural issues in the development and localization of speech-based interfaces has not been studied in great detail; however, some discussion is provided in Chapters 9 and 11.

2.4 Writing Systems

A writing system is defined as a set of notational symbols along with conventions for using them, designed for the purpose of communicating the writer's message in exact words. This definition excludes so-called **pictographic**, or **ideographic**, systems—that is, pictorial representations of concepts or ideas whose interpretation is subject to conventions but could be translated into a variety of different verbal expressions. In an ideographic system, the picture of a house could be translated into actual speech not just by *house* but also by *shed*, *dwelling*, *abode*, *hut*, or *mansion*. The writing systems described below are distinguished from ideographic systems in that they uniquely specify the words of the intended message.

The basic unit of a writing system is a **grapheme**. Writing systems can be subclassified according to the types of graphemes they employ—in particular, the relationship between grapheme, sound, and meaning. Following Sampson (1985), we draw a basic distinction between **logographic** systems, which are based on semantic units (units of meaning), and **phonographic** systems, which are based on sound units. The former can be subdivided into morphemic and polymorphemic systems, and the latter fall into syllabic, segmental, and featural subtypes. Figure 2.4 summarizes this classification, which is explained in detail below.

2.4.1 Types of Writing Systems

Sumerian cuneiform writing—the first writing system ever to be developed—was a logographic system (probably derived from an earlier pictographic system) in which each grapheme stood for a distinct word. Systems with a one-to-one correspondence between units of meaning and graphemes are termed **morphemic**. Those systems in which several units of meaning are combined into one grapheme are labeled **polymorphemic**. Their use is typically restricted because in general, the inventory of graphemes must be finite in order to permit the recombination of characters to form new words.

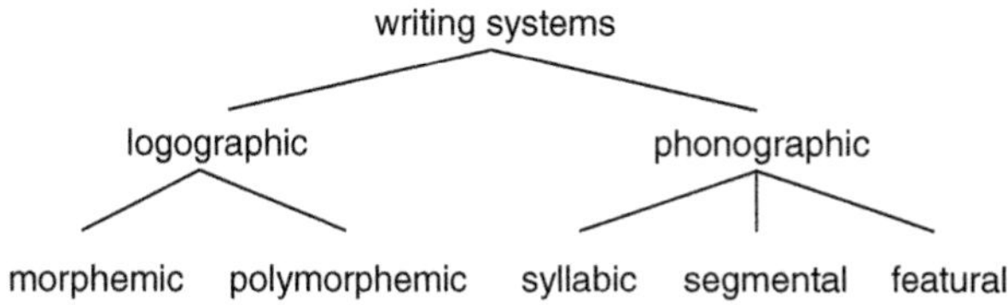

Figure 2.4: Classification of writing systems.

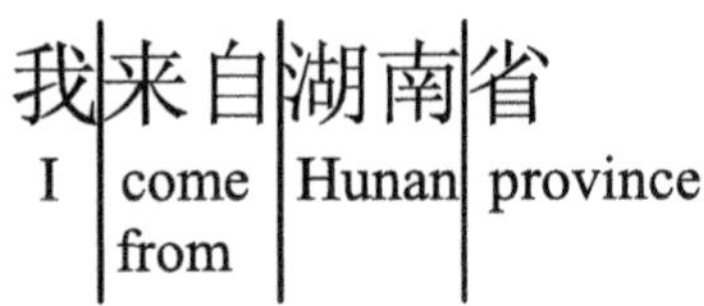
我来自湖南省
I
come from
Hunan
province

Figure 2.5: Representation of the sentence *I come from Hunan* in Chinese Hanzi, with indication of word segmentation.

The most well-known present-day logographic system is the Chinese Hanzi system, which represents sentences as sequences of characters that represent words or morphemes rather than alphabetic elements, as shown in Figure 2.5.

No whitespaces are used to delimit larger units of meaning (words). Native speakers typically agree on which characters form a semantic unit; however, this is not explicitly indicated in writing. The Hanzi system is often described as an ideographic or a pictographic system; however, this terminology is rejected by linguists. Although Chinese characters are historically derived from pictograms, the percentage of present-day characters that can be traced back to these origins is less than 3%. Many characters are instead used phonetically (some words are represented by the character for an identical-sounding though semantically different word) or in phonographic-logographic combinations.

In phonographic systems, graphemes represent sound units. All Western alphabets (Greek, Roman, Cyrillic), for example, are phonographic. More specifically, the graphemes in these systems roughly correspond to phonemes, thus belonging to the **segmental** subtype. However, the relationship between graphemes and phonemes is not necessarily one-to-one; English, for example, has numerous many-to-one grapheme-phoneme relations such as <ou> for /ʌ/ and <gh> for /f/ in *enough*. So-called **consonantal writing systems** are an important subtype of segmental writing systems in that they represent only consonants (and possibly long vowels) but omit short vowels. Consonantal writing systems were originally developed for Semitic languages, such as Arabic, Hebrew, and Akkadian, and are also sometimes called **abjads** (Daniels and Bright, 1996). The most well-known consonantal writing systems are those of Arabic and Hebrew; the former has also spread to other, non-Semitic languages (sometimes with slight modifications), such as Farsi, Hausa, and Urdu. In Arabic,

أُحِبُّ ٱلسَّفَرَ إِلَى ٱلْقَاهِرَة

آحب السفر إلى القاهرة

Figure 2.6: Arabic script representation of the sentence *I like to travel to Cairo*, with diacritics (top row) and without (bottom row).

only consonants and the long vowels /i:/, /a:/, and /u:/ are represented by the basic letters, whereas short vowels and certain other types of phonetic information such as consonant doubling can only be indicated by **diacritics**, which are short strokes placed above or below the preceding consonants. Figure 2.6 shows an example of Arabic script with and without diacritics. In everyday practice, diacritics are usually omitted, except in beginners' texts, in important religious writings, or to avoid ambiguities. In all other cases, readers infer the missing vowels from the context based on their knowledge of the language. Thus, consonantal scripts are defective in that they lack essential phonetic and word-distinguishing information.

In **syllabic** systems, graphemes stand for entire syllables; thus, the language in principle needs to have at least as many graphemes as syllables. A historic example of a syllabic script is Linear B, the writing system used for Ancient Greek. A present-day example of syllabic writing is the Japanese Kana system. Japanese uses three different types of graphemes: Kanji characters, which are logograms adopted from Chinese, plus Hiragana and Katakana. The latter two are collectively referred to as Kana, each of which consists of 46 graphemes representing all of the possible syllables in the language; Figure 2.7 shows examples of Hiragana characters. While it is possible to write a text entirely in Kana, a mixture of Kanji and Kana is used in practice, with Kana characters being used mainly for those words for which no Kanji characters exist, such as particles, foreign loan words, etc.

Many languages use a mixture of syllabic and segmental systems. The Southeast Asian languages Khmer, Thai, and Gujarati, for example, use related writing systems (all derived from the Devanagari script originally used for Sanskrit and for modern-day Hindi), in which graphemes mostly represent syllables. These systems contain basic characters for a consonant plus an inherent default vowel, and all other graphemes

a	あ	i	い	u	う	e	え	o	お
ka	か	ki	き	ku	く	ke	け	ko	こ
sa	さ	shi	し	su	す	se	せ	so	そ
ta	た	chi	ち	tsu	つ	te	て	to	と
na	な	ni	に	nu	ぬ	ne	ぬ	no	の
ha	は	hi	ひ	fu	ふ	he	へ	ho	ほ
ma	ま	mi	み	mu	む	me	め	mo	も
ya	や			yu	ゆ			yo	よ
ra	ら	ri	り	ru	る	re	れ	ro	ろ
Wa	わ	Wi	ゐ			we	ゑ	wo	を
								n	ん

Figure 2.7: Japanese Hiragana characters.

(either for consonants followed by other vowels or for bare consonants) are then derived from these by means of modifications, such as diacritic marks or rotation of the character. Such mixed writing systems are also termed **alphasyllabaries**, or **abugidas** (Daniels and Bright, 1996).

Featural writing systems use elements smaller than the phone, corresponding to phonetic or articulatory features. The most prominent of these is the Korean writing system. Korean was originally written using Chinese characters; however, in the 15th century, Hangul—an entirely original script—was developed for the Korean languages. The Hangul system contains 28 letters (though four of these are now obsolete) composed of elementary strokes representing the phonetic features of the corresponding sound. Consonantal features representing place and manner of articulation are expressed by stylized shapes evoking the shape and position of the active articulator (tongue or lips) during the production of the sound (see Figure 2.8). Vowels are expressed by vertical and horizontal bars that are combined with the consonant signs to form syllables. Syllables corresponding to words are then grouped perceptually by surrounding white spaces, which makes segmentation an easier task than in Chinese. In practice, Chinese characters are still sometimes used in combination with Hangul characters in present-day Korean.

IPA	p	t	c	k	p^h	t^h	c^h	k^h	r	h
Hangul	ㅂ	ㄷ	ㅈ	ㄱ	ㅍ	ㅌ	ㅊ	ㅋ	ㄹ	ㅎ
IPA	m	n	ŋ	s	b	d	ĵ	g̊	z	
Hangul	ㅁ	ㄴ	ㅇ	ㅅ	ㅃ	ㄸ	ㅉ	ㄲ	ㅆ	
IPA	i	e	ɛ	a	o	u	ʌ	w	wj	je
Hangul	ㅣ	ㅔ	ㅐ	ㅏ	ㅗ	ㅜ	ㅓ	ㅡ	ㅢ	ㅖ
IPA	jɛ	ja	jo	ju	jʌ	wi	we	wɛ	wa	wʌ
Hangul	ㅒ	ㅑ	ㅛ	ㅠ	ㅕ	ㅟ	ㅞ	ㅙ	ㅘ	ㅝ

Figure 2.8: Basic Korean Hangul characters.

2.4.2 Digital Representation and Input Methods

The rapid spread of modern information technology and mouse-keyboard-screen interfaces has led to advancements in new encodings and input methods for the multitude of different scripts used internationally. Speakers of languages that have not historically used a Western alphabet—in particular those using syllable-based or logographic writing systems—often face the problem of their writing systems containing many more characters than keys on a standard keyboard. As a result, two different trends have emerged: the first is the establishment or revision of existing **transliteration** schemes for representing script-based languages by means of the Roman (or some other Western) alphabet; this is sometimes called **romanization**. In the course of their history, many languages have switched from their original writing system to the Roman alphabet—for example, Turkish switched from a version of the Arabic script to the Roman alphabet in 1928, and Turkmen reverted from Cyrillic to Roman in 1991. Chinese has several systems for the transliteration of Standard Mandarin, including *zhùyīn* and several types of *pīnyīn*. Similarly, Japanese has the *Nihon-shiki* systems, and South Korea developed the Revised Romanization of Korean as a new standard for Korean transliteration in 2000. The second trend is the development of better encoding standards and keyboard input methods for a wide variety of languages. Until the mid-1990s, various encodings were used to represent character sets for different languages. No single encoding was able to encompass all languages, and sometimes different, conflicting encodings were used for the same language. Common encodings were established in the ISO standards, defined by the International Organization for Standardization, which include sixteen different encodings for the major European languages plus Arabic and Hebrew (ISO 8859-1 through ISO 8859-16). Others include the internal encodings (code pages) of the Microsoft DOS and Windows operating systems, or the Indian Script Code for Information Interchange (ISCII). In 1991, the **Unicode** standard was proposed as a means of simplifying the representation and exchange of plaintext in multilingual environments. Unicode (Allen and Becker, 2004) is a universal character encoding standard that is independent of platforms and operating systems. It provides a means of encoding all existing characters used in all of the world's languages (plus historical scripts, punctuation marks, diacritics, mathematical symbols, and so on) by mapping each character to a unique numeric value or character code. The encoding specifies code elements and their order whenever relevant, but the elements themselves are abstract and need to be interpreted by a Unicode-compatible software tool to be able to display or print actual characters of a specific size, shape, or style (glyphs). Unicode is increasingly gaining importance as the dominant character-encoding scheme, and a large number of software tools—such as Web browsers, e-mail readers, and text editors—now support the input and interpretation of Unicode.

2.4.3 Oral Languages

Whereas the task of transliteration is to translate the same text from one writing system into another, the task of **transcription** is to transform speech into a written representation, which can be either orthographic or phonetic. Written material is used to provide training data for language and acoustic models in speech recognition, language identification, speech translation, and other areas of multilingual speech processing, and therefore plays an essential role in developing speech technology. In the ideal case, the chosen representation (typically orthographic) provides as much information as possible about the spoken form. There are, however, many so-called oral languages—that is, languages that have no writing system. These not only include many indigenous languages without a literary tradition but also include many dialects that are used for everyday conversations but not for written communication, often because dialectal forms are stigmatized. Chinese and Arabic are examples of languages with a large number of unwritten regional dialects that depart significantly from the standard language. If a language has no standard writing system, the acquisition of linguistic resources is rendered much more difficult. In this case, the only way to collect text material is to record and transcribe spoken data in that language. This often requires defining a new writing system and orthographic standards, and training native speakers to use this standard consistently. In cases in which some written material can be found for the desired language, it will need to be normalized. For instance, in diglossia situations in which dialects and standard varieties coexist, speakers often use standard spellings and try to map them to dialectal pronunciations in inconsistent ways. The difficulties in undertaking data collection for oral languages are described in more detail in Chapter 3.

Grapheme-to-Phoneme Mapping

Another problem is the degree to which a word's written representation departs from its actual phonetic form. As mentioned previously, some writing systems are deficient in that they omit essential phonetic information; in those cases, the reader needs to use contextual information in order to infer the exact word identity. The prototypical examples of this case are the consonantal writing systems of Arabic and Hebrew, which do not provide information about the type and location of short vowels and other phonetic segments. This in turn makes it more difficult to train accurate acoustic models. One could argue that to some extent, even English does not provide complete phonetic information, since it omits prosodic marks such as word stress, which can be distinctive in such cases as *récord* (noun) versus *recórd* (verb). However, the occurrence of such examples in English is infrequent, and contextual information is sufficient to resolve these ambiguities. In Arabic or Hebrew, by contrast, the much larger degree of ambiguities significantly decreases the predictive power of language models. In other languages, the writing system may be phonetically complete but the relationship between orthography and pronunciation may be complex and nontransparent. Thus, while Spanish has a relatively simple mapping between graphemes and phonemes, English has many one-to-many correspondences between graphemes and phonemes and a large number of exceptions. The development of pronunciation dictionaries and techniques for predicting pronunciations from graphemic representations are discussed in Chapters 4, 5, and 7.

2.5 Languages and Speech Technology

What is the current status of speech technology for different languages? The availability of speech technology for different languages closely mirrors the levels of research effort and funding in different countries over the past two or three decades. English has probably had the largest share in speech research and development. This is not only due to the amount of research funding in English-speaking countries such as Canada, the United States and the United Kingdom, but also due to the fact that early efforts toward standardization of language resources have concentrated on English. As a result, English corpora such as Switchboard and the *Wall Street Journal* have become benchmark databases that are being used by researchers in non-English-speaking countries as well. Another factor is that annual benchmark evaluations in a range of different speech processing applications (automatic speech recognition, language identification, speaker identification) have been carried out by the National Institute for Standards and Technology (NIST) since the early nineties. These evaluations have had international impact and regularly involve non-U.S. participants. Significant research effort has also been spent on major European languages (French, Spanish, German, Dutch, Italian) and on Chinese and Japanese.

When trying to determine which languages are going to play a significant role in speech processing for the foreseeable future, the most important predictors are the current and future number of speakers, the economic and political potential of the countries where those languages are spoken, and the information-technology needs of the population. Clearly, the level of speech technology will never be the same for all languages; commercial developers of speech technology will only invest in a given language when there is the prospect of investment return, which in turn depends on market demand and cost-effectiveness. For instance, speech-based telephony systems will not sell well in a country where the cost of employing human phone operators is cheaper in the long run than that of developing, fielding, and maintaining an automated dialog system. Yet at the time of this writing, state-of-the-art speech technology is not available for even the twenty most widely spoken languages in the world (see Table 2.2)—in particular, the languages spoken in India.

Based on these considerations, it is probably safe to say that the languages listed above will continue to be of major interest to speech researchers and developers, with Asian and Southeast Asian languages becoming increasingly more important. However, speech technology research has much to contribute even to those languages that do not fall into this category. First, languages with a relatively small number of speakers and few linguistic resources may suddenly become of interest for military reasons or because of a catastrophic event requiring international rescue efforts, such as the 2004 tsunami in Southeast Asia. Second, as mentioned earlier, a large number of languages are in danger of becoming extinct, but there are ongoing projects for preserving them (E-MELD, 2005). If speech recognizers or speech-to-speech translation systems could be ported to new languages rapidly and at a reasonable cost, speech applications could be made available for a much larger number of languages. Language portability is a difficult goal. At present, the standard way of building a speech application for a new language is still to collect a sizable training corpus and to train statistical models for the new language from scratch. Considering the enormous number of languages and dialects in the world, this is clearly a suboptimal strategy, which highlights the need for more sophisticated modeling techniques. It would be desirable to develop models that can take advantage of similarities between dialects of the same language, or languages of the same type, and that can share data across different varieties, leading to true multilingual speech processing. The following chapters survey the state of the art of these approaches at various levels of linguistic modeling and with respect to a range of different applications.

Chapter 3

Linguistic Data Resources

Christopher Cieri and Mark Liberman,
Victoria Arranz and Khalid Choukri

3.1 Demands and Challenges of Multilingual Data-Collection Efforts

The development of technologies for spoken or written language is a resource-intensive venture. Furthermore, recent developments have blurred the lines between spoken and written language technologies with the result that the resource needs of many communities have increased in the types of data as well as in the volume of data and variety of languages. As Reynolds and colleagues (2003) explain, state-of-the-art speaker recognition systems focused on low-level features such as cepstra until recently, when higher level features such as lexical choice were proven useful. While older systems were trained and tested using speech samples attributed as to speaker, systems that exploit high-level features often require transcripts too. Similarly, speech-to-speech translation technology requires all of the resources typically prepared for speech recognition and speech synthesis in both languages plus the resources required for machine translation between the languages.

3.1.1 Use of Multilingual Speech Data

Although **corpus** is typically defined as a set of data collected and prepared for a specific use, linguistic resources are frequently exploited for purposes other than those for which they were designed. Perhaps the most convincing case of re-annotation and reuse of spoken data is the Switchboard corpus (Godfrey et al., 1992). Originally collected at Texas Instruments to support research into topic and speaker identification, Switchboard has been enhanced with orthographic annotations and transcripts, phrase-level time stamps, discourse and disfluency annotations, part-of-speech tagging, and syntactic bracketing. The enhanced data supports basic linguistic research as well as the development of technologies for automatic speech recognition, tagging, and parsing (Graff and Bird, 2000). Similarly, the CallFriend and Topic Detection and Tracking corpora have been re-annotated and reused for multiple purposes. This tendency toward reuse has implications for both corpus developers and evaluators.

As detailed below, much of the speech data developed in the United States supports research and technology development in one of several common task programs. Within such programs, multiple research organizations cooperate and compete in the creation of technologies that address a common problem, such as the need for systems to automatically transcribe the audio of news broadcasts. The programs are also characterized by centralized, objective, metrics-based evaluation and shared language resources. Research sites from around the world are attracted to such programs even though many may not be funded for their participation. In the parlance of common task programs, data may be designated in one of three ways. Language resources intended to provide statistical information and to support basic research and system development are designated as **training** data. To permit local testing of the effectiveness and generality of systems, common task programs generally provide **development/test** data. Finally, robust, objective, metrics-based, blind evaluation of technologies relies on **evaluation data**. Development/test and evaluation data are often grouped together under the term "test" data. The size of an evaluation data set depends on the number of conditions to be tested and the size of data in each combination of conditions necessary to give statistical significance. Development/test sets are generally similar in size to evaluation sets and are stratified in the same way. However, development/test and evaluation sets are often selected so as not to overlap in time. Training data may also be selected to differ in time epoch from test data. Within recent common task programs in the United States, it is generally accepted that system performance increases as more data becomes available. As a result, the primary factors determining the size of training data sets tend to be the time and funding allocated to create them.

3.1.2 Issues for New Languages

There are a number of issues that need to be considered when planning resource creation in a new language to support speech processing; several of the most important will be discussed here. Perhaps the most important issue is the availability of the speech data itself. Many recent technology development programs have focused on broadcast news not only because of its intrinsic interest and broad vocabulary coverage, but also because of its abundant availability. However, of the thousands of languages currently spoken on the planet, only a few hundred are represented in news broadcasts, and subsets of these are localized so that collecting broadcast news requires locating a collection system within the range of the broadcasts. Furthermore, data from a single source or a small number of sources may not provide adequate variability among speakers and broadcast conditions, and the resulting systems may lack generality. For other kinds of spoken data, such as conversational, read, or prompted speech, the ubiquity of the telephone somewhat improves the situation by allowing researchers to collect samples from a population that is not necessarily local. Unfortunately, the quality and variation characteristics of telephone networks and handsets make telephonic collection a limited solution. Telephone networks are bandwidth limited, and cellular telephony adds both the complications of compression and the artifacts of interference, drop-outs, and cell handoff. Furthermore, the cost of long-distance phone calls can still be prohibitive in large-scale collections—especially in today's world of voice mail and answering machines, where several attempts may be required to reach a subject. Even in cases where telephone networks are cost-effective and provide adequate quality, variation among them may conflate channel effects with other signal properties under study. Some speech processing communities in the United States have addressed this concern by constraining the location of subjects in their multilingual collections. Typically all callers participate from within the United States. Similar efforts have been devoted in Europe to the recording of databases, such as those from the SpeechDat family. Finally, multilingual collections that require subjects to be physically located in a special recording facility face the greatest logistical challenges when the multilingual subject pool is not local to the facility.

Assuming that one has solved the logistical problems of locating collection systems within reach of subject populations, one may still need to acquire permission to record. Although the principle of fair use, familiar to many readers of this volume, applies to much of the research and data described herein, that principle is not universal. Even where fair use is in effect, its applicability is subject to interpretation on a case-by-case basis. As a result of this uncertainty, data centers such as the **Linguistic Data Consortium (LDC)** and the **European Language Resources Association (ELRA)**[1] prefer to acquire explicit permission to use material such as broadcast audio and news text for linguistic research. In the case of broadcasts, a history of sharing footage both helps and complicates the situation. Broadcasters are accustomed to the idea of sharing and have procedures in place for doing so. On the other hand, their default model and costs are based on commercial reuse of footage. Research use may be unfamiliar to some broadcasters and is generally less lucrative for all. Still, the tremendous progress made in processing broadcast audio over the past several years would not have been possible without the generous support of dozens of news agencies and broadcasters. To alleviate the burden on both data creators and data users, LDC and ELRA act as intellectual property intermediaries. By maintaining common data provider and data user agreements, these centers reduce the number of agreements, and negotiations, necessary. For a set of x researchers who need y corpora, the number of agreements necessary is $x + y$ rather than the $x*y$ that would be necessary without this mediation. Whereas language resources include conversations or other speech collected directly from individual subjects, permission takes the form of informed consent rather than copyright. Within the United States, sponsored research projects involving humans must inform subjects of the purpose of the study and any associated risks and benefits. Subjects must give their consent affirmatively and researchers must document it via a procedure approved by a federally registered internal review board. Linguistic data collection has few risks compared to clinical or pharmaceutical trials;

[1] see http://www.ldc.upenn.edu and http://www.elra.info

the primary risk is loss of subject anonymity. Because requirements for the fair treatment of human subjects vary from place to place and because willingness to participate in language studies differs across cultures and within a single culture over time, permission to record remains a challenge for multilingual data collection.

The final issue to be discussed is sampling of subjects. In order to provide adequate representation of the variation that challenges speech technology, language resources need to cover the locally important categories related to sex, age, region, class, education, occupation and so on. The specific linguistic features that covary with any of these demographic factors differ from language to language and often from dialect to dialect; in addition, the boundaries between one category and the next along any demographic dimension may reflect cultural as well as linguistic constraints.

3.1.3 Transcription Approaches and Problems

Quick and Careful Transcription Specifications

To support automatic speech recognition (ASR) research, it is necessary to provide time-aligned transcripts of the speech data collected. However, normal rates for careful transcripts of conversational telephone speech average 20×RT (real time), or 20 hours for each hour of conversation per channel. Such a **transcription specification** is impractical for large-scale collections, in which transcription costs threaten to swamp all other costs. To address this problem, the EARS (Effective, Affordable, Reusable Speech-to-Text) research community recently developed a Quick Transcription (QTr) specification that requires only about six hours of effort for every hour of speech. One of these specifications was developed at LDC; the other was developed by BBN Technologies and based on that of a commercial transcription bureau: WordWave International (WordWave, 2005). The LDC variant relies on automatic segmentation of conversational audio into utterances of two to eight seconds in duration. Transcriptionists then perform a single pass over the audio, creating a verbatim record of the conversation. Although they make an effort to provide capitalization, punctuation, and special markers (for background noise, mispronunciations, nonce words, etc.), their prime directive is to complete the transcription using no more than five hours of effort for every hour of speech. This distinguishes QTr from careful transcription following the specifications developed by previous research thrusts, in which the prime directive was to create a (nearly) perfect transcript, and considerably more time and effort were allocated to approach that goal. The BBN/WordWave QTr variant transcribes the entire conversation without initial segmentation but then applies automatic (forced) alignment to yield time stamps for each pause group. Although these approaches were expected to yield lower quality transcripts it was anticipated that the much greater volume of available data would compensate for the loss in quality. In fact, experiments (Kimball, 2004) show that transcripts created this way are, for purposes of training ASR systems, of equal value to the considerably more expensive careful transcripts.

Transcription of Unwritten Varieties: The Case of Arabic Colloquials

Currently, technologies that perform automatic transcription of spoken language have intense requirements for language resources, including audio samples of speech, time-aligned transcripts, and pronunciation lexicons. Transcripts are generally produced in standard orthography while pronunciation lexicons map between surface orthographic forms appearing in the transcripts and their pronunciations. The pronunciations support acoustic modeling while the orthographic forms support language modeling. Because there are generally inadequate supplies of transcribed speech to support robust language modeling, written language is used as a proxy. This approach is recognized as an unfortunate but necessary compromise. There are cases, however, in which the approach is even more problematic. Many of the world's languages are either never written or written in a form that differs dramatically from their spoken form. In such cases, building language models from text to help select among hypotheses generated by an acoustic decoder will be challenging if not impossible. The colloquial Arabic dialects are regional varieties that are generally not written and that differ significantly from the variety that is written, Modern Standard Arabic (MSA).

Levantine Colloquial Arabic was one of the focus languages of the DARPA EARS (Effective, Affordable, Reusable Speech-to-Text) program a research program aimed at developing automatic speech-to-text transcription technology for conversational and broadcast news speech in English, Chinese, and Arabic. In order to develop automatic transcription systems for conversational telephone speech in Levantine, EARS research groups needed audio and transcripts in this variety. Because Levantine is not generally written, researchers at the Linguistic Data Consortium invented a writing system. The desiderata were (1) that the system be easy for native speakers to read and write, (2) that the same word be represented the same way each time it is uttered and that different words be represented differently, (3) that the representation not exaggerate differences between it and MSA, and (4) that the representation capture the phonetics of the dialect as well as possible given the other desiderata. The result is a two-layer transcription specification. The "green" layer represents Levantine words that are cognates of MSA words by using the MSA form, as long as the differences between the two can be explained by general rules of sound change. Levantine-specific words are also represented in Arabic orthography using the symbol-to-sound correspondences present in MSA, allowing for the regular sound changes that distinguish Levantine. The effects of these decisions are that native speakers quickly learn to transcribe Levantine and that the green layer bears a great similarity to written text—a useful feature for language modeling. The "yellow" layer refines the green layer by adding short vowels and noting significant changes in pronunciation, including added, deleted, and changed consonants and vowels. The green and yellow layers are aligned at the word level whenever possible and together support both acoustic and language modeling.

3.1.4 Cost and Dissemination

Cost factors in language-resource creation include staff to negotiate licenses; collect and annotate data; perform quality control on, publish, distribute, and track corpora; maintain information; and answer questions about resources. Costs also cover infrastructure for automatic collection and processing of resources to support human activities, as well as staff to maintain the infrastructure. Costs may be divided into those necessary to develop the resource in the first place and those necessary to maintain it and ensure its availability. Development costs are generally borne by a sponsoring organization that has an acute need for the resource. To the extent that resources are shared, production and maintenance costs may also be shared among subsequent users of the data. Data centers such as the LDC and ELRA tend to use this model. However, it is important to note that this model is far from universal. In some cases, corpus authors will seek to recover development costs via higher licensing fees. On the opposite end of the scale, production and maintenance costs are sometimes covered by sponsors who seek the broadest possible distribution for their corpora. A notable example is Talkbank, a five-year interdisciplinary research project funded by the U.S. National Science Foundation (NSF) (BCS-998009, KDI, SBE, and ITR 0324883) to advance the study of communicative behavior by providing not only data but tools and standards for the analysis and distribution of language resources. NSF subsidized the distribution of the first 50 or 100 copies of Talkbank corpora above and beyond those copies LDC distributes without cost to its members. The Talkbank corpora include field recordings in two Grassfields Bantu varieties, video recordings annotated for gesture kinematics, a finite-state lexical transducer for Korean and morphologically annotated Korean text, a corpus of sociolinguistic interviews in dialects of American English, the Santa Barbara corpora of spoken American English, and annotated field recordings of vervet monkey calls. Talkbank has also subsidized the second release of the American National Corpus.[2]

3.2 International Efforts and Cooperation

3.2.1 U.S.–European Collaboration

Increasing demand for language resources make international cooperation more than just a good idea. The Linguistic Data Consortium and the European Language Resources Association cooperate in a number of ways to reduce resource costs and increase availability. The **Networking Data Centers** project (Net-DC)—sponsored by the National Science Foundation and the European Commission (EC)—funded LDC and ELRA to advance transatlantic cooperation by collaborating on the development and distribution of a broadcast news corpus that would focus attention on differences in collection and annotation approaches, data formats, and distribution practices. As a result of this effort, LDC and ELRA have jointly released the Transnational English Database and negotiated data-sharing

[2]ANC Web site http://americannationalcorpus.org

arrangements, including the use of LDC data in the EC funded TC-STAR project, and the use of the EMILLE corpus, published by ELRA, for NSF-funded research on less commonly taught languages underway at LDC. LDC and ELRA have also collaborated in the development of the Open Language Archives,[3] a union catalog of linguistic resources held not only at these two organizations but also at two dozen other archives. Perhaps the most important outcome of Net-DC is a 40-hour Arabic broadcast news collection with time-aligned transcripts and a pronunciation lexicon that will be jointly published.

Another excellent example of international cooperation is the definition of a Basic Language Resource Kit, organized under the EU-funded ENABLER project, which joined teams from Europe but included input from the LDC and several groups from Asia. For many kinds of language technology, there now exist lists of required resources, including data types, processes, best practices, and formats.[4]

As described in Choukri and Mapelli (2003), technical, legal, and commercial prerequisites have to be taken into consideration for the production of language resources in a cooperative framework. To strengthen such cooperation, there is no doubt that an effort in coordinating this cooperation is required.

A coordinated operation was launched in the framework of speech language resources with the creation of an International Committee for the Coordination and Standardization of Speech Databases and Assessment Techniques, COCOSDA.[5] COCOSDA was established to encourage and promote international interaction and cooperation in the foundation areas of spoken language processing, especially for speech input/output. COCOSDA promotes the development of distinctive types of spoken language data corpora for the purpose of building and/or evaluating current or future spoken language technology, and offers coordination of projects and research efforts to improve their efficiency.

A new committee was also launched in 2003 in the field of written language resources—the ICWLRE (International Committee for Written Language Resources and Evaluation)—in agreement with the informal model of COCOSDA in the speech area. COCOSDA and ICWLRE have

[3] http://www.language-archives.org

[4] http://www.hltcentral.org/htmlengine.shtml?id=997

[5] http://www.cocosda.org

been working together in order to move toward a language resources and evaluation road map.

3.2.2 Collaboration within the European Union

On the European side, collaboration among member states has been strongly supported by the programs of the European Commission, which have launched calls for proposals for key players to interact under certain conditions. These conditions stated that at least 3 partners from 3 different countries participated in the submitted proposals. A principle behind these proposals was that of "coinvestment," which established that 50% of the incurred expenses would be covered by the Commission and the remaining 50% by the proposal applicants.

Collaboration among member states in the field of language resources is supported by the framework of the ERA-NET scheme of the sixth Framework Programme of the European Commission (2002–2006). Following the Fifth Framework Programme (FP5) of the European Community for research, technological development, and demonstration activities (1998–2002), the Sixth Framework Programme (FP6) has been aiming at strengthening the scientific and technological bases of industry and encouraging its international competitiveness while promoting research activities in support of other EU policies. With a budget of 17.5 billion euros for the years 2002–2006, it represents about 4 to 5% of the overall expenditure on research and technological development (RTD) in EU member states. The main objective of FP6 has been to contribute to the creation of the European Research Area (ERA) by improving integration and coordination of research in Europe.

The ERA-Net scheme is aimed at encouraging the coordination and cooperation of research at a national or regional level in the member and associated states through national and regional programs. It was implemented via an open call for proposals that took place in December 2002, welcoming proposals for coordination actions in any field of science and technology. The Commission pays all additional costs related to the coordination up to 100%. (Cf. Section 3.4 for further details on these funding schemes.)

Within this scheme, a number of government organizations have collaborated on a project proposal, whose objective is to coordinate European national programs in the field of human language technologies, addressing issues such as:

- Multilingual language resources identification (data and tools).
- Assesment of spoken and written language processing systems.
- Standards for language resources exchange.
- A language technology survey: programs, projects, products, actors, companies.

This collaboration has resulted in a proposal to the ERA-Net program termed "LangNet". The "LangNet" proposal is still under preparation and it takes into consideration four key issues:

- The availability of "portable" language resources to enable the development of usable applications: large corpora with sufficient coverage and quality; lexicons with morphological, syntactic, semantic, and pronunciation information; and so on.
- The assessment or evaluation of the quality of the developed technologies.
- The availability of standards, norms, and best practices to allow the interoperability and reusability of resources.
- The identification of existing national programs and developed technologies, and the promotion of awareness through a language technology portal.

These issues will be addressed at the same pace within all European countries to avoid a two-speed Europe, from a linguistic point of view. To achieve this goal, a collaboration will be established between member states and the European Commission through member-state funding of the production of language resources and general coordination of the action and infrastructure by the European Commission.

As presented in Choukri and Mapelli (2003), a number of national and cross-national initiatives are already in place. Contacts have been made with the corresponding representatives in several countries, and a number of them have confirmed their willingness to coordinate their national efforts and to share their experiences with others.

3.3 Data Collection Efforts in the United States

The past decade has seen significant effort devoted toward multilingual speech processing in the United States. Much of this effort has been associated with annual metrics-based technology evaluation projects administered by the National Institute for Standards and Technology (NIST) and sponsored by the Defense Advanced Research Projects Agency (DARPA). In the following sections some of the larger efforts—including the CallHome, CallFriend, Switchboard, Fisher, and Mixer collections—will be discussed.

Space limitations prevent us from giving adequate treatment to the efforts of several groups that have created important linguistics resources. The Center for Spoken Language Understanding (CSLU) at the Oregon Graduate Institute of Science and Technology has published 20 different corpora since 1992. Two are collections of telephone speech involving multiple languages: Eastern Arabic, Cantonese, Czech, English, Farsi, French, German, Hindi, Hungarian, Japanese, Korean, Malay, Mandarin, Italian, Polish, Portuguese, Russian, Spanish, Swedish, Swahili, Tamil and Vietnamese. The Johns Hopkins Center for Language and Speech Processing (CLSP) typically develops one or more databases each year through its annual summer workshops. The Institute for Signal and Information Processing (ISIP) of the Mississippi State University has contributed several important language resources, including software for echo cancellation, a speech recognition tool kit, a corpus of southern accented speech, the resegmentation of the Switchboard corpus, and JEIDA—a collection of prompted Japanese speech. ISIP resources are available from its home page or the Linguistic Data Consortium. Last but not least, the U.S. Military Academy's Center for Technology Enhanced Language Learning (CTELL) has created speech corpora in Arabic, Russian, Portuguese, American English, Spanish, Croatian, Korean, German, and French along the way to creating recognition systems to support language learning at West Point.

The Linguistic Data Consortium addresses the needs of researchers in speech and written language processing by licensing, collecting, creating, annotating, and sharing linguistic resources, including data, tools, standards, and best practices. Since its creation in 1992, the LDC has distributed more than 25,000 copies of more than 300 titles and otherwise shared data with 1,820 organizations in 93 countries. The LDC often serves as data coordinator for NIST technology evaluation and DARPA common task programs. This role is not assigned by fiat but is decided on the basis of competition. Although the LDC is located within the United States, its membership is open to researchers around the world. More than half of all LDC shipments have had destinations outside the United States. The LDC catalog lists all of the corpora LDC has released and may be searched by the languages of the corpus, types and sources of data included, recommended uses, research programs for which it is relevant, and release year.[6]

3.3.1 CallHome

The CallHome collections supported the large-vocabulary conversational speech recognition (LVCSR) program (NIST, 1997) in which researchers built systems to automatically transcribe large vocabulary, continuous speech, specifically conversational telephone speech. They represent perhaps the earliest effort to define and create a basic resource kit for developers of a specific linguistic technology. For each CallHome language—English, Mandarin Chinese, Egyptian Colloquial Arabic, Spanish, Japanese, and German—at least 200 international telephone conversations, 20 to 30 minutes in duration, were collected. Subjects participated in a single call, speaking to partners about topics of their choosing. All calls originated within and terminated outside of the United States.

Parts of each of the calls in each language were transcribed orthographically and verbatim. If a call was designated as training or development test data, 10 minutes were transcribed. If a call was designated as evaluation data, only 5 minutes were transcribed. For English, Spanish, and German, verbatim, orthographic transcription was a straightforward, if challenging, task. However, Mandarin Chinese, Egyptian Colloquial Arabic, and Japanese presented additional challenges. Within the Japanese transcripts, word segmentation was performed by hand. The Mandarin Chinese transcripts were automatically segmented using software developed at the Linguistic Data Consortium. Egyptian Colloquial Arabic offered perhaps the largest challenge since it is not generally written. Researchers at the Linguistic Data Consortium needed to invent a writing system for this variety and then train native speakers to use it consistently before transcription could begin. The verbatim transcripts include every word and

[6] See http://www.ldc.upenn.edu/catalog

speech sound produced: acronymns, disfluencies, nonlexemes, repetitions, truncated words, restarts, coughs, lip smacks, and so on.

The CallHome data sets also include a pronunciation lexicon for each language. These lexicons contain an entry for each form that appears in the transcripts. The lexical entries include the surface form and its pronunciation, part-of-speech, and frequency within the transcripts. For languages with complex morphology, the lexical entry also contains a morphological analysis of the attested form. Where possible, the frequency information includes counts in other genres, for example, in news text.

The first CallHome corpora were published in 1996. By 1997, speech, transcripts, and lexicons for all six languages were available. The initial releases included 100 or 120 calls designated as training material. The remainder of the calls were held in reserve and used to construct test sets for NIST's annual LVCSR/HUB-5 technology evaluations. Once exposed, the test sets were released for general use between 1998 and 2003. The Egyptian Colloquial Arabic lexicon was updated in 1999. In 2002, the LDC released supplements to the Egyptian speech and text corpora.

3.3.2 CallFriend

The goal of the CallFriend project was to support research on language identification (LID) (NIST, 2003). CallFriend Phase 1 calls are similar to those of CallHome in that subjects chose both their partners and their topics, and conversations could last as long as 30 minutes. However, there are several major differences between the two collections, which reflect their different purposes. First, CallFriend targeted a smaller number of calls, 100, in a larger variety of languages: Arabic, Canadian French, American English from both the North and South of the United States, Farsi, German, Hindi, Japanese, Korean, Mandarin, Caribbean and South American Spanish, Tamil, and Vietnamese. Second, CallFriend calls all originated and terminated within North America so that the phonetic characteristics of conversations that might be used to distinguish, for example, English from Mandarin, would not be confused with differences in the telephone networks of the United States and China. CallFriend calls were not initially transcribed, and no pronouncing lexicons were created. However, over time, some of the Korean and Farsi CallFriend calls have been transcribed for other purposes, and Russian has been both collected and transcribed. All calls were audited to verify dialect and gender and to indicate noise, echo, or other distortion present in the call.

A second phase of CallFriend collection began in the fall of 2004. In many ways, CallFriend Phase 2 echoes the conditions under which the original CallFriend was collected. Calls are again limited to the continental United States. The collection targets 100 short telephone conversations, typically 10 minutes long, for each language, pairing subjects who know each other to discuss topics of their choosing. Phase 2 data will also be used in conjunction with the NIST LID technology evaluation program.

Phase 2 changes the focus, however, from supporting simple language identification to supporting language, dialect, and variety identification. The collection effort is divided into two tasks: language identification and dialect identification. The targets for the former are Farsi, Georgian, Hindi, Italian, Japanese, Korean, Punjabi, Russian, Tamil, and Vietnamese, seven of which appeared in the first phase.

The dialect identification collection focuses on varieties that are typically considered dialects in common parlance. Geographically conditioned dialect variants of English, such as those spoken in the United States, Britain, India, and Australia, will be given the same status as the Arabic "colloquials" or the Chinese dialects of Standard Mandarin (Putonghua), Southern Min (Minnan Hua in mainland China and Taiwanese in Taiwan), Cantonese (Yueyu or Guangdong Hua), and Shanghai Wu (Shanghai Hua). Clearly there are varying degrees of mutual intelligibility among the Chinese, Arabic, and English dialects. However, without knowing in advance how differences in mutual intelligibility affect language identification system performance scores, the LID community will treat them similarly on a first attempt.

The CallFriend Phase 1 corpora were published via LDC in 1996. Similarly, Phase 2 corpora will be published once they have been used in the NIST language recognition evaluations.

3.3.3 Switchboard-II

Following the success of the original Switchboard collection, the speaker recognition research community decided to conduct additional Switchboard-style collections. Switchboard-II was primarily collected at the Linguistic Data Consortium in five phases—three phases characterized by geographical diversity and two phases characterized by diversity of transmission type. Switchboard-II Phases 1, 2, and 3 focused on the Mid-Atlantic, Mid-Western, and Southern regions of the United States, respectively. Switchboard Cellular Phases 1 and 2 focused, respectively, on GSM and non-GSM cell phones used in the United States. In all of the Switchboard-II collections, subjects were required to participate in multiple phone calls, speaking to subjects whom they generally did not know about assigned topics. In all phases, subjects were asked to participate in 10 conversations and to vary the locations, phones, and handset types they used as much as possible. The first three collections were evaluated by the number of subjects enrolled such that the average number of calls per subject was 10. The final two phases established minimum numbers of subjects to have completed at least 10 calls. All five phases have been published through the LDC and together comprise 14,167 conversations involving 2,649 subjects. Although Switchboard-II conversations were conducted exclusively in English, the Switchboard protocol was an important source of influence in the creation of other protocols, and the data has helped advanced speaker-recognition technologies around the world through its use in NIST's speaker recognition evaluations, which bring together researchers from Europe, Africa, the Middle East, and Australia.

3.3.4 Fisher

The DARPA EARS program[7] sought to develop technologies that create accurate, readable transcripts of broadcast news (BN) and conversational telephone speech (CTS) in English, Chinese, and Arabic (DARPA, 2005). EARS research focused on three areas: novel approaches, speech-to-text (STT), and metadata extraction (MDE). MDE research sought to produce transcripts that indicated who said what when, identified disfluent speech, and tagged discourse structure so that the resulting rich transcripts could be either formatted for human use or processed by downstream systems. A combination of challenging annual goals and multidisciplinary approaches drove the program and requisite data collection forward aggressively. EARS researchers required raw audio, particularly of conversational telephone speech but also of broadcast news speech in

[7] DARPA (2005), EARS home page, http://www.darpa.nil/ipto/programs/ears

the three EARS languages, with accompanying time-aligned transcripts and metadata annotation in volumes not previously available.

The Fisher telephone-conversation collection protocol was created at the Linguistic Data Consortium to address the needs of the EARS research community. Previous collection protocols, such as CallHome and Switchboard-II, and the resulting corpora have been adapted to the needs of speech-to-text research but were in fact created for other technologies. CallFriend was originally designed to support language identification while Switchboard was meant to support both topic and speaker identification. Although the CallHome protocol and corpora were developed to support speech-to-text technology, they feature relatively small numbers of speakers making a single telephone call of relatively long duration to a friend or family member about topics of their choosing. This approach yields a relatively narrow vocabulary across the collection. Furthermore, CallHome conversations are challengingly natural and intimate. In contrast, under the Fisher protocol, a very large number of participants make a few calls of short duration to other participants whom they typically do not know about assigned topics. This maximizes inter-speaker variation and vocabulary breadth although it also increases formality relative to CallHome and CallFriend. Whereas protocols such as CallHome, CallFriend, and Switchboard relied on participant initiative, the Fisher protocol relies on the robot operator to drive the study. In the earlier protocols, the robot operator waited for incoming calls from participants. The CallHome and CallFriend robot operator used a single telephone line to call a single previously identified subject in response to an incoming call, and, in the case of Switchboard, the robot operator responded to an inbound call by serially calling any subjects who indicated availability at that time. In the Fisher protocol, the robot operator simultaneously accepted calls and, using twelve to eighteen lines, initiated calls to all subjects who indicated availability at a given hour, pairing them as soon as they responded. Fisher subjects needed only to answer their phones at the times they specified when registering for the study. Subjects were permitted to initiate calls when it was convenient for them.

Fisher subjects were given a topic to discuss, which changed daily in order to ensure broad vocabulary coverage and to give purpose to a conversation that otherwise joins two strangers with no especial reason to talk. Nothing prevented subjects from diverging from the topic though most chose to stay with it.

Fisher data collection began in December 2002 and continued through 2004. The Fisher English collections sought to recruit a balanced pool of subjects from a variety of demographic categories, including males and females from three age groups and four dialect regions—both native speakers of English and fluent nonnatives. Subjects also indicated their level of education, occupation, and competence in languages other than English. The study began with the assumption that subjects would make from one to three calls each. However, in order to facilitate the selection of evaluation test sets that included subjects not seen elsewhere in the study, there was a period of nearly four months in which the Fisher robot operator only placed calls to unique subjects. Once the initial evaluation sets were developed, this limitation was removed, and the rate of collection more than tripled from an average of 15 to an average of 54 calls per day. The second phase of Fisher collection emphasized the cost per hour of speech collected by removing the constraint on demographic balance and allowing subjects to participate in up to 20 calls each.

In order to recruit the volume and diversity of subjects desired for Fisher Phase I, it was necessary to automate recruitment and registration. The study was announced in newsgroups and Google banners related to linguistics, speech technologies, and job opportunities. LDC placed print advertisements in the largest markets of each of the major dialect regions. E-mail lists devoted to finding money-making opportunities for their members eventually discovered and publicized the study independently of LDC.

Subjects registered primarily via the Internet by completing an electronic form, though a small percentage called LDC's toll-free number. Prior to registering, subjects learned that their conversations would be collected for purposes of education, research, and technology development, that their identities would be kept confidential; and that they would be compensated per full-length call.

The initial pool of 40 topics was supplemented in November 2003 with 60 new topics that were used throughout November and December 2003 and all of the second phase of collection. Thus, each of the first set of topics was used eight or nine times. Each of the second set appeared once in Phase I and multiple times in Phase II.

One of the greatest challenges in the Phase I collection was recruiting enough subjects to keep up with the collection platform. Because Fisher subjects typically completed only one to three calls, the platform accepted and then retired subjects as fast as the recruiters could register them. Even though registration was almost completely automatic and recruiter teams could easily process up to 500 registrations per day, the number of actual registrations was consistently well below what the platform could handle. In Phase II, in which subjects could complete up to 20 calls, recruitment ceased to be the problem it was in Phase I.

During its first cycle of operation, the Fisher protocol produced as much data as had been collected at LDC in the previous 12 years combined. In just over 11 months, LDC collected 16,454 calls averaging 10 minutes in duration and totaling 2,742 hours of audio, that is 37% more than the most aggressive target set by sites and sponsors.

Fisher sought to collect CTS data from a representative sample of the U.S. population. The first parameter such studies generally attempt to balance is gender. Previous studies found that in the United States, females join more frequently and participate more fully, and so have struggled to attain a 60/40 female to male ratio. In Fisher Phase I, females made just 53% of all calls. Age variation was also an important variable in Fisher Phase I. Whereas previous collections of conversational telephone speech focused on college-student populations, in Fisher Phase I, 38% of subjects were aged 16–29, 45% were aged 30–49, and 17% were over 50.

In contrast with previous, regionally based Switchboard collections, Fisher subjects represent a variety of pronunciations, including U.S. regional pronunciations, non-U.S. varieties of English, and foreign-accented English. Using the major dialect boundaries drawn by William Labov's Phonological Atlas of North America (Labov, 2004), the Fisher project recruited subjects from the four major United States dialect regions: North, Midland, South, and West. However, the study also admitted speakers from Canada, non-native speakers of English, and speakers of other national varieties of English outside the United States.

While Fisher Phases I and II continued in the United States, complementary collections of Mandarin Chinese and Levantine Colloquial Arabic were underway. For the Arabic collection, some 13,100 Levantine subjects were recruited. Most were Lebanese, though more than 1,000 were Jordanian, and a small number of Syrian and Palestinian subjects were also represented. Levantine Colloquial Arabic may be further divided into Northern, Southern, and Bedouin dialects, all of which were represented.

The Levantine collection used the same Fisher protocol as the American English collection, and the robot operator was located in the United States. Levantine topics were not simply translated from the English topics but were modified for the subject population, with some topics removed and a few added. Levantine subjects were permitted to make up to three calls. About 25% of recruited subjects succeeded in participating in one or more calls of 5–10 minutes duration. The study produced more than 300 hours of conversational telephone speech, of which 250 hours have been transcribed.

The Mandarin Chinese collection was managed by researchers at the Hong Kong University of Science and Technology (HKUST) and differed from Fisher English and Arabic in a few important respects. Naturally, topics for the Mandarin collection were adapted from the English topics. Although subjects were permitted to do more, most participated in only one call. Finally, Mandarin calls were arranged by a human operator whereas the English and Arabic calls were managed by completely automatic means. The HKUST collection has yielded some 350 hours of conversational telephone speech, all transcribed, in which eight major regions of China are represented. The data has been used in EARS technology evaluations and will be published through the LDC.

Due to the success of the previous Fisher collections, LDC is now conducting a Fisher Spanish collection of up to 1,200 conversations in both Caribbean and non-Caribbean varieties of Spanish using the Fisher collection protocol and subjects living in the United States.

3.3.5 Mixer

Where the Fisher collections were designed to support research and development of speech-to-text systems, the Mixer collections were intended to support speaker recognition research and technology development with an emphasis on forensic needs. Two goals of such a system are text independence and channel independence. The definition of channel independence has recently broadened, and requirements have increased to include language independence as well. The NIST-sponsored technology evaluation programs SRE2004 and SRE2005 have embraced the new requirements for speaker recognition systems. Responding to these new needs, the Linguistic Data Consortium created the Mixer corpora of multilingual, cross-channel speech.

Mixer Phase I was a collection of telephone conversations targeting 600 speakers participating in up to 20 calls of at least six minutes duration. Like previous speaker recognition corpora, the calls feature a multitude of speakers conversing on different topics and using a variety of handset types. Unlike previous studies, many of the subjects were bilingual, conducting their conversations in Arabic, Mandarin, Russian, and Spanish as well as English. Some telephone calls were also recorded simultaneously via a multichannel recorder using a variety of microphones. Mixer relies on the Fisher protocol in which a robot operator initiates 18 calls at a time, leaving six lines open to receive calls, and pairs any two subjects who agree to participate at the same time. Mixer adds one refinement to the Fisher protocol, ordering its calls to increase the probability of pairing speakers of the same native language. Considering the pool of subjects who indicated availability in a given hour of the day, the Mixer robot operator called all speakers of Arabic first, Mandarin second, Russian third, Spanish fourth, and English last.

In previous call-collection projects of this kind, about half of all recruits failed to participate in the study and about 70% of those who did participate achieved 80% of the stated goals. To compensate for shortfalls in participation, more than 2,000 subjects were recruited, and performance goals were set 20–25% higher than needed. In addition to per-call compensation, incentives such as completion bonuses and lotteries successfully encouraged subjects to complete calls in foreign languages, via unique handsets or LDC's multimodal recording device. Candidates registered, via the Internet or phone, providing demographic data and an availability schedule, and describing the handsets on which they would receive calls. The personal information candidates provided for payment purposes is kept confidential and is not delivered with the research data.

In order to encourage meaningful conversation among subjects who did not know each other, 70 topics were developed after considering which had been most successful in previous studies. Topics ranged in breadth from "Fashionably late or reasonably early" to "Felon re-emancipation." Since Mixer required bilinguals, topics were balanced between those of domestic interest and those having international appeal. The robot operator selected one topic each day. Subjects had the ability to decline calls after hearing the topic of the day. Once a pair of subjects were connected, the robot operator described the topic of the day fully and began recording. Although subjects were encouraged to discuss the topic of the day, there was no penalty for conversations that strayed from the assigned topic. All calls were audited shortly after collection to ensure that the speaker associated with each unique identification number was consistent within and across calls; to log the language of the call; and to indicate the levels of background noise, distortion, and echo observed.

All call activity in Mixer contributed to a core collection in which the goal was 10 calls from each of 600 subjects. To support evaluations of the effect of volume of training data on system performance, it was desirable to have a group of subjects who completed not 10 but 20 calls. Subjects who appeared so inclined were encouraged to complete 25–30 calls through per-call compensation and a bonus for exceeding 25. Subjects who completed 30 calls were immediately deactivated from the study and compensated for their considerable effort. To support the development and evaluation of systems that recognize multilingual speakers regardless of the language they speak, LDC recruited subjects bilingual in English and either Arabic, Mandarin, Russian, or Spanish. These subjects were encouraged to make at least five calls in English and five in their other language.

Although the robot operator clustered its outbound calls by the native language of the subjects, the persistence of the robot operator coupled with the preponderance of subjects who did not speak the same non-English language allowed the core collection to race ahead of the foreign-language collection. "Language-only" days, in which the robot operator only allowed calls among speakers of the day's target language, helped guarantee that the study reached its foreign-language goals.

To support an enhanced definition of channel independence, one side of a series of Mixer conversations was collected using a variety of sensors. The sensors were chosen to represent a real-world environment, such as conference rooms, courtrooms, and cell phone use. Participants placed calls to the Mixer robot operator while being recorded simultaneously on the cross-channel recorder. Participants living near one of the three cross-channel collection centers—at LDC, Institute of Signal Processing (ISIP), and the International Computer Science Institute (ICSI), Berkeley—made at least five separate cross-channel recordings.

The recording system consisted of a laptop, a multichannel audio interface, two FireWire attached hard drives, a set of eight microphones/sensors, and a simple eight-channel recording application. The multichannel audio interface (MOTU 896HD) connected to the laptop via FireWire and handled eight balanced microphone connections, sampling each channel at

48 kHz with 16-bit samples. The multichannel sensors were: a side-address studio microphone (Audio Technica^{TM} 3035), a gooseneck/podium microphone (Shure^{TM} MX418S), a hanging microphone (Audio Technica^{TM} Pro 45), a PZM microphone (Crown Soundgrabber^{TM} II), a dictaphone (Olympus Pearlcorder^{TM} 725S), a computer microphone (Radio Shack^{TM} Desktop Computer Microphone), and two cellular phone headsets (Jabra^{TM} Earboom and Motorola^{TM} Earbud).

Mixer call collection began in October 2003, after approximately 200 participants had been recruited, and continued through May 2005. In all, 4,818 subjects were recruited, of which 63% were female and 37% male. About 47% of all recruits actually completed at least one full-length, on-topic call—a typical participation rate for telephone speech studies. Subjects completed 14,442 total conversational sides (7,221 calls), of which 58% contain female speakers and 42% contain male speakers.

For Mixer Phases I and II, some 1,150 subjects completed 10 or more calls, 611 completed 20 or more calls, and 382 completed 30 calls. For the foreign-language goal, in which 100 speakers of each language were required to complete four or more calls in one of the Mixer languages other than English, 129 Arabic speakers, 115 Mandarin speakers, 107 Russian speakers, and 100 Spanish speakers met or exceeded the requirement. Two hundred subjects completed four or more cross-channel calls.

The Mixer corpora are the first large-scale, publicly available resources to support research and evaluation on multilingual and bilingual speaker recognition and channel independence, including a broad variety of microphone types. All of the data resulting from the Mixer collections will be used to support evaluations of speaker recognition systems as well as in 2004, 2005, and possibly 2006, and will be published via the Linguistic Data Consortium.

3.4 Data Collection Efforts in Europe

3.4.1 Funding Structures in the European Union

Within the European Union (EU), the production of language resources takes place both at the national and at the supranational level. The funding of supranational activities has a long tradition. Technical and scientific collaboration across national boundaries was established as a major goal by

Table 3.1 European research programs funding speech and language technology.

Program	Funding	Duration
ESPRIT 0	115 Million ECU	1983
ESPRIT 1	750 Million ECU	1984–1988
ESPRIT 2	1,600 Million ECU	1987–1992
ESPRIT 3	1,532 Million ECU	1991–1994
ESPRIT 4	2,084 Million ECU	1994–1997
1st Framework Programme	3,750 Million ECU	1984–1987
2nd Framework Programme	5,396 Million ECU	1987–1991
3rd Framework Programme	6,600 Million ECU	1990–1994
4th Framework Programme	11,879 Million ECU	1994–1998
5th Framework Programme	14,960 Million EURO	1999–2002
6th Framework Programme	17,500 Million EURO	2002–2006

the 1951 European Coal and Steel Community (ECSC) Treaty. The 1957 Euratom Treaty established the Joint Research Centre (JRC), the concept of cost-sharing contract research programs, and procedures for the coordination of national research projects. In 1983, the Council of the European Union approved the creation of so-called framework programs, that is four-year broad research programs with a number of subdivisions ranging from agricultural productivity development to research on new technologies. The underlying goal of framework programs is to fund projects submitted by multiple European partner institutions by sharing the cost between those institutions and the European Commission. Framework programs are currently in their sixth phase; each phase has included a certain level of funding for language resource-related activities. In addition, a number of precursor programs (the ESPRIT programs 0 to 4) addressed topics of interest for language engineering. Table 3.1 lists these programs along with their duration and level of funding.

3.4.2 Language Engineering Related Programs

Specific programs addressing language engineering issues have been funded within all programs listed above.

A well-known project within the second framework program was the EUROTRA project, which aimed at creating an advanced machine-translation system intended to deal with all official languages of the European Community by developing a limited prototype which could later be produced on an industrial scale. The project lasted about 10 years (1982–1992).

Within the third framework program, the so-called Language Resources and Evaluation (LRE) program focused on the development of linguistic resources and computational tools (software tools, grammars, dictionaries, terminological collections, text corpora); support for the formulation of standards and guidelines for the encoding and interchange of linguistic data); pilot and demonstration projects (machine translation; document abstracting and indexing; aids for mono- and multilingual document generation, storage, and retrieval; man-machine communication; construction of knowledge bases from natural language text; computer-aided instruction); and so on. The official objective was to develop basic linguistic technology that would be incorporated by the European industrials into a large number of computer applications, in which natural language is an essential ingredient.

Within the fourth framework program, the Language Engineering program was put into place, which focused on the integration of new oral and written language processing methods and general linguistic research aimed at improving possibilities for communicating in European languages.

The largest program within the Fifth Framework was the Information Society Technologies (IST) program, which included a Human Language Technology (HLT) component. Funding for HLT related research focused on the usability and accessibility of digital content and services while supporting linguistic diversity in Europe.

Finally, the Sixth Framework has introduced priority themes in the areas of software, computing and user interfaces that are directed at facilitating the interpretation of speech, vision, and gestures within multimodal human-computer communication.

3.4.3 National Projects Supporting Language Technologies

In addition to European initiatives, a number of national organizations within Europe have been working toward creating a significant body of language resources.

A selection of these is listed below.

France

In 1999, the Conseil Supérieur de la Langue Française appointed a committee dedicated to language processing called CTIL (Comité pour le Traitement Informatique de la Langue) whose task it is to issue guidelines regarding the development and usage of French language processing.

Based on CTIL reports, several research networks and specific language engineering projects were launched, the most important of which is the Technolangue program.[8] The main objective of this program is to provide language resources, standards, and evaluation methodologies that meet the needs of the current information society and language technology markets. An important action behind this program is the maintenance of its portal, where all results and information regarding the program are made public for the Human Language Technology (HLT) community.

Germany

Germany has funded several long-term R&D projects that have resulted in a number of language resources. The most well-known projects are Verbmobil and SmartKom. Verbmobil focused on machine translation of speech (German to Japanese, with English as a bridge language) while Smartkom has aimed at developing multimodal, multilingual human-human interaction. One of the resources resulting from these projects is the multimodal corpus Smartkom, produced by the Bavarian Archive for Speech Signals (BAS). The release SKP 2.0 of this corpus contains recordings of 96 German speakers recorded in public places (cinema and restaurant) in the SmartKom scenario, which is a traditional public phone booth but equipped with additional intelligent communication devices. The recorded modalities are the audio in 10 channels, the video of the face, the video of the upper body from the left, the infrared video of the display area (to capture the 2-D gestures) as input to the SIVIT device (Siemens gesture recognizer), the video of the graphical user interface, the coordinates of graphic tableau (when the pen was used), and the coordinates of the SIVIT device (when fingers/hands were used). Annotation files are also included with transliteration, 2-D gesture, user states in three modalities, and turn segmentation.

[8] http://www.technolangue.net

Italy

In Italy, two national projects have been carried out within two different programs aimed at extending core resources developed in the EU projects and creating new language resources and tools. Specific objectives include the development of a natural language processing platform and technology transfer to small and medium enterprises (SMEs). SI-TAL (Sistema Integrato per il Trattamento Automatico del Linguaggio), with 13 partners, including both research centers and SMEs; and LCRMM (Linguistica Computazionale: Ricerche Monolingui e Multilingui), with 16 partners made up of private and public organizations. Among the language resources produced within these projects, are the following speech data: the Spoken Italian Corpus of 100 hours of speech, annotated dialogues for speech interfaces (human-human and human-machine interactions), and software for the fast integration of speech technology into telephone and Web applications.

Netherlands and Belgium

The Dutch HLT Platform comprises all organizations involved in HLT in the Dutch speaking area of Europe. This platform aimed at strengthening the position of HLT in the Netherlands and the Flemish part of Belgium and was instigated by the intergovernmental organization NTU. It involves the following Flemish and Dutch partners: the Ministry of the Flemish Community/AWI, the Flemish Institute for the Promotion of Scientific-Technological Research in Industry (IWT), the Fund for Scientific Research (FWO-Flanders), the Dutch Ministry of Education, Culture and Sciences (OcenW), the Dutch Ministry of Economic Affairs (EZ), the Netherlands Organization for Scientific Research (NOW), and Senter (an agency of the Dutch Ministry of Economic Affairs). Important outcomes of this platform are both the spoken Dutch corpus and the HLT agency devoted to the extension and maintenance of the Dutch LRS.

Norway

Other efforts worth mentioning are the Norwegian Language Bank, containing spoken data, text, and lexical resources. This project was initiated by the Norwegian Ministry of Culture and Religious Affairs and the Ministry of Commerce and Industry, and coordinated by the Norwegian Language Council.

Spain

In Spain, a number of programs are in progress, currently under the organization and guidance of the Ministerio de Educación y Ciencia, the Ministerio de Industria, Turismo y Comercio and the Ministerio de Fomento, as well as the Centro para el Desarrollo Tecnológico Industrial. These include projects such as RILE (Spanish Language Resources Server), BASURDE (Spontaneus-Speech Dialogue System in Limited Domains), PETRA (Speech Interfaces for Advanced Unified Messaging Applications), ALIADO (Speech and Language Technologies for a Personal Assistant), and OPENTRAD (Open Source Machine Translation for the Languages of Spain).

3.4.4 European Language Resources Association (ELRA)

ELRA's Background, Mission, and Activities

Research and development efforts in language technologies rely heavily on the availability of large-scale Language Resources (LRs), together with appropriate standards, methodologies, and tools. In order to define a broad organizational framework aiming at the creation, maintenance, and dissemination of such LRs, the European Commission launched the project RELATOR. The major outcome of this project was the decision to set up an independent, permanent, centralized nonprofit membership-driven association. In 1995 this association was founded under the name of "The European Language Resources Association (ELRA)." ELRA's board consists of twelve members, who are elected by an open vote of all ELRA members. The board defines the association's goals, the policy is implemented by the chief executive officer and his staff.

Since its foundation, ELRA broadened its mission to include the production and commissioning of production of LRs as well as the evaluation of language engineering tools technology.

Thus, ELRA's mission could now be outlined as follows: "The mission of the association is to promote language resources and evaluation for the HLT sector in all their forms and their uses, in a European context. Consequently, the goals are to coordinate and carry out identification, production, validation, distribution, standardization of LRs, as well as support for evaluations of systems, products, tools, etc., related to language resources."

ELRA's technical activities are directly linked to its mission within language technology. They can be broken down into the following subtasks and strategic activities:

1. Issues concerning storage and providing of language resources for the HLT community.
   - Maintenance of ELRA's catalogue
   - Development of the Universal Catalogue
   - Development of the Basic LAnguage Resource Kit (BLARK)
   - Development of the Extended LAnguage Resource Kit (ELARK)
   - Identification of useful LRs
   - Validation and quality assessment
   - Production or commissioning of the production of LRs
2. Administrative and legal issues regarding language resources.
   - Handling of legal issues related to the availability of LRs
   - Distribution of LRs and pricing policy
3. Involvement in the evaluation of human language technologies.
   - Evaluation projects with (Evaluation and Language resources Distribution Agency).[9]
4. Promotion of the language technology field.
   - Information dissemination, promotion and awareness, market watch, and analysis

3.4.5 Identification of Useful LRs

The efforts regarding the identification of LRs have been considerably increased at ELRA. In addition to the catalog of ready-for-distribution LRs, intensive work is taking place in the identification of all existing

[9] http://www.elda.org

LRs so as to compile them in a universal catalogue and provide all possible information to the language technology R&D community. The methods and procedures that are already used by ELDA for this identification task are:

- "LR watch" through ads, and announcements in mailing lists, such as Linguist list, Corpora, LN, and ELSNET.
- Hiring the service of internal and external experts with the mission of identifying LRs.
- Cooperation with institutions; participation in national, European, and international projects related to LR creation.
- Development of expertise on national, European, and international research programs and their relevant projects, including:
    - *European Community Initiatives and Programs*: INTAS, CORDIS, IST, COST, EUREKA
    - EU/International Networks and Associations: AFNLP, ALTA, ATILF, Dutch Language Union, ENABLER, ELSNET, EUROMAP, HLTCentral, LT World, NEMLAR, OntoWeb, SEPLN
    - LRs distribution agencies: LDC, CCC, ChineseLDC, GSK, Indian LDC, OGI, TELRI
    - *Information Portals*: COLLATE, ENABLER Survey
    - *Project Consortia*: National & International
    - *Metadata Infrastructure*: IMDI, INTERA, OLAC
    - *Standards*: ISLE, ISO/TC 37/SC 4
    - *International Coordination Committees*: ALRC, ICCWLRE, COCOSDA
- Participation in HLT events and conferences.
- Search for information in technical documentation and publications.
- Information about the existence of LRs by ELDA contacts.
- Maintaining ELRA's and ELDA's Web sites.
- Mouth-to-ear information.

3.4.6 Administrative and Legal Issues

One of ELRA's priority tasks is to simplify the relationship between producers/providers and users of LRs. In order to encourage producers

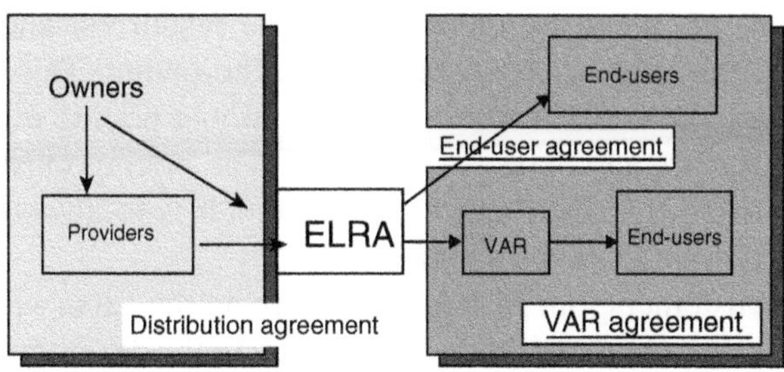

Figure 3.1: Contract model followed by ELRA.

and/or providers of LRs to make such data available to others, ELRA (in collaboration with legal advisors) has drafted generic contracts defining the responsibilities and obligations of both parties. Figure 3.1 shows the model on which these contracts are based. ELRA's effort to distribute existing LRs has certainly contributed toward an easier path for both academic and industrial partners to either distribute or acquire LRs. Furthermore, this effort has represented an active battle against the wasting of developed LRs that remain forgotten and are never used or shared.

ELRA contracts establish, among other things, how LRs can be used, for example, whether they can be used for both research and technology/product development or only for research purposes. Providers are protected by requiring that the user shall not copy or redistribute the LRs. The production and distribution of such data licenses is one of ELRA's contributions to the development of LR brokerage. These licenses are available on ELDA's Web site[10] and all interested actors are encouraged to use them.

The pricing policy is another issue requiring careful attention. Essentially, LRs are nowadays being traded like any other commodity, and their price is therefore subject to market demands as well as provider requirements. Market knowledge and contacts with potential providers allow ELRA to always have reliable and useful information on the demands and needs of the market. The ELRA approach is to simplify the price setting, to clarify possible uses of LRs, and to reduce the restrictions imposed by the producer.

The prerequisite for acting as a broker is that each purchase renders a payment, covering the compensation claimed by the owner of the resource.

[10] http://www.elda.org/article1.html

In general, ELRA is not the owner of the resources, and can therefore only set a fair price in cooperation with the owner. This cooperation in setting the price is often based on conventional pricing methods, such as production costs and expected revenues. The pricing must also take into account the ELRA distribution policy, which is to always try to offer a discounted price to its members.

Most customers join ELRA before buying the LRs (which is enforced by the pricing policy). ELRA's contribution to the development of research activities has seen considerable growth, and its involvement in research and commercial development is balanced and shows a substantial increase in the items distributed for R&D.

3.5 Overview of Existing Language Resources in Europe

3.5.1 European Projects

A number of projects in Europe have been working toward the production of multilingual speech and language resources, many of which have become key databases for the HLT community. Details on some of these speech projects follow.

The SpeechDat Family:

The SpeechDat projects[11] are a set of speech data-collection efforts funded by the European Commission with the aim of establishing databases for the development of voice-operated teleservices and speech interfaces. Most of the resulting databases are available via ELRA. These projects include:

- OrienTel: This project focused on the development of language resources for speech-based telephone applications for the Mediterranean and the Middle East, roughly spanning the area between Morocco and the Gulf States and including several variants of local German, French, English, Arabic, Cypriote Greek, Turkish, and Hebrew.

[11] http://www.speechdat.org

- SALA (SpeechDat Across Latin America) which can be divided into two further projects:
    - SALA-I: Fixed telephone network in Latin America: Speech databases were created for the purpose of training speech recognizers that performed well in any Latin American country. The databases covered all dialectal regions of Latin America that were representative in terms of Spanish and Portuguese language variants.
    - SALA-II: Mobile/cellular telephone network in Latin America, the United States, and Canada: Speech databases were created to train speech recognition systems for various cellphone applications in the Americas. The databases cover all dialectal variants of English, French, Portuguese, and Spanish languages represented in North and Latin America.

- SpeechDat-Car: This project focused on the development of ten in-vehicle and mobile telephone network databases, each of which contains 600 recording sessions. The ten languages covered were Danish, British English, Finnish, Flemish/Dutch, French, German, Greek, Italian, Spanish, and American English.
- SpeechDat(E): This project aimed to provide hitherto nonexistent resources for SpeechDat scenarios in Eastern European languages, such as Czech, Hungarian, Polish, Russian, and Slovak.
- SpeechDat(II): Twenty-five fixed and mobile telephone network databases and three speaker-verification databases were developed for Danish, Dutch, English, Flemish, Finnish, French, Belgian French, Luxemburgian French, Swiss French, German, and Luxemburgian German.
- SpeechDat(M): Eight fixed telephone network databases and one mobile telephone network database were developed for Danish, English, French, Swiss French, German, Italian, Portugese, and Spanish.

Further projects related to the SpeechDat family are:

- LILA: The goal of this recently started project is to collect a large number of spoken databases for training automatic speech recognition systems for the Asian-Pacific languages, such as those found in

Australia, China, India (including English), Indonesia, Japan, Korea, Malaysia, New Zealand, the Philippines, Taiwan, Thailand, Vietnam, etc. The data will be collected via mobile phone networks.
- SPEECON: The overall goal of SPEECON was to enable each partner of the consortium to produce voice-driven interfaces for consumer applications for a wide variety of languages and acoustic environments. The languages covered by the project are Cantonese (China, Hong Kong), Czech, Danish, Dutch (Belgium and Netherlands), Finnish, Flemish, French, German, Hebrew, Hungarian, Italian, Japanese, Korean, Mandarin Chinese, Russian, Spanish (American), Swedish, Swiss-German, Turkish, U.K.-English, and U.S.-English.

Other Key European and Collaborative Projects Producing LRs

A number of projects have targeted not only the development of LRs per se but also the development of multilingual technology for human-human or human-machine communication. Some of these are:

- C-STAR (Consortium for Speech Translation Advanced Research):[12] This is a voluntary international consortium of laboratories devoted to the development of spoken translation systems. The consortium became official in 1991 and since then has undergone three phases of collaborative research and development.
- CHIL (Computers in the Human Interaction Loop):[13] This project focuses on multimodal human interaction supported by computer services that are delivered to people in an implicit, indirect, and unobtrusive way. ELDA is involved in the activities related to LRs and evaluation.
- FAME (Facilitating Agents in Multicultural Exchange):[14] FAME pursued innovation in the areas of augmented reality, perception of human activities, and multiparty conversation modeling and understanding. In the context of the latter, language resources have been collected for Catalan and Spanish language technology, information retrieval, conversational speech understanding, robust multilingual

[12] http://www.c-star.org/

[13] http://chil.server.de/servlet/is/101

[14] http://isl.ira.uka.de/fame/

spontaneous speech recognition, and speech recognition using distant microphones.

- LC-STAR (Lexica and Corpora for Speech-to-Speech Translation Components):[15] This recently finished project focused on creating language resources for speech-to-speech translation (SST) components. The project has created lexica for thirteen languages (Catalan, Classical Arabic, Finnish, German, Greek, Hebrew, Italian, Mandarin Chinese, Russian, Slovenian, Spanish, Turkish, and U.S.-English) and text corpora and a demonstration system for three languages (Catalan, Spanish, and U.S.-English). These resources will be available shortly through ELRA.
- NESPOLE! (Negotiating through Spoken Language in E-Commerce):[16] This project, finished in 2002, aimed at providing a system capable of supporting advanced needs in e-commerce and e-service by means of speech-to-speech translation.
- TC-STAR (Technology and Corpora for Speech to Speech Translation):[17] The TC-STAR project is aimed at realistic speech-to-speech translation technology. ELDA is involved in the activities related to LRs and evaluation, including all components of a speech-to-speech translation system (speech recognition, machine translation, and speech synthesis). The project targets a selection of unconstrained conversational speech domains (political speech and broadcast news) and three languages: European English, European Spanish, and Mandarin Chinese.
- VERBMOBIL:[18] This German project targeted the development of a speech translation interface for human-human communication and involved the English, German, and Japanese languages.

3.5.2 Language Resources Archived and Provided by ELRA

As mentioned in Section 3.4.4, one of ELRA/ELDA's activities concerning the storage and providing of Language Resources for the HLT community is the maintenance of its catalog,[19] which provides detailed

[15]http://www.lc-star.com/
[16]http://nespole.itc.it/
[17]http://www.tc-star.org
[18]http://vermobil.dfki.de/
[19]http://www.elda.org/sommaire.php

information about all the resources available though ELRA. The catalog is provided with a search utility that allows the users to specify their needs or preferences, in terms of resource type and/or language, among other search possibilities. Alternatively, the interface also allows for the listing of the full catalog, organized by resource type. These resource types categorize the Language Resources within ELRA's catalogue into Multimodal/Multimedia, Speech, Written and Terminology LRs. The full list of available resources comprises the following:

Multimodal/Multimedia	Speech	Written	Terminology
3	284	238	278

where the 3 Multimodal/Multimedia LRs are included in the 284 Speech LRs since they contain speech data as well.

The Speech resources cover a wide variety of languages, such as Basque, Bulgarian, Chinese, Czech, Danish, Dutch, English, Estonian, Finnish, Flemish, French, German, Greek, Hebrew, Hungarian, Japanese, Korean, Italian, Latvian, Mandarin, Norwegian, Persian, Polish, Portuguese, Romanian, Russian, Slovakian, Slovenian, Spanish, Swedish, Turkish, and Ukrainian. Among these languages, the following language varieties are also considered: Cantonese and Mandarin Chinese; American and British English; English spoken by Dutch speakers; Belgian French; Swiss French; Luxembourgian French; Austrian German; Swiss German; Finnish Swedish; Mandarin from Shanghai; Spanish from Chile, Colombia, Mexico, and Venezuela or German spoken by Turkish speakers, etc.

Regarding ELRA's written resources, the following items can be found in the catalog, also with information on their production, description, languages covered, etc.:

Written Corpora	Monolingual Lexicons	Multilingual Lexicons
45	71	122

The languages covered by these written resources also aim to cover as wide a variety as possible. Written-language LRs are available for Arabic, Assamese, Bengali, Bosnian, Bulgarian, Czech, Chinese, Croatian, Danish, Dutch, English, Estonian, Finnish, French, German, Greek, Gujarati, Hindi, Hungarian, Icelandic, Italian, Irish, Japanese, Kannada, Kashmiri, Korean, Malay, Malayalam, Marathi, Oriya, Polish, Portuguese, Punjabi, Romanian, Russian, Serbian, Sinhala, Spanish, Swedish, Tamil, Telegu, Turkish, and Urdu. These also consider language varieties, such as Arabic from Lebanon, Mandarin Chinese, or Brazilian and European Portuguese.

In relation to the last broad resource type considered, ELRA distributes terminological resources for languages like Catalan, Danish, English, French, German, Italian, Korean, Latin, Polish, Portuguese, Spanish, and Turkish.

In addition to ELRA/ELDA's work on the maintenance of its catalog, work is also taking place toward the development of archiving tools with a different objective: the Universal Catalogue, the Basic Language Resource Kit (BLARK), and the Extended Language Resources Kit (ELARK):

- *The Universal Catalogue*: ELRA's emphasis and increased activity in the identification of LRs has also led to the development of the Universal Catalogue. Currently under development, it will be a repository for all identified LRs, where users will have access to relevant information about as many existing language resources as possible. The objective is to provide the HLT community with a source of information about existing LRs and those under development, which will help us avoid double investment in the development of the same LRs. The relationship between the Universal Catalogue and ELRA's Catalogue will be one of self-information, that is, the existing resources included in the Universal Catalogue will serve as a guide toward potential negotiations and acquisitions for ELRA's Catalogue.
- *The Basic Language Resource Kit (BLARK)*: Closely related to the Universal Catalogue is the concept of a Basic Language Resource Kit. The BLARK concept was defined by Steven Krauwer (Krauwer, 1998) and was first launched in the Netherlands in 1999, through ELSNET (*European Network of Excellence in Language and Speech*), supported by an initiative called Dutch Human Language Technologies Platform. Then, in the framework of the ENABLER thematic network (*European National Activities for Basic Language Resources – Action Line: IST-2000-3-5-1*), ELDA published a report defining a (minimal) set of LRs to be made available for as many languages as possible and identifying existing gaps to be filled in order to meet the needs of the HLT field. A good illustration of the BLARK concept is the work on Arabic resources within the NEMLAR project[20] (Krauwer et al., 2004).

[20]http://www.nemlar.org/

- *The Extended Language Resource Kit (ELARK)*: This is an extended version of the BLARK. The ELARK will comprise Language Resources for those languages that require, and are able to support more advanced processing.

3.5.3 Language Resources Produced by ELRA

In addition to its role in the identification, archiving, and distribution of LRs, ELRA also plays an active role in the production and commissioning of language resources. For example, ELRA's involvement in the production of spoken language resources includes the following projects:

- CHIL: production of LRs appropriate for evaluation tasks: transcription of English-speaking seminars and annotations of video recordings.
- C-ORAL-ROM: preparation of the specifications of the validation criteria for the C-Oral-Rom database package and annotations of video recordings.
- OrienTel: handling the recordings for Morocco and Tunisia, including three linguistic variants: French, Modern Standard Arabic (MSA), and Modern Colloquial Arabic (MCA) and annotations of video recordings.
- Neologos: creation of two speech databases for the French language; a children-voice telephone DB called PAIDIALOGOS and an adult-voice telephone DB called IDIOLOGOS and annotations of video recordings.
- Speecon: production and supervision of production of the speech DBS for French, Italian, and Swedish languages and annotations of video recordings.
- TC-STAR: audiovisual material has been recorded in the European Parliament and agreements are being negotiated to obtain the usage rights of the recordings of the European and Spanish Parliaments. Speech data have also been transcribed.
- ESTER (evaluation campaign): recordings from French radio stations were carried out within the evaluation campaign organized within the framework of Technolangue and annotations of video recordings.
- MEDIA (evaluation campaign): recording and transcription of a speech database containing 250 adult speakers in a Wizard of Oz scenario.

Chapter 4

Multilingual Acoustic Modeling

Tanja Schultz

4.1 Introduction

The aim of an **automatic speech recognition (ASR)** system is to convert speech into a sequence of written words $W = w_1 w_2 \ldots w_n$. A typical system consists of four major components, as illustrated in Figure 4.1. These are (1) the **acoustic model** $P(\mathbf{X}/W)$, which models the sound units of a language based on speech features **X** extracted from the speech signal beforehand; (2) the pronunciation dictionary, which usually describes the pronunciation of words as a concatenation of the modeled sound units; (3) the language model, which estimates the probability of word sequences $P(W)$, and (4) the decoder, which efficiently searches the huge number of possible word sequences and selects the most likely word sequence W^* given the spoken input: $W^* = \underset{W}{\operatorname{argmax}}\, P(\mathbf{X}|W) \cdot P(W)$.

In the last decades, the fundamental technologies of speech processing have been dominated by data-driven approaches that use statistical algorithms to automatically extract knowledge from data. For language

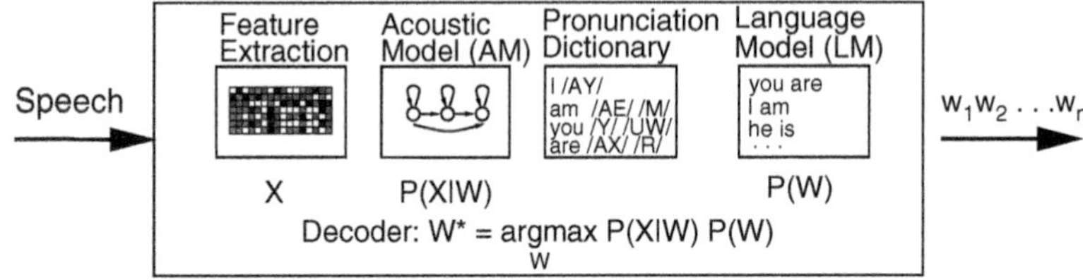

Figure 4.1: Automatic speech recognition (ASR).

modeling, the most successful techniques rely on statistical N-gram language modeling (as discussed in Chapter 6); for acoustic modeling, the predominant paradigm are Hidden Markov Models.

Hidden Markov Models (HMMs)

The purpose of an HMM λ is to model the probability $P(\mathbf{X}|W)$. This is done by two embedded processes: one stochastic process that produces a state sequence and a second stochastic process that produces a sequence of output observation symbols according to a probabilistic function associated with each state. While the latter process is directly observable, the former underlying stochastic process cannot be observed, hence the name Hidden Markov Model. Figure 4.2 depicts a three-state left-to-right HMM typically used in speech recognition, and shows the generation of an observation sequence $O = o_1 o_2 \dots o_T$.

An HMM λ is defined by five components:

- A set $S := \{S_1, S_2, .., S_N\}$ of N HMM states
- A probability distribution π that assigns a probability to each state S_i to be the initial state q_1 of a state sequence $\pi_i = P(q_1 = S_i),\ i = 1 \dots N$
- A matrix $\mathbf{A} = (a_{ij})$ of state-transition probabilities, where $a_{ij} = P(q_t = S_j \mid q_{t-1} = S_i),\ i,j = 1 \dots N$ describes the probability to transition from state S_i to S_j
- A set of K observation symbols $V := \{v_l, v_2, \dots, v_K\}$ to be emitted per time frame by the observable stochastic process
- A matrix $\mathbf{B} = (b_j(k))$ of emission probabilities, where $b_j(k) = P(o_t = v_k \mid q_t = S_j), j = 1 \dots N, k = 1 \dots K$ is the probability of emitting the observation $o_t = v_k$ in state S_j

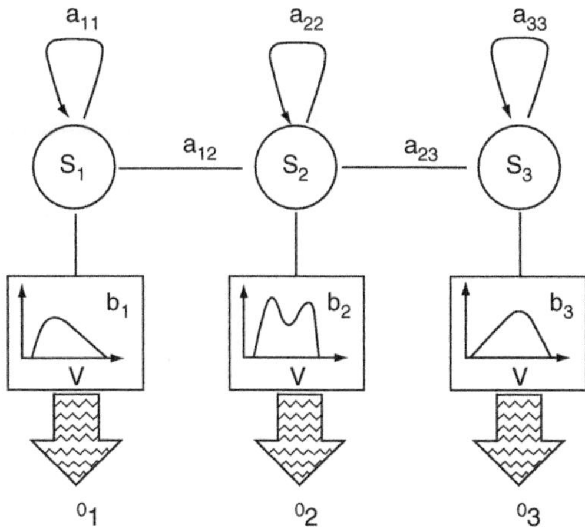

Figure 4.2: Generation of an observation sequence $O = o_1 o_2 \ldots o_T$ with a three-state left-to-right Hidden Markov Model.

In **discrete HMMs**, the observations come from a finite set of symbols. In this case, the emission probabilities $b_j(k)$ are discrete output distributions. In a **continuous HMM**, the observations are drawn from a continuous space. The output probabilities are therefore continuous probability density functions $b_j(x)$, usually defined by multivariate Gaussian mixture density functions

$$b_j(\boldsymbol{x}) = \sum_{l=1}^{L_j} c_{jl} \cdot N(\boldsymbol{x} | \mu_{jl}, \Sigma_{jl}), \quad \sum_{l=1}^{L_j} c_{jl} = 1 \tag{4.1.1}$$

where L_j denotes the number of Gaussian mixture components and c_{jl} is the weight of the lth mixture in state S_j. A single **Gaussian density function** $N(\boldsymbol{x}|\mu, \Sigma)$ is defined by a mean vector μ and the covariance matrix Σ:

$$N(\boldsymbol{x}|\mu, \Sigma) = \frac{1}{\sqrt{(2\pi)^d |\Sigma|}} \cdot e^{\frac{1}{2}(\boldsymbol{x}-\mu)^T \Sigma_i^{-1} (\boldsymbol{x}-\mu)} \tag{4.1.2}$$

The most commonly used HMM in speech processing is the first-order HMM. Here the assumption is made that the probability of being in state

$q_t = S_i$ only depends on the preceding state $q_{t-1} = S_j$. The second assumption of HMMs is the output-independence assumption. This means that the probability of emitting a particular symbol at time t only depends on state $q_t = S_i$ and is conditionally independent of the past observations. These two assumptions allow the formulation of very efficient algorithms to solve the three problems of acoustic modeling, which has resulted in the enormous success of HMMs in speech recognition. These three problems are the evaluation, decoding, and learning problem for given acoustic models and observed speech feature sequences.

1. The Evaluation Problem: What is the probability $P(O|\lambda)$ to generate an observation sequence $O = o_1 o_2 \ldots o_T$ for a given HMM $\lambda = (\mathbf{A}, \mathbf{B}, \pi)$?
2. The Decoding Problem: What is the state sequence q^* that most likely generated the observation sequence $O = o_1 o_2 \ldots o_T$ for a given HMM $\lambda = (\mathbf{A}, \mathbf{B}, \pi)$?
3. The Learning Problem: For a given HMM $\lambda = (\mathbf{A}, \mathbf{B}, \pi)$ and a set of observations O, how can we adjust the model parameters that maximize the probability to observe $O : \lambda^* = \underset{\lambda}{\operatorname{argmax}} P(O|\lambda)$?

HMMs have been extensively studied and described in the context of speech processing. The interested reader is referred to more comprehensive introductions such a Rabiner and Juang (1993) and Huang et al. (2001).

Context-Dependent Acoustic Modeling

Among the well-known factors determining the accuracy of speech recognition systems are variations due to environmental conditions, speaking style, speaker idiosyncrasies, and context. The main purpose of an accurate acoustic model $P(\mathbf{X}|W)$ is to cope with these variations. Accuracy and robustness are the ultimate performance criteria. The modeling units to represent acoustic information have to be chosen such that they are accurate, trainable, and generalizable. Therefore, the recognition of large vocabularies requires that words are decomposed into sequences of subwords. Conventionally, these subwords are phones. The decompositions are described in the pronunciation dictionary (see Chapter 5). Consequently, $P(\mathbf{X}|W)$ needs to consider speaker variations, pronunciation variations, context-dependent phonetic coarticulation effects, and environmental variations.

It is well known that recognition accuracy improves significantly if phones are modeled depending on the context. This context usually refers to the left and right neighborhood of phones. A **triphone model** is a phonetic model that takes both the left and the right neighboring phones into consideration. A **quinphone** considers two neighboring phones to the left and two to the right. A **polyphone** is a model that considers the left and right context of arbitrary length. Since context-dependent models increase the number of parameters, their trainability becomes a limiting factor.

In order to balance the trainability and accuracy, parameter-sharing techniques are applied. For this sharing, we take advantage of the fact that many phones have similar effects on their neighboring phones. Consequently, the goal of *context clustering* is to find similar contexts and merge them. This can be further generalized to share parameters at the subphonetic level. Here, the state in phonetic HMMs is taken as the basic subphonetic unit. While one aspect of model sharing is the parameter reduction to improve the trainability and efficiency, another aspect is to map unseen polyphones of the test set to representative polyphones that are appropriately trained. Conventionally, this is done by **decision trees** that classify an object by asking binary questions. The question set usually consists of linguistic questions, for instance, "Is the left context a plosive?" The decision tree can be automatically built by selecting questions for each node that best split the node. Usually entropy reduction or likelihood increase is chosen as a splitting criterion. In most cases, the number of splits is determined by the amount of training data. After the building process, this decision tree can be applied to unseen contexts by traversing the tree from top to bottom along the linguistic questions and identifying the best matching leaf node.

Figure 4.3 shows the first two steps in the generation of a decision tree for the second (middle) state of a quinphone HMM for phone /k/. In the beginning (left-hand side of Figure 4.3), all quinphones found for /k/ in the training material are collected in the root node of the tree. In this example, we have four quinphones coming from the contexts /l f k o n/, /l f k a u/, /w e k l a/, and /j a k r a/. In the first step (middle part of Figure 4.3), the question that results in the highest entropy reduction is selected from the question set. In our example, it is the question of whether the left context of /k/ is a vowel. The tree node is split into two successor nodes. The contexts /l f k o n/ and /l f k a u/ are assigned to the

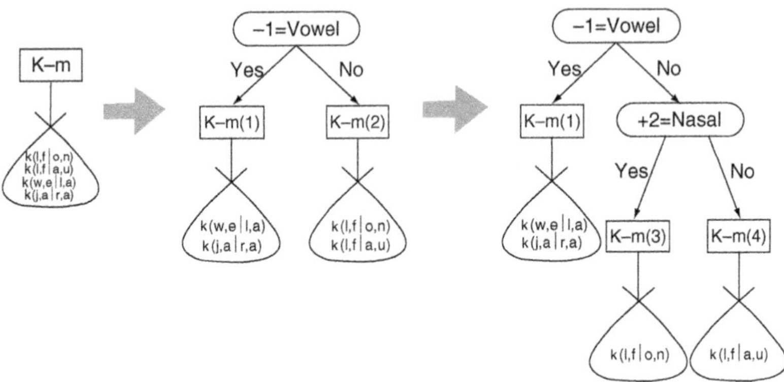

Figure 4.3: Context decision tree for the middle state of a quinphone HMM for phone /k/.

"no" successor node (left context of /k/ is not a vowel) and the contexts /w e k l a/ and /j a k r a/ are assigned to the "yes" node. In the second step (right-hand side of Figure 4.3), the question about the second right context gives the highest entropy reduction, and the quinphones are separated accordingly. During decoding, the decision tree is traversed from top to bottom, finding the most appropriate leaf node. If, for instance, the unseen context /o f k a n/ occurs in testing, K-m(3) will be taken to model this particular quinphone of /k/ (left context of k is not a vowel, second context to the right is nasal).

Typically, the implementation of the core algorithms such as HMM modeling or decision-tree clustering are independent of the language to be recognized. When a recognition system is developed for a new language, all core technology components remain untouched, only the knowledge components are customized to the language in question—that is, the acoustic model, the pronunciation dictionary, and the language model trained on a particular language. Many groups showed that speech recognition systems originally developed for one language have been successfully ported to several other languages, including IBM (Cohen et al., 1997), Dragon (Barnett et al., 1995), BBN (Billa et al., 1997), Cambridge University (Young et al., 1997), Philips (Dugast et al., 1995), MIT (Glass et al., 1995), LIMSI (Lamel et al., 1995), and ISL (Schultz and Waibel, 2001). The transformation of English recognition systems to such diverse languages as French, Japanese, and Mandarin Chinese indicates that the core speech technology components generalize across languages and that similar modeling assumptions hold for various languages.

4.1.1 Multiple Language Systems versus Multilingual Systems

The success of sharing the technology components across languages developed into the idea to share knowledge components across languages as well. As can be seen from Figure 4.4, this knowledge sharing can happen on three levels: the acoustic model, the pronunciation dictionary, and the language model. This chapter focuses on the acoustic model aspect. Issues with respect to multilingual pronunciation dictionaries are discussed in Chapter 5, and multilingual language modeling is addressed in Chapter 6.

We refer to a system that is designed to recognize speech from one particular language at a time as a **monolingual recognizer**. In such a system, the acoustic model is trained solely on data from that language. By training and evaluating monolingual recognizers in multiple languages, we can investigate the differences between languages on levels such as phonetics, phonology, or morphology. We can furthermore highlight resulting challenges for speech recognition in multiple (even less familiar) languages, and extrapolate the amount of effort necessary to port a speech recognition system to a new target language.

While in a monolingual recognizer all knowledge sources are specialized to one particular language, a **multilingual recognizer** is capable of simultaneously recognizing several languages by sharing knowledge

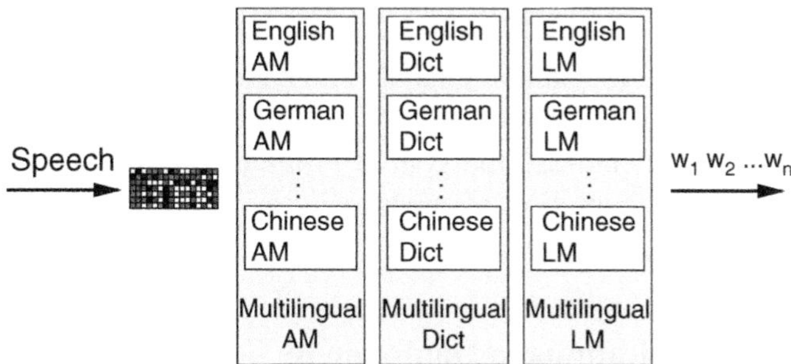

Figure 4.4: Multilingual ASR system.

sources across languages, as shown in Figure 4.4. Multilingual recognition systems that have a multilingual acoustic model consist of either a collection of coexisting subsets of **language dependent** acoustic models each addressing one language or a combination of **language independent** acoustic models that were trained by sharing data across languages. The latter requires the definition of a global unit set that describes which data can be shared across languages.

4.1.2 Compact and Maintainable Systems

The conception of a multilingual speech recognizer has several benefits. First, it allows several languages to be recognized simultaneously, which means the system can be exposed to speech from various languages, such as in public information kiosk systems. If the set of languages is closed but the identity of the actual spoken language is unknown, the system performs an implicit language identification. Another typical application is voice dialing, in which proper names of several languages need to be recognized, and merging the acoustic models of the languages in question can lead to significant improvements (see Chapter 9). Multilingual recognizers furthermore enable code-switching, which is when a bilingual speaker switches from one language to another. Second, a multilingual system eases the burden of system maintenance since only one acoustic model component has to be maintained independent of the number of languages involved. A third benefit, which has recently become very important in the context of mobile personal devices such as PDAs and cell phones, is that a multilingual system may save CPU and memory resources. This is accomplished by significantly reducing the overall parameter size of the acoustic models via sharing models across the languages.

In summary, the major goal of language independent modeling for acoustic model combination targets compact and maintainable multilingual systems. In contrast, the goal of **language adaptive** modeling is the efficient, rapid adaptation of preexisting acoustic models toward a new target language.

4.1.3 Rapid Language Deployment

Although speech recognition systems have been successfully ported to several languages, the majority of extensions has so far been limited to the most widespread languages from among the most extensively studied language groups, where huge amounts of spoken and written data have been accumulated over time. Building a large vocabulary speech recognition system usually requires data that include (1) dozens of hours of recorded and transcribed speech in the targeted language to train the acoustic models; (2) a large vocabulary, on the order of 100,000 words; (3) a matching pronunciation dictionary to guide the decoding procedure; and (4) a huge corpus of appropriate text data for language modeling, containing hundreds of millions of words. The lack of such data is one of the major bottlenecks in the development process of speech recognizers. Furthermore, since the collection of these data is an extremely time- and cost-consuming process, it also hampers the rapid deployment of speech recognition systems in highly demanded languages.

Unfortunately the assumption that large speech databases can be provided on demand does not hold, for several reasons. As described in Chapter 2 more than 6,900 languages exist in the world. Due to the tremendous costs (see Chapter 3), only a very small selection of languages would ever be able to be collected. However, the selection criteria may rapidly change with the political and economic situation. Furthermore, robust speech driven applications are built for various tasks, domains, languages, and accents. Therefore, combinatorial reasons prohibit data collections that satisfy all requirements.

As a consequence, recent research focuses on methods and techniques to rapidly deploy speech recognition systems to new languages using a considerably reduced amount of data resources. Examples for such research projects are Avenue (Lavie et al., 2003), which focuses on low-density languages such as Mapudungun; Babylon and TransTac (BABYLON, 2003), which push for faster development cycles and field trials in new languages; and Spice (SPICE, 2005; Schultz, 2004), which develops tools to enable rapid development of speech processing systems in new languages.

4.2 Problems and Challenges

The development of a multilingual speech recognizer applying multilingual acoustic models poses several problems and challenges. One reason derives from the nature of the task, namely to merge knowledge sources across multiple languages. This either has to be done by the close cooperation of various language and technology experts, which often results in rather heterogeneous solutions, or is performed by a single expert who then has to familiarize him/herself with all languages in question.

Another reason is the large variety of language properties which make the comparison and merging process very challenging. These properties affect all linguistic characteristics of languages, namely the written form, the sound system, prosodic and phonological features, the relation between letters and sounds, the presence or absence of a segmentation of the written text into useful units, and the morphology of the language. All these factors have a significant impact on the task of multilingual speech processing, and most of them, on multilingual acoustic modeling.

4.2.1 Lack of Resources: Uniform Data in Multiple Languages

Techniques to create multilingual acoustic models require multilingual speech and text databases that cover many languages and that are uniform across languages. Uniformity here refers to the quantity as well as to the quality of data. Quantitative uniformity means that the total amount of text and audio should be balanced across languages. The audio data should cover a reasonable number of speakers with a representative distribution in demographics, such as gender, age, place of origin, and education. The amount of text data should allow for reliable estimates for language modeling or grammar development. The quality of data refers to aspects such as recording conditions (noise, channel, environment), recording equipment (microphone, soundcard), the collection scenario (task, setup), the speaking style (read, planned, conversational, sloppy, formal), and transcription conventions.

For an ideal multilingual acoustic model, the data should provide a good representation of the languages of the world. Unfortunately, the majority of languages are spoken by less than 100,000 speakers, and only about 150 languages (3%) have more than 1 million speakers. This makes it unreasonable to consider all languages, but how do we come up with a good language selection? What are good selection criteria? Of course, this depends on the research target or the application. Relevant criteria for academic and commercial purposes could be the size of the speaker population, the economical or political relevance of a country or language, the desire to preserve and foster minority languages, the necessity to reach out to less connected and/or rapidly developing countries and communities, or simply the geographical coverage in the world.

For the development of language independent sound unit sets, it is desirable to cover the phonetical, phonological, and morphological aspects of languages. For text-related processing, it is important to cover orthographic variations and segmentation issues. Ideally, we would have a database that covers the most relevant languages of the world, with full coverage of the phonetic inventory that human beings use. The collected speech would come from a representative group of speakers and would be fully transcribed on at least a (segmented) word level. Per language, about 10,000 utterances, or 100,000 words spoken by 100 different speakers, are considered to be the minimal amount to guarantee a relatively robust speaker-independent acoustic model. The text material should comprise millions of words to allow a good estimate for language model probabilities. For comparison across languages, it is desirable to get semantically equivalent text corpora that are comparable with respect to vocabulary size and domain.

In the late 1990s, it became obvious that research in the area of multilingual speech processing was considerably impeded by the lack of multilingual text and speech databases (Lamel et al., 1996; Young et al., 1997; Adda-Decker, 1999). As a consequence, research institutions and data consortiums, such as the Linguistic Data Consortium (LDC) in the United States (LDC, 2000) and the Evaluations and Language Resources Distribution Agency (ELDA) in Europe (ELDA, 2005) initiated and coordinated several multilingual data collections. An overview of available data and ongoing data-collection efforts is given in Chapter 3.

The GlobalPhone Database

One database collection effort with the special emphasis on fostering multilingual speech recognition research is the GlobalPhone project (Schultz and Waibel, 1997b; Schultz, 2002). The GlobalPhone corpus provides transcribed speech data for the development and evaluation of **large vocabulary continuous speech recognition** systems (LVCSR) in the most widespread languages of the world. GlobalPhone is designed to be uniform across languages with respect to the amount of text and audio per language, the audio data quality (microphone, noise, channel), the collection scenario (task, setup, speaking style), and the transcription conventions. As a consequence, GlobalPhone supplies an excellent basis for research in the areas of (1) multilingual speech recognition, (2) rapid deployment of speech processing systems to new languages, (3) language and speaker identification tasks, (4) monolingual speech recognition in a large variety of languages, and (5) comparisons across major languages based on text and speech data.

To date, the GlobalPhone corpus covers 17 languages: Modern Standard Arabic (AR), Chinese-Mandarin (CH), Chinese-Shanghai (WU), Croatian (KR), Czech (CZ), French (FR), German (GE), Japanese (JA), Korean (KO), Brazilian Portuguese (PO), Polish (PL), Russian (RU), Spanish (SP), Swedish (SW), Tamil (TA), Thai (TH), and Turkish (TU). This selection covers a broad variety of language characteristics relevant for speech and language research and development. It comprises widespread languages (Arabic, Chinese, Spanish), contains economically and politically important languages (Korean, Japanese, Arabic), and spans over wide geographical areas (Europe, America, Asia). The speech covers a wide selection of phonetic characteristics, such as tonal sounds (Mandarin, Shanghai, Thai), pharyngeal sounds (Arabic), consonantal clusters (German), nasals (French, Portuguese), and palatalized sounds (Russian). The written language covers all types of writing systems, including logographic scripts (Chinese Hanzi and Japanese Kanji), phonographic segmented scripts (Roman, Cyrillic), phonographic consonantal scripts (Arabic), phonographic syllabic scripts (Japanese Kana, Thai), and phonographic featural scripts (Korean Hangul). The languages cover many morphological variations, such as agglutinative languages (Turkish, Korean), compounding languages (German), and scripts that completely lack word segmentation (Chinese, Thai).

The data acquisition was performed in countries where the language is officially spoken. In each language, about 100 adult native speakers were asked to read 100 sentences. The read texts were selected from national newspaper articles available from the Web to cover a wide domain and a large vocabulary. The articles report national and international political news, as well as economic news mostly from the years 1995–1998. The speech data was recorded with a Sennheiser 440-6 close-speaking microphone and is available in identical characteristics for all languages: PCM encoding, mono quality, 16-bit quantization, and 16 kHz sampling rate. The transcriptions are available in the original script of the corresponding language. In addition, all transcriptions have been romanized, that is, transformed into Roman script applying customized mapping algorithms. The transcripts are validated and supplemented by special markers for spontaneous effects like stuttering, false starts, and nonverbal effects, such as breathing, laughing, and hesitations.

Speaker information, such as age, gender, occupation, etc., as well as information about the recording setup complement the database. The entire GlobalPhone corpus contains over 300 hours of speech spoken by more than 1,500 native adult speakers. The data is divided into speaker disjoint sets for training, development, and evaluation (80:10:10) and are organized by languages and speakers (Schultz, 2002).

The GlobalPhone corpus has been extensively used for studies comparing monolingual speech recognition systems for multiple languages, multilingual acoustic model combination, rapid language adaptation, experiments on multilingual and cross-lingual articulatory features, nonverbal cues identification based on multilingual phone recognizers (speaker identification, language identification, accent identification), non-native speech recognition, grapheme-based speech recognition, as well as multilingual speech synthesis, and is distributed by ELDA (2005).

4.2.2 Language Peculiarities

The characteristics of a language such as the letter-to-sound relation, the sound system, or phonological and morphological features have a large impact on the process of building a multilingual speech processing system. This section describes some of the resulting challenges, especially with respect to multilingual acoustic modeling.

Written Script and Relation to Spoken Speech

According to Weingarten (2003), there exist hundreds of different writing systems in the world, but a surprisingly large number of languages do not have a written form at all. For these languages, a textual representation has to be defined if they are targeted for speech recognition. Such new written standards have recently been defined for languages and dialects like Pashto (PASHTO, 2005), Egyptian Arabic (Karins, 1997), and Iraqi Arabic within the context of research project such as CallHome, Babylon or TransTac, which address rapid deployment of portable speech translation device. For those languages in the world that do have a written form, many different scripts are used, as described in Chapter 2. Table 4.1 shows the writing systems together with the number of graphemes for 12 languages of the GlobalPhone corpus. Clearly visible is the large variation in numbers of graphemes. Languages such as Korean and Japanese use several writing systems in parallel. It can also be seen that even among those languages using the Roman script, the number of defined graphemes can vary greatly (capitalized and lowercase letters are counted as two separate graphemes). When relating the number of graphemes to the number of phonemes of a language (see Figure 4.5) it becomes clear that some languages, such as Turkish and Croatian, have as many phonemes as grapheme representations (ignore capitalization) while others have more graphemes as for instance, German and English. However, the ratio between graphemes and phonemes does not reveal much about the mapping between them. These can vary from close one-to-one mappings to many-to-one or one-to-many mappings. Furthermore, many scripts have grapheme grouping (e.g., German "sch" is always pronounced /ʃ/).

Table 4.1 Writing systems and number of graphemes for twelve languages.

Language	Script	Category	# Graphemes
Chinese	Hanzi	Logographic	>10,000
Korean	Hangul / Gulja	Phonographic (featural)	40 / 5601
	Chinese characters	Logographic	<10,000
Japanese	Hiragana / Katagana	Phonographic (featural)	46 / 46
	Kanji	Logographic	6,000–7,000
Arabic	Arabic Alphabet	Phonographic (consonantal)	100
Thai	Devanagari	Phonographic (syllabic)	63
Russian	Cyrillic	Phonographic (segmental)	57
Portuguese	Roman Alphabet	Phonographic (segmental)	80 (52 + 28 special)
Spanish	Roman Alphabet	Phonographic (segmental)	66 (52 + 14 special)
Croatian	Roman Alphabet	Phonographic (segmental)	62 (52 + 10 special)
German	Roman Alphabet	Phonographic (segmental)	59 (52 + 7 special)
Swedish	Roman Alphabet	Phonographic (segmental)	58 (52 + 6 special)
Turkish	Roman Alphabet	Phonographic (segmental)	58 (51 + 7 special)
English	Roman Alphabet	Phonographic (segmental)	52

As described above, a speech processing system requires a pronunciation dictionary to guide the recognition process. Since manually building a pronunciation is a very time- and cost-consuming process, it is desirable (especially in large-vocabulary speech recognition) to automatically generate these pronunciations from the written form. While this is not possible for most logographic scripts, the phonographic scripts allow the application of (rule-based) letter-to-sound algorithms. However, among the languages, the letter-to-sound relation varies considerably. It ranges from nearly a one-to-one relation—such as for Spanish, Finnish, or Turkish—to languages with a relatively straightforward relationship—such as German and Thai—to languages like English, which require complex rules and have many exceptions. As a consequence, the prediction quality of letter-to-sound tools highly depends on the language. Since the quality of a pronunciation dictionary highly effects the performance of speech recognition and speech synthesis, a poor letter-to-sound relation needs to be counterbalanced by manual corrections, which considerably increases development time and costs. Issues related to pronunciation dictionaries will be discussed in greater detail in Chapter 5, the consequences for speech synthesis will be highlighted in Chapter 7.

Sound Systems

Across the world's languages, the sound inventories vary considerably. A comparison of the sound inventories of nine GlobalPhone languages is given in Figure 4.5. It shows the ratio between consonants and vowels for the phone set, for the pronunciation dictionary (large vocabulary, 30,000–60,000), and for read speech. The total number of phonemes ranges from 29 for Turkish to 48 for Chinese. The ratio between consonants and vowels in the inventory varies from 4:1 in the case of the Croatian language versus 0.8:1 for Portuguese. In spoken speech, Portuguese has the lowest consonant-to-vowel ratio (50%), and German has the highest (60%). These numbers confirm the often cited perception of German speech sounding harsh and being rich in consonantal clusters.

The plot on the right of Figure 4.5 compares the acoustic difficulties of languages. For this purpose, phoneme recognizers for each language have been evaluated. The recognizers are purely based on acoustic knowledge—that is, no language model constraints have been applied. The results indicate significant language differences in acoustic confusability, ranging

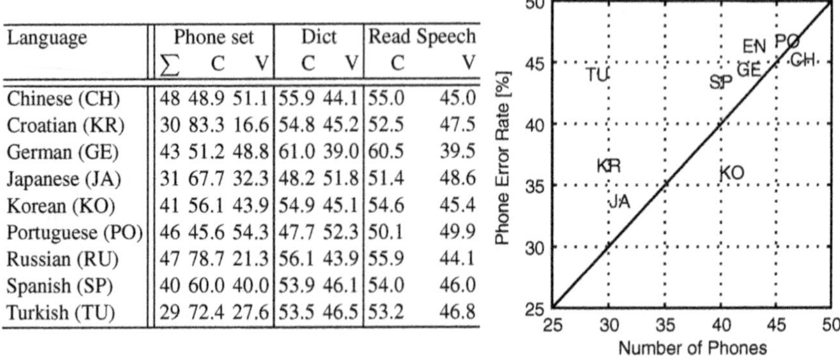

Language	Phone set			Dict		Read Speech	
	Σ	C	V	C	V	C	V
Chinese (CH)	48	48.9	51.1	55.9	44.1	55.0	45.0
Croatian (KR)	30	83.3	16.6	54.8	45.2	52.5	47.5
German (GE)	43	51.2	48.8	61.0	39.0	60.5	39.5
Japanese (JA)	31	67.7	32.3	48.2	51.8	51.4	48.6
Korean (KO)	41	56.1	43.9	54.9	45.1	54.6	45.4
Portuguese (PO)	46	45.6	54.3	47.7	52.3	50.1	49.9
Russian (RU)	47	78.7	21.3	56.1	43.9	55.9	44.1
Spanish (SP)	40	60.0	40.0	53.9	46.1	54.0	46.0
Turkish (TU)	29	72.4	27.6	53.5	46.5	53.2	46.8

Figure 4.5: Consonant (C) to Vowel (V) ratio (in %) and phone-based error rates for nine languages.

from a 33.8% to a 46.4% phone error rate. It can be seen that the phone error rate correlates with the number of phonemes used to model a language. This makes sense, since the existence of a larger number of model units increases the chances of confusing them with each other. Turkish seems to be an exception to this finding. The error analysis revealed that this is due to a very high substitution rate between the closed front vowels [e], [i], and [y].

Prosody and Phonology

In tonal languages, the lexical items are distinguished by contrasts in pitch contour or pitch level on a single syllable, such as in Mandarin Chinese and Thai, which both differentiate between five tones. In pitch languages like Japanese, pitch contrasts are not drawn between syllables but between polysyllabic words. In stress languages, individual syllables are stressed. In fixed-stress languages like Turkish, stress patterns always occur in the same position within a word (Turkish has in general word-final stress). Fixed-stress languages are easier to model than lexical stress languages such as English and German, in which the stress position varies across words.

Languages with syllable structure, such as the Japanese mora-syllables, have a very restricted phonology (consonant–vowel) with a small number of polyphones, while other languages, such as German, consist of many consonantal clusters, which are more difficult to recognize and result in many more polyphone types and higher phoneme perplexities.

Figure 4.6 shows the number of polyphone types for context width 0 to ±6 for nine GlobalPhone languages. Only the first phone of an adjacent word to the right and the last phone of the word to the left are considered when calculating the polyphones in order to mimic the common strategy in state-of-the-art speech recognition decoding. As a result, the amount of polyphone types depends not only on the phonotactic rules of a language but also on the length of word units. This becomes obvious for Korean, in which the number of polyphones saturates after context width ±2 due to the segmentation into short word units. The saturation for Chinese mainly results from the restricted syllable structure CV(C), while Japanese polyphones are limited due to the mora-syllable structure. Turkish, French, and English have a rather low number of polyphones, whereas German has a very high number of polyphones. Consequently, assuming a comparable size of acoustic models per language, the German models are forced to capture a broader variety of contexts per center phone than in other languages.

Segmentation and Morphology

Some languages, such as English and Spanish, provide a natural segmentation of the written form into word units that can be conveniently used

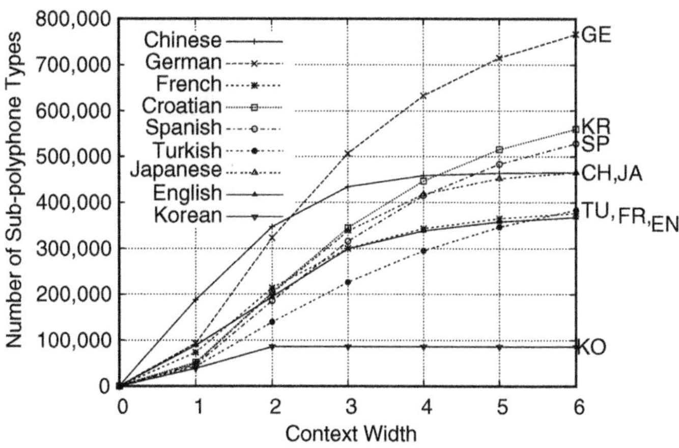

Figure 4.6: Number of polyphones over context width for nine languages.

as dictionary units for speech recognition. The words are long enough to be easily discriminated but short enough to limit the vocabulary growth rate. However, many languages lack a natural segmentation like Chinese, Japanese, or Thai, in which sentences are written in character strings without any spacing between adjacent words. In between those two extremes are many languages that provide a natural segmentation but whose number of derivative forms (i.e., inflected forms) per word stem is very large. This is due to their morphologic structure and leads to rapid vocabulary growth and a high rate of Out-of-Vocabulary (OOV) words. OOV words are those that a speech processing system is expected to handle but have never been observed in training data (e.g., unseen inflections of words that are important for the application). Words that do not occur in training data are usually not included in the search vocabulary of the speech recognition system. Thus, they cannot be recognized by the system. Consequently, each OOV word leads to at least one recognition error. As a word error can reduce the effective reach of the language model, we usually observe 1.2–1.5 word errors per OOV word.

Table 4.2 gives the size of vocabulary and resulting OOV rates for 10 GlobalPhone languages. The OOV rates differ significantly between these languages. For English we observed the lowest OOV rate—that of 0.3% with a 64,000-word vocabulary. This number increases to 4.4% for German, 13.5% for Turkish, and 34% for Korean if the natural segmentation is

Table 4.2 Out-of-vocabulary rates for ten languages.

Language	Vocabulary	OOV Rate
Korean	64K	34.0%
Turkish	64K	13.5%
German	61K	4.4%
Portuguese	60K	4.3%
English	64K	0.3%
Korean (segmented)	64K	0.2%
Chinese (segmented)	60K	0%
Croatian	31K	13.6%
Spanish	30K	5.2%
French	30K	4.7%
Japanese (segmented)	22K	3.0%

considered. For Chinese and Korean, the OOV rates drop down to 0% and 0.2% after applying automatic segmentation algorithms.

In sum, we can identify three groups of different behaviors. Low OOV rates can be expected from English as well as from segmented Chinese and Korean; moderate OOV rates are observed for German, French, Spanish, Portuguese, and segmented Japanese; and high OOV rates occur in high inflecting and agglutinating languages, such as Croatian, Turkish, and Korean.

Semantics and Pragmatics

In this section, the compactness of languages is investigated. Compactness refers to the amount of word types (vocabulary) and word tokens (running words in a text) in a language that is needed to convey a certain meaning. For the calculation of compactness, we used text material from the "European Corpus Initiative Multilingual Corpus I (ECI/MCI)," made available by ELSNET (1994). The experiments were carried out on "mul06," a parallel corpus of nine European languages based on the announcement text of the EC Esprit program. The corpus is relatively small, consisting of roughly 20,000–25,000 words per language; however, the specific feature of this corpus is that it gives a parallel nine-lingual text, meaning that for all languages, each text is supposed to convey exactly the same meaning. Since the data consist of only one sample per language (each text is translated only once, and the translations into the different languages are most likely done by different translators), the findings should be interpreted cautiously. Figure 4.7 plots the number of word tokens over the word types (left), and the number of grapheme tokens over word types (right).

English uses by far the smallest vocabulary needed to convey a meaning, which is due to the low number of inflection forms. However, many constructs require auxiliary verbs, which make the sentences longer. This explains the average rank in compactness. Dutch, French, Portuguese, Spanish, and Danish have an average-size vocabulary but Danish is by far the most compact. Spanish needs 20% more words to convey the same meaning. German, Italian, and Greek need a larger vocabulary, thus a higher OOV rate can be expected for those languages. When it comes to grapheme tokens, English becomes more compact, which is a reflection of the short word units, while German moves down in the ranking, since it uses long compound words. Danish is surprisingly compact in both word

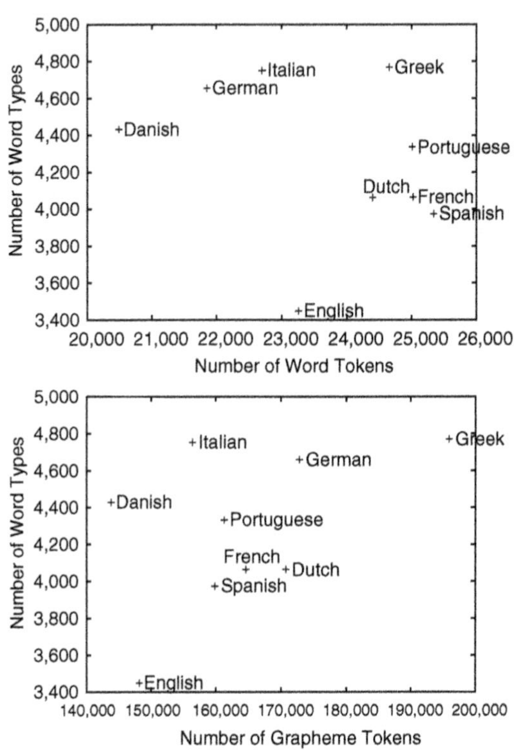

Figure 4.7: Compactness for nine EU-languages: Word Type (vocabulary) vs. Word Tokens (top) and vs. Grapheme Tokens (bottom).

and grapheme tokens—that is, the words are rather short, and few words are needed to convey the same meaning.

The language dissimilarities described above have a significant impact on multilingual acoustic modeling and make it is almost impossible to compare word error rates across languages. It has also been observed that languages with a rather simple structure at one level are more complicated on another. For instance, the Japanese phonetic and phonologic system is very restricted and easy to recognize (compare Figure 4.5), while on the morphological and semantic levels, it is known to be very challenging (no segmentation, strong inflection). For English it is just the opposite; the morphological structure is very simple (see Table 4.2 and Figure 4.7), but the phonetic structure is challenging (Figure 4.5).

4.2.3 Gap between Technology and Language Expertise

In addition to the large variety of language peculiarities, the coordination between experts across language boundaries is challenging. As mentioned previously, the task of multilingual acoustic modeling can be performed either by a group of technology and language experts, or by a single expert. In the first case, coordination and uniformity are the biggest concerns. Therefore it is often a more feasible approach to carry out the task by a single technology expert. However, most humans are knowledgeable in only a small number of languages. Consequently, the technology experts who work on multiple languages at one time are likely to be unfamiliar with most of the language they work with. This is a very difficult situation, as it is not feasible to learn language or even various scripts in a short time. Therefore, to facilitate access and error analysis, it might be more appropriate to convert the script into another form (e.g., romanization), which in itself is not a trivial task.

Ideally, the technology expert is supported by a language expert who has some knowledge about his or her language and, in the best of all worlds, some understanding of speech processing technologies. However, it is often very difficult to find such a person, particularly for low density languages or languages from developing countries. The lack of those experts dramatically slows down or even prevents the rapid development of speech processing technologies. The project SPICE (SPICE, 2005; Schultz, 2004) aims to bridge the gap between technology and language expertise by developing tools that support naïve users in building speech processing components in their language without the need for understanding the underlying technology. Knowledge of the language in question, which is important for the development of speech processing technologies, is solicited automatically from the user.

4.3 Language Independent Sound Inventories and Representations

The idea of sharing data across languages to train language independent and language adaptive acoustic models is based on the assumption that the articulatory representations of human speech sounds are so similar across languages that sound units can be considered to be independent from the underlying language. Based on this assumption, the language specific sound inventories of N languages $(L_1, \ldots, L_N)$ can be unified into one global set $(\Upsilon = \Upsilon_{L_1} \cup \Upsilon_{L_2} \cup \ldots \cup \Upsilon_{L_N})$.

Since the speech production process is limited by the anatomical prerequisite of the human articulatory apparatus and by physical laws, the assumption seems to be appropriate. However, particular production mechanisms may be dominant in one language and nonexistent in others, which makes the assumption a rather coarse approximation of the reality and raises the question of which sound units are appropriate as base units for this kind of approach.

4.3.1 Phonetic Sound Inventories

To date, by far the most widely used approach to model human speech in speech processing systems is based on *phonemes*. A phoneme is defined as the smallest sound unit of a language that discriminates between a minimal word pair. Phonemes as sound units have been quite successful in speech processing, since they are well defined and readily available for many languages. Furthermore, describing the pronunciation of a word as a concatenation of phonemes is straightforward (while finding an appropriate sequence is not).

Phonemes are defined in the context of a particular language as can be derived from the definition given above. Consequently, two sounds—for example, [l] and [r]—can be two different phonemes in one language (e.g., in English, with the minimal word pair /law/, /raw/) but allophones (variants of the same phoneme) in another language. For instance, Japanese speakers are famous for confusing /l/ and /r/ in English, because in their language [l] and [r] are allophones of the same phoneme.

The International Phonetic Alphabet (IPA) (see Chapter 2) assigns an IPA symbol to each identified phoneme of the known languages. Phonemes that are regarded as the same across languages are assigned the same IPA symbol. IPA was adopted by the speech processing community to describe similarities between phonemes across languages. Andersen and Dalsgaard, coined the term *polyphoneme* in the context of language identification, to describe the group of language independent phonemes (Υ_{LI}) occurring in more than one language, and N remaining groups of language dependent *monophonemes* ($\Upsilon_{LD_{L_1}}, \ldots \Upsilon_{LD_{L_N}}$). The set $\Upsilon_{LD_{Lm}}$ contains phonemes only occurring in language L_m, thus $|\Upsilon_{LD_{Lm}}| = 0$, if each phoneme of language L_m has a counterpart in at least one of the remaining $N - 1$ languages (Andersen et al., 1993).

Based on IPA (in its 1993 version, cf. Chapter 2), a **global unit set** for 12 languages has been defined (Schultz and Waibel, 2001). Sounds of different languages, which are represented by the same IPA symbol share one common unit (tonal variations of Mandarin vowels are treated as one vowel). Table 4.3 summarizes the polyphonemes and monophonemes for all 12 languages. For each polyphoneme, the upper half of Table 4.3 displays the number of languages that share it. The lower half of Table 4.3 contains the number and type of monophonemes for each language. The 485 language dependent phonemes of the 12 languages results in a set of 162 global units, of which the majority are shared among at least two languages.

4.3.2 Alternative Representations

The performance of phoneme-based speech recognition improved significantly over the last 20 years, mainly as a result of larger training corpora and better algorithms. However, speech recognition systems are still, fundamentally, template-based matching systems that model words as a sequence of phonemic or subphonemic units, sometimes referred to as the "beads-on-a-string" model (Ostendorf, 1999). The blind, data-driven principles using unstructured phonemes as base units are limited by the ability of their models to generalize to previously unseen conditions (Deng, 1997a). These limitations become more prominent as the robustness of speech recognition systems is challenged by varying effects, such as channel, speakers, speaking style, or domains, causing mismatched training and test conditions. Different languages can be viewed as extreme instances of effects causing mismatched conditions. Some argue that multilingual acoustic models based on phonemes are questionable, not due to their limited ability to generalize but also because phonemes are defined based on one language (Williams et al., 1998).

Deng et al. (1994) interprets the limited generalization as a results of the inappropriate mimicking of the speech production process. In his view,

Table 4.3 Global unit set for twelve languages.

Shared by	#	Modeled Phonemes (IPA symbols)	
	83	Polyphonemes shared across ≥ 2 languages	
		Consonants	Vowels
All	4	m,n,s,l	-
11	7	p,b,t,d,k,g,f	-
10	3	-	i,u,e
9	6	ŋ,v,z,j	a,o
8	1	ʃ	-
7	3	r,h,ʧ	-
6	1	-	ɛ
5	9	ɲ,ʒ,x,ʦ,ʤ	iː,y,ɘ,ɔ
4	4	-	ɨ,ø,ɑ,ei
3	11	ʎ,w,ç	ɪ,uː,eː,œ,oː,æ,ai,aʊ
2	34	p^h,t^h,d^j,k^h,g^j,ʁ,ɾ$_r$, θ,ð,s^j,z^j,ʂ̥,ɻ,ʦ^h,ʧ^j	ˈi,yː,ɯ,ʊ,ˈe,ɛː,øː,aː,ˈa,ɑː, ˈu,ˈo,aɪ,au,ia,io,eu,oi,oʊ
	79	Monophonemes belonging to *one* language	
		Consonants	Vowels
CH	15	tʂ̥,t^{h}ʂ̥,cç,cç^h	iʊ,iɛ,ua,uɛ,uɔ,ya,yɛ, iao,uɛi,uai,ioʊ
EN	5	ɾ$_d$	ʌ,ɝ,ɔi,ɚ
FR	5	ɥ	ɛ̃,œ̃,ɑ̃,ɔ̃
GE	3	-	ɐ,ʏ,ɔʏ
JA	2	ʔ	ɯː
KO	14	p̚,p',t̚,t',k̚,k',s',c'^{h}	ie,iɘ,iu,ɨi,oa,uɘ
KR	1	ʤ^j	-
PO	8	-	ĩ,ũ,ẽ,õ,ɐ̃,ew,ow,aw
RU	15	p^j,b^j,t^j,m^j,r^j,v^j, ʃ^j,ʒ^j,l^j,ʃtʃ,ʃtʃ^j	ja,jɛ,jɔ,ju
SP	2	β,ɣ	-
SW	9	ʈ,ɖ,ɳ,ɭ,ks	œː,æː,ʉː,ɵ
TU	0	-	-
∑	162	Silence and noises shared across languages	

speech variability cannot be well expressed in terms of general properties. Ostendorf (1999) argues that the major problem lies in the poor description of pronunciation variants by phoneme substitutions, deletions, or insertions. In fact, the context-dependent acoustic models of the phonetic units, which supposedly capture the pronunciation variability by providing different pronunciation models for the same phone in different contexts, are generated by data-driven clustering techniques, as described in Section 4.1. These models are forced to be very broad, since factors such as speaker dialect, speech rate, gender, and noise conditions also influence the acoustic representation of a sound. Including these factors into the clustering process fragments the data even more, such that every model only receives little training data while the resulting combinatorial explosion makes it hard to compute useful shared models.

As a consequence, scientists are looking for alternative representations of sound units. Subsegmental elements have been proposed that are expected to allow the description of all sounds of the world's languages through a combination of distinctive features. Williams et al. (1998) observe that distinctive features can be recognized more reliably than phonemes. However, modeling their temporal relations turns out to be difficult. Deng (1997b) introduced a set of rules and a finite state automaton to describe the temporal relation of features. He applied an English model set to Mandarin and found that the model parameter set had to be extended only slightly. From these experiments, he concludes that only a small number of languages need to be seen in order to cover all features.

Kirchhoff (1998) proposed an alternative representation that approximates the articulatory state of the speaker's vocal tract when speaking a sound, such as [voiced] or [velar]. A phone is described by a vector of such **articulatory features**. Speech recognition is based on recognition of the state sequence in the parallel feature streams. Since not all features have to change synchronously between phones, this approach allows speech to be described at a finer level than the phonetic segmentation. Since articulatory features are common to multiple phonemes, training material can be shared more efficiently. Furthermore, fewer classes have to be distinguished (e.g., binary features), which makes the training of statistical models for articulatory features more robust and consequently recognition rates for articulatory features are better than for phonemes. Stüker and Schultz (2003) extended the concept of a global unit set to articulatory features and defined a global feature unit set for five languages, as given in

Table 4.4 Global feature unit set for five languages.

	Feature	Languages
CONSONANT		CH GE EN JA SP
	VOICED	CH GE EN JA SP
	UNVOICED	CH GE EN JA SP
	ASPIRATED	CH EN
	PLOSIVE	CH GE EN JA SP
	NASAL	CH GE EN JA SP
Manner	TRILL	GE SP
	FLAP	EN SP
	FRICATIVE	CH GE EN JA SP
	AFFRICATE	CH GE EN JA SP
	APPROXIMANT	CH GE EN JA SP
	LATERAL-APPROXIMANT	CH GE EN JA SP
	BILABIAL	CH GE EN JA SP
	LABIODENTAL	CH GE EN JA SP
	DENTAL	EN SP
	ALVEOLAR	CH GE EN JA SP
Place	POSTALVEOLAR	GE EN JA SP
	RETROFLEX	CH EN
	PALATAL	CH GE EN JA SP
	VELAR	CH GE EN JA SP
	UVULAR	JA
	GLOTTAL	GE EN JA
VOWEL		CH GE EN JA SP
	ROUND	CH GE EN JA SP
	UNROUND	CH GE EN JA SP
	TONAL1-5	CH
Tongue Position		
	CLOSE	CH GE EN JA SP
Vertical	CLOSE-MID	GE EN JA SP
	OPEN	CH GE EN JA SP
	OPEN-MID	CH GE EN
	FRONT	CH GE EN JA SP
Horizontal	CENTRAL	GE EN
	BACK	CH GE EN JA SP

Table 4.4. The resulting set has a much smaller total number of units compared to the phoneme-based global unit set (37 feature-based units, 204 phoneme-based units). The feature units are tightly shared; 21 articulatory feature units occur in all languages. Only the features [uvular] and [tonal] are not shared across languages. This demonstrates that training data can indeed be shared more effectively.

In the next section, we introduce a measure for the efficiency of unit sharing and compare the phoneme-based and feature-based unit sets. Section 4.4 presents recognition results for both unit sets.

4.3.3 Sound Unit Sharing across Languages

To measure the relation between language specific units and the size of the global unit set, Schultz et al. (2001) defined the **share factor** sf_N for a set of N languages—that is, sf_N gives the average number of languages sharing the units of a global unit set:

$$\mathrm{sf}_N = \frac{\sum_{i=1}^{N} |\Upsilon_{L_i}|}{|\Upsilon|}, \quad |\Upsilon| = |\Upsilon_{LI}| + \sum_{i=1}^{N} |\Upsilon_{LD_{L_i}}| \tag{4.3.1}$$

The share factor is one if no polyphonemes exist at all, and N if each of the N languages uses the identical phonetic inventory, $1 \leq \mathrm{sf}_N \leq \mathrm{N}$.

In the case of the 12 languages as showed in Table 4.3, we have 485 language specific phonemes, resulting in

$$\mathrm{sf}_{12} = \frac{|\Upsilon_{ch}| + |\Upsilon_{kr}| + \cdots + |\Upsilon_{tu}|}{|\Upsilon|} = \frac{485}{162} = 2.99 \tag{4.3.2}$$

In other words, each phoneme of the global unit set is shared on average by 3 languages. Figure 4.8 plots the average share factor over all possible k-tuples ($k = 1, \ldots, 12$) of 12 languages $\binom{12}{k}$. Obviously, the share factor increases with the number of languages involved, however the increase rate is lower than expected. Furthermore, the share factor has a large variance, since the phoneme inventory overlap is impacted by the selection of languages. Overall, using the global unit set allows for a significant unit reduction but the efficiency of data sharing leaves room for improvement.

Figure 4.8 also shows the average and variance of the share factor based on the articulatory feature unit set for the five languages presented in Table 4.4. The share factor for the feature-based unit set clearly outperforms the share factor of the phoneme-based set. In addition, it grows almost linearly over the number of involved languages. We can therefore expect that the training of language independent articulatory features is going to make better use of the training data of different languages than in the phoneme-based case.

Articulatory feature detectors have been applied to speech recognition in order to improve robustness toward noise and reverberation. Kirchhoff (1998) compared individual acoustic, articulatory, and combined speech recognition systems. She found that under very noisy conditions, a recognition system relying solely on articulatory features performs better than a system that uses phonetic models. The output combination of phoneme models and feature models even performs better under clean, reverberant, and low-noise conditions. Eide (2001) includes the output of articulatory feature classifiers into the front-end of a standard phone based system and shows performance improvements in noisy conditions. Metze (2002) introduced a flexible stream-based architecture to fuse the information given by

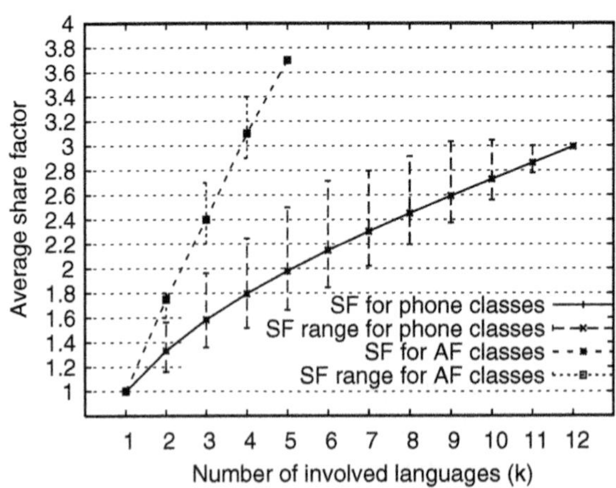

Figure 4.8: Average and range of the share factor for phoneme based and articulatory feature based units over the number of $\binom{12}{k}$ and $\binom{5}{l}$ involved languages, respectively, with $k = 1, \ldots, 12$ and $l = 1, \ldots, 5$.

articulatory feature detectors with a continuous density HMM by computing a weighted sum of the corresponding log likelihoods. Section 4.4 will present speech recognition results following from this approach. Deng et al. (1994) proposed a speech recognition framework that makes sole use of articulatory features using the concept of overlapping features.

Although revisions to classical phoneme-based approach are often discussed, and the use of articulatory features may intuitively be more appropriate—especially in the context of multilingual acoustic modeling—most of current recognition systems still rely on phoneme-based acoustic models. Experimental results indicate that alternative features outperform phoneme-based systems in adverse conditions, and that combining alternative features with phoneme-based models leads to improvements. However, no results have yet been published showing that a recognition system solely based on alternative sound units outperforms the standard phoneme-based approach.

4.3.4 Phonetic Coverage across Languages

The share factor indicates the percentage of data that can be shared to train acoustic models across known languages. For the purpose of adaptation to new languages, another aspect becomes interesting, namely how many sound units of the new language are expected to be covered by the global unit set. For this purpose, we define the **coverage coefficient $\mathrm{cc_N}(L_T)$** of the target language L_T to be:

$$\mathrm{cc_N}(L_T) = \frac{|\Upsilon_{L_T} \cap \Upsilon|}{|\Upsilon_{L_T}|} = 1 - \frac{|\Upsilon_{LD_{L_T}}|}{|\Upsilon_{L_T}|} \tag{4.3.3}$$

While the share factor *sf* measures the average sharing of all phonemes in the global unit set over all languages, the coverage coefficient *cc* indicates the portion of phonemes in the target language L_T, which are covered by phonemes of the global unit set. The coverage coefficient is zero if no phoneme of the target language L_T has a counterpart in the global unit set, and one if each phoneme is covered, $0 \leq \mathrm{cc}(L_T) \leq 1$.

This can naturally be extended to models of various context width. Equation (4.3.3) can be used to measure monophone coverage, triphone coverage, and in general polyphone coverage. We further distinguish

Table 4.5 Triphone coverage matrix for ten GlobalPhone languages; two numbers are given for each matrix entry (*i*, *j*), meaning that language *i* is covered by language *j* with triphone types (upper number) and triphone tokens (lower number).

	CH	EN	FR	GE	JA	KO	KR	PO	SP	TU
CH	100	0.3 6.8	0.1 5.8	0.1 5.3	0.1 4.2	0.0 5.3	0.1 4.2	0.1 5.4	0.1 5.3	0.2 4.9
EN	0.6 5.2	100	6.5 18.6	5.4 18.1	1.8 8.9	3.4 11.6	1.5 7.7	0.9 6.6	1.3 6.6	3.8 9.2
FR	0.1 3.9	9.7 16.4	100	29.0 53.3	10.2 22.7	11.2 28.7	25.8 45.5	18.4 36.4	17.4 41.3	23.1 35.6
GE	0.1 3.9	5.5 19.6	19.8 41.6	100	9.3 19.5	7.2 18.2	18.6 34.9	13.6 28.0	12.9 28.3	12.9 26.1
JA	0.2 2.5	4.5 9.9	16.8 37.4	22.3 33.6	100	9.8 25.6	16.0 29.2	11.0 27.6	13.6 31.2	25.9 52.5
KO	0.1 4.1	4.9 16.1	10.9 35.0	10.3 36.3	5.8 24.9	100	10.2 38.6	8.0 30.8	9.3 38.4	9.1 26.1
KR	0.2 1.8	3.2 5.0	37.0 64.7	39.0 68.8	14.0 28.2	15.0 34.5	100	31.0 63.0	34.3 61.8	31.5 50.4
PO	0.4 2.3	2.0 4.6	28.0 49.5	30.2 57.9	10.2 26.7	12.5 37.5	32.9 62.5	100	33.5 57.5	19.8 39.9
SP	0.2 2.5	2.7 5.6	23.5 60.1	25.4 60.2	11.2 34.0	12.9 40.1	32.2 64.2	29.7 58.2	100	17.5 41.0
TU	0.8 5.4	8.9 18.3	36.3 52.0	29.6 46.0	24.8 46.1	14.6 33.0	34.4 50.1	20.4 38.6	20.3 39.6	100

between the coverage of polyphone types and polyphone tokens to reflect the fact that coverage of frequent polyphones is more important than coverage of less frequent ones with respect to recognition performance.

Table 4.5 summarizes the triphone coverage for ten GlobalPhone languages. The coverage of triphone types is given in the upper row, and coverage of triphone tokens in the lower row. The table shows that 33.6% of Japanese triphone tokens are covered by German triphones, and that 22.3% of the triphone types are responsible for this coverage rate. On the other hand, only 19.5% of all German triphone tokens are covered by Japanese triphones. This effect is due to Japanese phonotactics, which allow consonant vowel combinations but no consonant clusters.

The polyphone coverage indicates how well a generic polyphone decision tree fits to a new target language. To calculate the coverage, a very simple greedy procedure can be applied. For illustration purposes, Portuguese will serve as the example of a new target language. First, the language among all pool languages that achieves the highest coverage for Portuguese is selected. Second, this language is removed from the pool, and the coverage between Portuguese and each language pair resulting from the combination of removed language plus remaining pool language is calculated. The procedure is repeated for triples and so forth until the pool of languages is exhausted. Thus, in each step, the language that maximally complements the polyphone set is determined. The percentage coverage (cc(Po) $\times$ 100) is plotted in Figure 4.9 for context width ± 0 (monophones), ± 1 (triphones), and ± 2 (quinphones).

From Figure 4.9 we see that the coverage decreases dramatically for wider contexts. With a nine-language pool from GlobalPhone, the coverage of Portuguese monophones achieves 91%, then drops to 73% for triphones and to 47% for quinphones. After incorporating the three main contributing languages, the coverage for monophones cannot be increased any further. When enlarging the context width to one, the coverage saturates after four languages. For a context width of two, at least five languages contribute to the quinphone coverage rate, therefore, it is expected that increasing the context width will require more languages. We experimented with removing the main contribution languages from the pool—Spanish, Croatian, and French. Removing Spanish could nearly be compensated by German

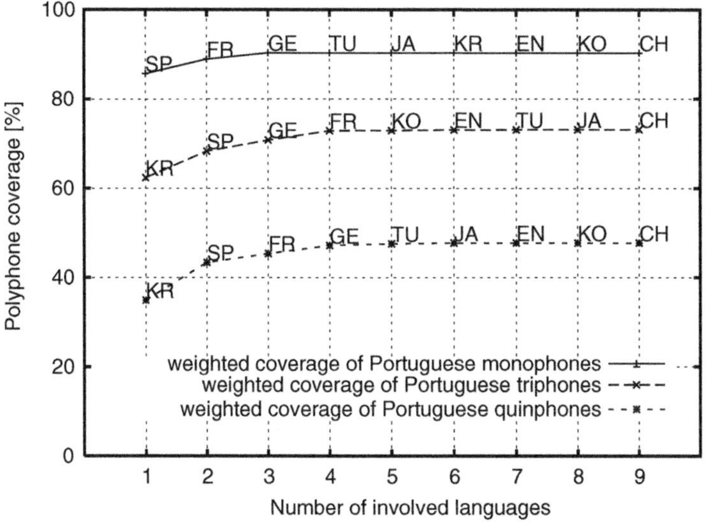

Figure 4.9: Portuguese polyphone coverage by nine languages.

plus Croatian, and vice versa. This indicates that these three languages cover similar portions of the Portuguese polyphone set. It is not possible to compensate for the removal of French by including other languages, since French provides unique polyphones not found elsewhere. In this case, the missing phonemes are nasal vowels, which are frequent in Portuguese. We can conclude from this observation that when designing a language pool for adaptation purposes, it is more critical to find a complementary set of languages than to cover a large number of languages.

By analyzing the coverage in Figure 4.9 and Table 4.5, we infer that a polyphone decision tree, even build on several languages, cannot be successfully applied to a new language without adaptation. We will see in the next section how the context mismatch between languages effects the multilingual acoustic model combination.

4.4 Acoustic Model Combination

In this section, we introduce common methods and technologies to combine acoustic models across languages. As described earlier, there are two major purposes for multilingual acoustic model combination: (1) truly multilingual applications that can handle multiple languages simultaneously, and (2) rapid language adaptation. In the first case, the goal might be to build a system that can handle several languages at a time, to get a more compact system with a smaller number of total parameters and thus reduced memory footprint, or to get a system that is easier to maintain. In the second case, the goal is to cover as many sound characteristics as possible to be prepared for rapid adaptation to future languages of interest. Since the targets are different, the acoustic model combination methods differ as well, and will therefore be discussed separately. We will first describe various methods to combine models, and then discuss special issues related to rapid language adaptation. The discussion will further discriminate between phoneme-based and articulatory-feature-based combination experiments.

4.4.1 Language Independent Acoustic Modeling

The idea to share phoneme models across languages was first formulated by Dalsgaard, Andersen, and Barry (1992) and was motivated by the task of language identification. Dalsgaard et al. used monophonemes as well as polyphonemes to identify the spoken language of an utterance, while others concentrated on monophonemes emphasizing the language-discriminating information inherent in monophonems to identify a language (Berkling et al., 1994; Zissman and Singer, 1995). The concept of language independent acoustic models was applied successfully in several other studies on language identification (Corredor-Ardoy et al., 1997; Kwan and Hirose, 1997) and fueled the idea that language independent acoustic models could be useful for speech recognition purposes as well.

Three major approaches to combine acoustic models across languages are discriminated:

- Heuristic model combination based on linguistic knowledge
  - Phonetic/articulatory (Dalsgaard and Andersen, 1992; Cohen et al., 1997; Ward et al., 1998; Weng et al., 1997b)
  - IPA-based (Köhler, 1997, 1998, 1999; Schultz and Waibel, 1998c) or Sampa-based (Ackermann et al., 1996, 1997; Übler et al., 1998)
- Purely data-driven model combination
  - A phoneme confusion matrix provides similarity measure between phonemes (Andersen et al., 1993; Andersen et al., 1994; Dalsgaard and Andersen, 1994; Imperl, 1999)
  - A combination of distance measures is used to calculate similarity between phonemes (Bonaventura et al., 1997; Micca et al., 1999)
  - Agglomerative clustering procedures based on:
    - ∗ Likelihood distances (Andersen and Dalsgaard, 1997, Köhler, 1999)
    - ∗ A-posteriori distances (Corredor-Ardoy et al., 1997)
- Hierarchical combination of both heuristic and data-driven methods
  - Step 1: Heuristic grouping of phonemes into classes
  - Step 2: Data-driven clustering within the classes defined by Step 1 (Köhler, 1999, 1996; Weng et al., 1997b; Cohen et al., 1997; Ward et al., 1998; Schultz and Waibel, 1998c, b)

Heuristic Model Combination

Several different heuristics have been proposed to assign phonemes to classes. Dalsgaard and Andersen (1992) used auditive phonetic criteria to assign phonemes to classes. This approach was adopted by Cohen et al. (1997) and Ward et al. (1998), while others applied broader articulatory classes (Weng et al., 1997b) or reference schemes—such as IPA (Köhler, 1997, 1998, 1999) or Sampa (Ackermann et al., 1996, 1997; Übler et al., 1998)—to decide the assignment of phonemes.

The heuristic mapping of phonemes into a class is followed by training one common acoustic model per class. The class model is trained by sharing the data of those languages that occur to be in this class (data-sharing principle). When these heuristically combined multilingual acoustic models are applied to speech recognition on the training languages, the results are usually worse than those of monolingual acoustic models. Multilingual models outperform monolingual ones in two scenarios—either when the training data in the target language is very small (Andersen et al., 1993, 1994; Dalsgaard and Andersen, 1994), or when speech from truly bilingual speakers or slightly accented speech from nearly bilingual speakers is recognized (Ackermann et al., 1996, 1997; Übler et al., 1998) (see also Chapter 9).

Data-Driven Model Combination

Data-driven acoustic model combination requires a combination scheme and a similarity measure to decide which acoustic models should be combined. Various schemes and measures have been proposed: Andersen Andersen et al. (1993) calculated a phoneme confusion matrix and merged the most similar phonemes into classes; Corredor-Ardoy (1997) introduced an agglomerative clustering scheme that uses similarities calculated on the a-posteriori probabilities between phonemes; Köhler (1996, 1999) applied likelihood-based distances to agglomerative clustering; and Bonaventura et al. (1997) introduced a distance measure that combines five different similarity measures based on Gaussian mixture models. Interestingly, the latter three studies found the same phonemes to be shared among languages (up to six languages were investigated at a time): the voiceless plosives /p/, /t/, and /k/; the fricatives /s/ and /f/; and the nasals /n/ and /m/ (Corredor-Ardoy et al., 1997; Köhler, 1996; Bonaventura et al., 1997).

In addition, Bonaventura et al. found the vowel clusters /o/, /e/, and /a/. Imperl and Horvat (1999) extended the data-driven clustering approach to context-dependent acoustic models by defining the similarity between two triphones of different languages to be the weighted sum of the similarity between the center phones and the similarities between the right and left phones, respectively.

The resulting combined models are applied to various recognition experiments and mostly result in severe recognition performance degradation. The performance gap becomes larger as the number of shared languages grows (Bonaventura et al., 1997). However, multilingual models are found to significantly outperform monolingual models when the latter become unreliable due to lack of data. In such a scenario, multilingual acoustic models benefit from increasing the amount of data per model by sharing it across languages (Andersen et al., 1993, 1994; Dalsgaard and Andersen, 1994). Overall, the data-driven approach outperforms the purely heuristic one (Köhler, 1999), but still cannot achieve better results than the monolingual acoustic models.

The data-driven approach blindly clusters similar phonemes across languages. Besides the computational costs, purely data-driven approaches have two major drawbacks: first, sounds of the same language might be clustered into one common class due to data artifacts, as observed by Andersen et al. (1993) and Köhler (1996). As a consequence, the model granularity of languages, whose phoneme sets are collapsed, can decrease to a critical point, resulting in a major performance degradation for the language. Second, the resulting phone clusters are in many cases not linguistically meaningful—as found by Andersen and Dalsgaard (1997) and Dalsgaard et al. (1998)—and therefore may not allow a unique mapping from the phone clusters to phonemes of a new target language. Therefore, it becomes difficult to use these phone clusters as seed models for rapid model bootstrapping across languages.

Hierarchical Model Combination

The hierarchical model combination intends to overcome these limitations by applying a two-step approach: in the first step, phone categories are defined; in the second step, the acoustic models of these heuristically defined phone categories are clustered in a data-driven fashion.

Heuristic phone categories are based on either phonetic (Weng et al., 1997b) or IPA criteria (Köhler, 1999). The data sharing is applied to the resulting categories. Weng et al. (1997b) built semicontinuous HMM-based models, in which all phones of the same category share one Gaussian mixture model. Köhler (1999) clustered phones within IPA-based classes in a bottom-up fashion. Weng (Weng et al., 1997b) investigated the effect of data sharing by comparing models shared within one language with models shared across two languages. Sharing across languages resulted in performance reduction of 5% compared to sharing within languages. Part of the performance loss can be avoided by using language questions in a top-down clustering procedure with phone decision trees (Cohen et al., 1997; Ward et al., 1998; Schultz and Waibel, 1998b). Language questions are added to the phonetically motivated context questions. Phones that contain language questions at the tree root are modeled as monophonemes, while the others are modeled as polyphonemes.

Among the combination methods described above, the purely heuristic IPA-based approach performs worst, and the hierarchical one performs best. However, all three methods are outperformed by monolingual acoustic models when tested on the training languages (Köhler, 1999). Cohen et al. found a relative performance degradation of 5–9% compared to monolingual models. This includes the loss due to language model combination (Cohen et al., 1997). Schultz found on average 5.8% relative performance degradation on five languages. Reducing the parameter set to 40% leads to another 4.3% relative loss (Schultz and Waibel, 1998c).

Combination of Context-Dependent Acoustic Models

The above described studies were mostly restricted to context-independent acoustic models. Schultz and Waibel (2000, 2001) extended the hierarchical combination method to the case of context-dependent acoustic modeling. They drastically reduced the amount of model parameter and improved the model robustness for the purpose of rapid language adaptation. Three different methods for acoustic model combination are introduced and compared to each other with respect to performance on the three tasks of monolingual recognition, simultaneous multilingual recognition, and rapid language portation. The three methods are the language separate *ML-sep* method, the language mixed *ML-mix* method, and the language

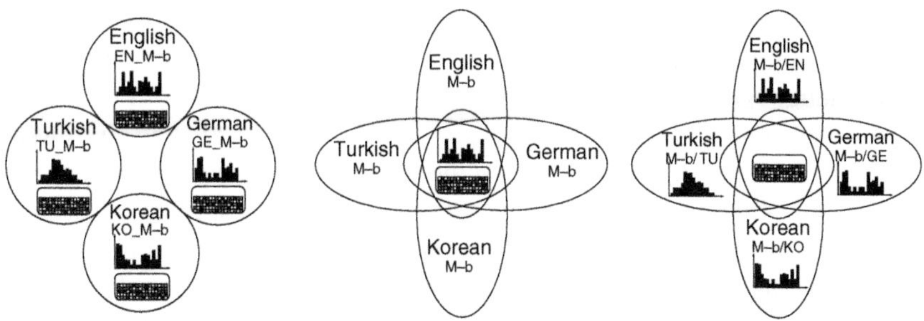

Figure 4.10: Acoustic model combination ML-sep, ML-mix, ML-tag (from left to right).

tagged *ML-tag* combination method, as illustrated in Figure 4.10. The acoustic modeling is HMM-based; the probability $p(x|s_i)$ to emit x in state s_i is modeled by a mixture of K_i Gaussian components $p(x|s_i) = \sum_{k=1}^{K_i} c_{s_i k} N(x|\mu_{s_i k}, \Sigma_{s_i k})$, as described in Section 4.1. In Figure 4.10, the mixture weights c are symbolized as distributions, and the Gaussian components $N(x|\mu, \Sigma)$ are symbolized as rounded boxes. The figure describes the acoustic modeling combination for the begin state of the model of phoneme /m/, thus "M-b."

In the *ML-sep* combination method, each language specific phoneme is trained solely with data from its own language—that is, no data are shared across languages to train the acoustic models. The multilingual component of *ML-sep* is the feature extraction, since one global Linear Discriminant Analysis (LDA) matrix is calculated to reduce the dimensionality of the feature vectors. This global matrix uses all language specific phoneme models as classes for the LD criterion (Fukunaga, 1972). Context-dependent models are created by applying an entropy-based decision tree clustering procedure. Using the Gaussian mixture based approach for modeling HMM emission probabilities, as described above, the *ML-sep* combination method can be described as:

$$\text{ML-SEP}: \begin{cases} c_{s_i} \neq c_{s_j} & , \quad \forall i \neq j \\ \mu_{s_i,k} \neq \mu_{s_j,k}, & \forall i \neq j \\ \Sigma_{s_i,k} \neq \Sigma_{s_j,k}, & \forall i \neq j \end{cases}$$

In the *ML-mix* combination method, data are shared across different languages to train the acoustic models of polyphonemes—the phonemes of different languages, which belong to the same IPA unit defined in our global unit set (see Section 4.2.2). During training, no information about the language is preserved. In other words, for each IPA unit of the global unit set, one Gaussian mixture per state is initialized, and the model of this IPA unit is trained by sharing the data of all languages belonging to the IPA unit. Context-dependent models are created by applying the aforementioned clustering procedure. The questions for the decision tree clustering process are linguistically motivated and are derived from the IPA reference scheme. With the function ipa(s_i), which returns the IPA unit to which s_i belongs, the *ML-mix* model combination method can be described as:

$$\text{ML-MIX}: \begin{cases} c_{s_i} = c_{s_j} & , \quad \forall i,j : \text{ipa}(s_i) = \text{ipa}(s_j) \\ \mu_{s_i,k} = \mu_{s_j,k} , & \forall i,j : \text{ipa}(s_i) = \text{ipa}(s_j) \\ \Sigma_{s_i,k} = \Sigma_{s_j,k} , & \forall i,j : \text{ipa}(s_i) = \text{ipa}(s_j) \end{cases}$$

Another way to share phoneme models across languages is performed in the model combination method *ML-tag*. Here, each phoneme receives a language tag, attached to preserve information about the language to which the phoneme belongs. *ML-tag* is similar to *ML-mix* in the sense that they both share all the training data and use the same clustering procedure. However, for *ML-mix*, the training data are only labeled by phoneme identity, whereas for *ML-tag*, the training data are labeled by both phoneme and language identity. The clustering procedure is extended by introducing questions about the language and language groups to which a phoneme belongs. The Gaussian components are shared across languages as in the *ML-mix* method, but the mixture weights are trained separately. Therefore, the relative importance of phonetic context and language membership is resolved during the clustering procedure by a data-driven method. The *ML-tag* combination method can be described as:

$$\text{ML-TAG}: \begin{cases} c_{s_i} \neq c_{s_j} & , \quad \forall i \neq j \\ \mu_{s_i,k} = \mu_{s_j,k} , & \forall i,j : \text{ipa}(s_i) = \text{ipa}(s_j) \\ \Sigma_{s_i,k} = \Sigma_{s_j,k} , & \forall i,j : \text{ipa}(s_i) = \text{ipa}(s_j) \end{cases}$$

These three modeling approaches have been evaluated for five GlobalPhone languages: Croatian, Japanese, Korean, Spanish, and Turkish (Schultz and Waibel, 1998c). Speech recognition performance is compared between a truly multilingual speech recognition system (simultaneously recognizing five languages) and five monolingual speech recognition system. All knowledge components except for the acoustic model are language dependent (see Section 4.1). When comparing the monolingual systems to the *ML-sep* system, whose only difference lies in the multilingual LDA, the LDA increases the word error rate slightly but not significantly. The *ML-tag* system outperforms the mixed system *ML-mix* in all languages by an average error rate of 5.3% (3.1–8.7%). Since the collection of the GlobalPhone speech data is uniform in terms of recording and channel conditions the conclusion can be drawn that preserving the language information achieves better results with respect to simultaneous recognition. Two *ML-tag* systems were trained, one with the same parameter size as five monolingual systems, and one compact system that reduces the model size to 40% compared to the monolingual case. The compact system results in a 3.1% performance degradation on average (1.2–5.0%) compared to the monolinguial case. However, not all of the degradation can be explained by the reduction of parameters, this can be derived from the performance gap of 1.1% (0.3–2.4%) between the monolingual systems and the full size *ML-tag* system.

To investigate the degree of language sharing coded in the context-dependent acoustic models, the polyphone decision tree of the five-languages *ML-tag* was analyzed. For this purpose, the language distribution of each tree node was computed by calculating the language entropy gain when traversing the given multilingual tree. The gain D_H is calculated as

$$D_H = p(n_{yes})H_{yes} + p(n_{no})H_{no} - (p(n_{yes}) + p(n_{no}))H_{org}$$

where H_n is the entropy of the language distributions in node $n \in \{org, yes, no\}$ defined as $H_n = \sum_{i=1}^{l} p_n(i) \log_2 p_n(i)$ und $l = 5$. The resulting sum of the entropy gain D_H is plotted over the number of clustered subpolyphones in Figure 4.11. This procedure allows analyzing the ratio of language questions compared to phonetic questions. The curve *sum of all questions* gives the overall language entropy gain of all questions, whereas the curve *phonetic context questions* shows the entropy gain belonging to non-language questions. The large gap between both curves indicates that major parts of the entropy gain result from language questions. The remaining five curves indicate the contribution of questions belonging to only one language. It is shown that questions about Korean and Turkish are more important than questions about other languages, especially in the

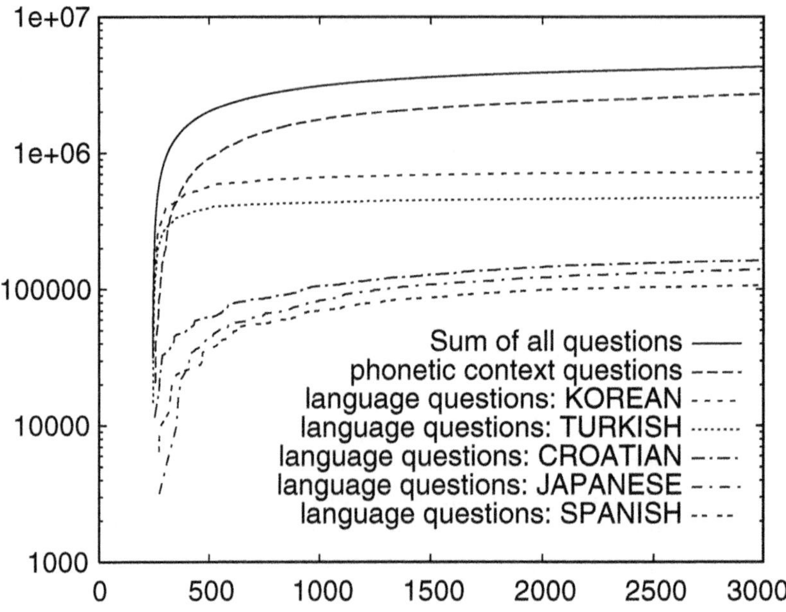

Figure 4.11: Entropy gain over number of subpolyphones for a five-lingual system.

initial stages of the clustering process. This indicates that sounds in those two languages seem to be different from the rest. Both results demonstrate that language questions are frequently asked and are more important in the beginning of the splitting process than questions about the phonetic context of a phoneme. After about 3,000 splits, the main parts of language information are clustered out, which implies that a multilingual system *ML-tag* with more than 3,000 polyphone clusters is composed largely of language dependent acoustic models.

4.4.2 Language Adaptive Acoustic Modeling

The major cost factor for developing speech recognition systems in new languages is the large amount of transcribed audio data that is required for the training of accurate acoustic models. To accommodate the variations in available data, three issues are addressed in language adaptive acoustic modeling:

1. No training data available: cross-language transfer
2. Limited training data available: language adaptation technique
3. Large amounts of training data available: bootstrapping approach

The term **cross-language transfer** refers to the technique of using a recognition system on a new language without having ever seen any training data of the language in question. Research in this area investigates whether cross-language transfer between two languages of the same family performs better than transfer across family borders (Constantinescu and Chollet, 1997), and whether the number of languages used for training the original acoustic transfer models influences the performance on the target language (Gokcen and Gokcen, 1997; Schultz and Waibel, 1998a). Some results indicate a relation between language similarity and cross-language performance (Bub et al., 1997; Constantinescu and Chollet, 1997). Other experiments have clearly shown that multilingual transfer models outperform monolingual ones (Schultz and Waibel, 1998c; Bub et al., 1997).

In the **language adaptation technique**, an existing recognizer is adapted to the new target language using very limited training data. Ongoing research (Wheatley et al., 1994; Köhler, 1998; Schultz and Waibel, 1998b) concentrates on two issues: (1) amount of adaptation data needed to get reasonable results, and (2) finding suitable acoustic models to start from. As expected, language adaptation performance is strongly related to the amount of data used for adaptation (Schultz and Waibel, 2000). It has also been demonstrated that the number of training speakers is more critical than the number of training utterances (Wheatley et al., 1994). Multilingual acoustic models have proven to be more effective than monolingual ones (Schultz and Waibel, 1998b; Köhler, 1998).

The key idea in the **bootstrapping approach** is to initialize the acoustic models of the target language recognizer using seed models developed for other languages. After this initialization step, the resulting system is completely rebuilt using large amounts of training data from the target language. This approach has been successfully applied in earlier studies (Osterholtz et al., 1992) to bootstrap a German recognizer from English. Experiments have proven that cross-language seed models achieve lower word error rates than flat starts or random models (Wheatley et al., 1994). The usefulness of the described global unit set and multilingual acoustic models as seed models has been shown in Schultz and Waibel (1997a).

Polyphone Decision Tree Specialization

Previous approaches to language adaptation have been limited to context-independent acoustic models. Since in the language dependent case, wider contexts increase recognition performance significantly, we investigated whether such improvements extend to the multilingual setting. The use of wider context windows raises the problem of phonetic context mismatch between source and target languages, as described in Section 4.3.4. In order to overcome the problem of the observed mismatch between represented context in the multilingual polyphone decision tree and the observed polyphones in the new target language, a Polyphone Decision Tree Specialization (PDTS) procedure has been proposed (Schultz and Waibel, 2000).

The PDTS procedure works in three steps. First, a multilingual polyphone decision tree is trained by sharing the training data across several languages. Second, the tree is reduced to those phonemes occurring in the target language. As we saw from the context mismatches across languages, the tree is likely to only poorly estimate all polyphones of the target language, since the context questions may not reflect the contexts in which the target language phonemes occur. Therefore, this tree is adapted in a third step by restarting the decision tree growing process using the limited adaptation data available from the target language. The regrowing process is completed after reaching a predefined number of new leaf nodes depending on the amount of training data (Schultz and Waibel, 2000).

4.4.3 Approaches Based on Alternative Sound Units

As noted in Section 4.3.2, the introduction of articulatory features has improved the performance of speech recognition systems under various conditions, especially when the detectors for articulatory features are combined with standard phoneme models (Eide, 2001; Metze and Waibel, 2002). Experimental results have been carried out on monolingual speech recognition applying monolingual articulatory feature models. Stüker and

Schultz extended this concept to multilingual articulatory features in which the models of the articulatory features are shared across languages (Stüker et al., 2003a, 2003b). In order to show that articulatory features are robust to interlanguage variability, a global set of features for five languages from the GlobalPhone global phoneme set was derived (as described in Section 4.3). Then an IPA-based mapping from the global phoneme set to articulatory features was applied. The articulatory features therefore describe abstract classes that focus on the main aspect of the articulators (e.g. [voiced], [lip-rounding], [bilabial]) when a human sound is articulated, rather than the exact position of the articulators during the speech process.

Marked positions of continuous features are modeled by binary features—that is, rather than having a continuous feature for the horizontal positions of the dorsum, three discrete values ([frontal]), [central], and [back]) are used, each seen as a binary feature that is either absent or present. In analogy to the phoneme-based case, it is assumed that these binary features are so similar across languages that they can be considered to be language independent (Stüker et al., 2003). As a consequence, the models are not only trained monolingually but also trained multilingually by sharing the training data of the models across the five languages: Chinese, English, German, Japanese, and Spanish. As shown in Figure 4.8, the resulting sharing factor sf for articulatory features is 3.7, which clearly outperforms the phone-based sharing factor (phone-based sf is below 2 for these five languages). Therefore, it can be expected that articulatory features make better use of the training data. The resulting monolingual and multilingual feature detectors are evaluated by calculating their classification accuracy on the training language (monolingual and multilingual evaluation). In addition, the multilingual feature detectors are also evaluated on a language that has not been presented during training (cross-lingual evaluation).

Figure 4.12 shows the performance of the individual monolingual feature detectors from five languages calculated on English test data. The solid line shows the classification accuracy of the English detectors. The additional data points show the classification accuracy of the detectors trained on the other four languages. Data points that appear above the solid line indicate a detector that classifies an English feature more accurately than the corresponding English articulatory feature detector. This shows the potential to reliably detect articulatory features across languages. Selecting the best detectors of the set of all languages improves the overall classification accuracy. In the case of English, the classification accuracy averaged

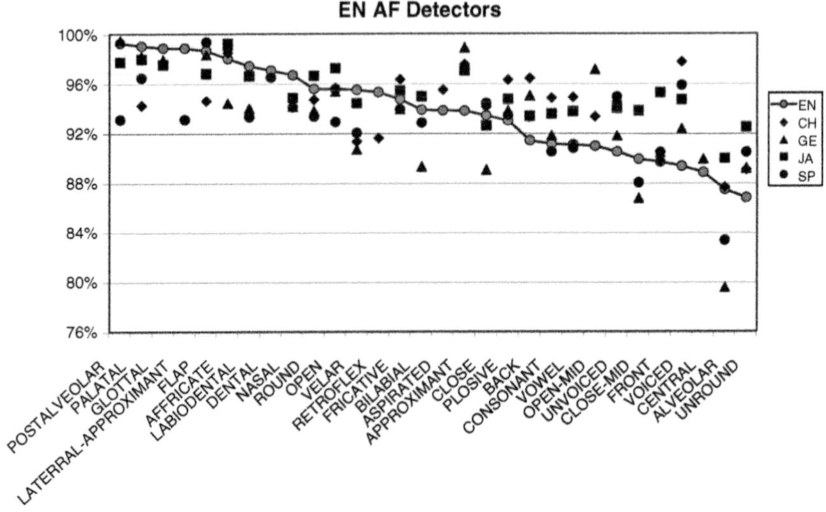

Figure 4.12: Classification accuracy of articulatory feature detectors from five languages on English test data.

over all English features gives 93.83% compared to 96.13% when selecting the best detectors out of five languages. If the best detectors trained on all but the English language are selected, the classification performance achieves 95.61%. This indicates that it is possible to reliably detect articulatory features on new languages without the need to train detectors on those particular languages.

Multilingual feature detectors were trained applying the *ML-mix* acoustic model combination method as previously described. The set of shared feature detectors trained on the five languages (MM5) was then compared to the monolingually trained feature detectors. Table 4.6 shows the result: The monolingual models clearly outperform the multilingual ones when tested on the training languages. On the other hand, the multilingual models always outperform monolingual models when applied across languages—for example, MM5 achieves 90.40% on English while the monolingual detectors trained on Chinese, German, Japanese, or Spanish range between 86.36% and 87.90%. Leaving out the training language when building four-lingual models (MM4) gives better results than the monolingual ones across languages in most cases. All these effects must result from more global models, since the amount of training material was controlled to

Table 4.6 Comparison between monolingual and multilingual articulatory feature detectors; monolingual detectors are trained on one particular language, MM5 are trained by sharing the data of all five languages, and MM4 detectors are trained on all but the test language.

	Test Language (TID)				
	CH	EN	GE	JA	SP
Monolingual					
CH	**93.52%**	87.42%	88.23%	86.45%	83.22%
EN	87.74%	**93.83%**	89.17%	88.41%	87.90%
GE	88.57%	87.90%	**92.94%**	86.46%	82.68%
JA	87.11%	87.65%	86.77%	**95.22%**	87.39%
SP	84.76%	86.36%	83.31%	87.76%	**93.46%**
Multilingual					
MM5	90.56%	90.40%	88.94%	90.90%	88.71%
MM4 (w/o TID)	89.51%	88.27%	88.04%	88.02%	87.06%

be the same throughout the experiments. We conclude that multilingual acoustic modeling of articulatory features leads to similar results as in the phoneme-based case. When tested on the training language, multilingual models are outperformed by language specific monolingual ones, but when trained on one set of languages and tested on another, multilingual models are better than the monolingual models.

These articulatory feature detectors can be combined with an HMM-based speech recognition system to significantly improve the overall system performance. A flexible stream-based architecture proposed by Metze (Metze and Waibel, 2002) is applied that merges articulatory feature information with standard continuous density HMMs by computing the weighted sum of the corresponding log likelihoods. This approach was shown to improve performance on several monolingual large vocabulary tasks. As shown in Table 4.7, by selecting and weighting English

Table 4.7 Word error rates for English when decoding with articulatory feature detectors as additional stream.

		Monolingual			Multilingual	
	Baseline	EN	GE	Best of 5	MM4	MM5
English	13.1%	11.7%	11.9%	11.5%	11.8%	11.9%

feature detectors in a discriminative way, a relative gain of 10.8% on an English test set could be achieved. Very similar gains result when using the language independent feature detectors from MM4 and MM5, while a best selection from monolingual articulatory features out of five languages even outperform the multilingual and the English detectors (Stüker et al., 2003).

Grapheme-Based Recognition

The use of phonemes or subphonetic units as acoustic model units requires a dictionary that describes the pronunciation of each vocabulary entry in terms of these units. The best speech recognition results are usually achieved with hand-crafted dictionaries. However, this manual approach is very time and cost consuming especially for large-vocabulary speech recognition. Moreover, as applications become interactive, the demand for on-the-fly dictionary expansion increases. Consequently, methods to automatically create dictionaries are necessary in all those cases where no language expert knowledge is available or time and cost limitations prohibit the manual creation. Several methods have been introduced in the past, especially in the context of text-to-speech processing. Here, methods are mostly based on finding rules for converting the written form of a word into its phonetic transcription by either applying rules (Black et al., 1998a) or using statistical approaches (Besling, 1994). In speech recognition, only very few approaches have been investigated so far (Singh et al., 2002). Recently, the use of graphemes as modeling units for speech recognition has been proposed (Kanthak and Ney, 2002). The idea of using graphemes as model units—that is, speech recognition based on the orthography of a word—is very appealing, especially in the context of rapid portability to new languages, since it makes the generation of a pronunciation dictionary a very straightforward task. However, it requires that the orthographic representation of a word is given and that the relation between the written and the spoken form is reasonably close. Killer and Schultz investigated the potential of the grapheme-based approach in the context of rapid language deployment (Killer et al., 2003).

According to Weingarten, the majority of the languages writing systems are phonographic scripts, and the most widely used script is the Roman script (2003). Due to its widespread use, languages without written forms are likely to adopt some variation of the Roman script. Consequently, it is reasonable to assume that the grapheme-based approach can be applied to a very large number of languages. Furthermore, as the results in this section will show, the grapheme-based approach is also feasible for other phonographic scripts, such as Cyrillic and Thai.

As already discussed in Section 4.2.2, the grapheme-to-phoneme relationship varies widely across languages. To investigate the potential of the grapheme-based modeling approach, a variety of languages with different degrees of letter-to-sound relation were selected including English, German, and Spanish as examples of the Roman script, in which English shows the weakest grapheme-to-phoneme correspondence, Spanish shows the strongest, and German lies somewhere in between. Languages written with other than Roman scripts were also examined, namely Russian and Thai.

As described previously, the acoustic units are modeled using polyphones—that is, phonemes in the context of neighboring phonemes. Since the number of polyphones for even a very small context width is too large to allow robust model-parameter estimation, context-dependent models are usually clustered into classes using decision tree-based state tying (Young et al., 1994). Due to computational and memory constraints, these cluster trees are grown for each phoneme substate. However, this scheme prohibits parameter sharing across polyphones of different center phonemes. To cope with the fact that depending on the context, the same grapheme might be pronounced in different ways and different graphemes might be pronounced the same way, we applied a flexible tree tying (FTT) scheme (Yu and Schultz, 2003). This clustering scheme allows parameter sharing across polyphones of different center phones by constructing a single decision tree for all substates of all phonemes.

The results of grapheme-based speech recognition with and without flexible tree clustering are given in Table 4.8. The first two rows compare the performance of phoneme-based and grapheme-based speech recognizers

Table 4.8 Phoneme versus grapheme-based speech recognition [word error rates] for five languages.

System	SP	GE	EN	RU	TH
Phoneme (standard)	24.5	15.6	11.5	33.0	16.0
Grapheme (standard)	26.8	14.0	19.2	36.4	26.4
Grapheme (w FTT)	-	12.7	18.4	32.8	18.3

for five languages. The third column shows the results on the flexible tree tying scheme. All settings and components of the recognition systems as well as parameter sizes are the same except for the acoustic model and dictionary. The results show that grapheme-based systems perform significantly worse for languages with poor grapheme-to-phoneme relation such as English (Mimer et al., 2004), but achieve comparable results for closer relations such as Spanish (Killer et al., 2003), Russian (Stüker and Schultz, 2004), and Thai (Schultz et al., 2005). For German, we even see a gain by using graphemes over phonemes, which is most likely due to the more consistent dictionary. With the enhanced tree clustering, grapheme-based speech recognition outperforms the phoneme-based approach in the case of German and Russian, significantly improves the performance for Thai, and closes the gap for English. The absolute performance differences across the languages are due to a variety of factors, such as system maturity, different out-of-vocabulary rates due to morphology and/or vocabulary size, and language model training corpus size.

Additionally, we built language independent grapheme models by reassembling the work on language independent phoneme acoustic models and investigated the potential of rapid deployment to new languages (Killer et al., 2003). The results have so far shown limited success—mainly due to the fact that writing systems are relatively consistent within one language but not across languages.

4.5 Insights and Open Problems

In this chapter, we addressed the concept of multilingual acoustic modeling in the context of a large variety of languages. More specifically, we introduced the terms language dependent, language independent, and language adaptive acoustic models. We highlighted language similarities and dissimilarities based on many different languages, and discussed the resulting challenges for multilingual speech processing. Several methods and techniques to combine language dependent acoustic models across languages to form a language independent model set were investigated. We showed how multilingual acoustic models can be applied to simultaneous speech recognition in a compact language independent LVCSR system. Furthermore, we introduced the development of language adaptive acoustic models that allow the rapid porting of acoustic models to a new target language by borrowing models and data from various languages but using only a limited amount of adaptation data from the target language. In addition, the effectiveness of these language adaptive acoustic models was explored with respect to context-dependent modeling in combination with polyphone decision tree specialization methods.

Multilingual acoustic modeling serves many different goals. The comparison of acoustic models across a large variety of languages offers insights into similarities and differences between languages. This allows us to extrapolate the expected performance when moving to unseen languages. In addition, multilingual acoustic models can often be applied to other tasks. The existence of phone recognizers in multiple languages has proven to be very useful for the identification of nonverbal cues, such as language identification (Zissman and Singer, 1995), accent identification, and speaker identification (Jin et al., 2003). By combining the output of multiple streams, we found significant improvements in all three tasks—especially in the case of mismatching training conditions (Schultz et al., 2002).

The combination of language dependent acoustic models to shared language independent models allows the simultaneous recognition of several languages. This includes implicit language identification (see Chapter 8), since during decoding, the acoustic models compete with each other and the resulting recognizer output indicates the spoken language. This procedure is required in public information kiosk systems that are exposed to a large (open) set of languages. Of course, such a system could also be built by a preceding explicit language identification module; however, the implicit strategy is usually more efficient (Aretoulaki et al., 1998) and more accurate due to higher-level knowledge sources (Schultz et al., 1996b). Furthermore, a language independent acoustic model together with a strategy for transparent dictionaries and language models results in multilingual interfaces, which enable code-switching—a seamless switching of the language in the middle or between utterances (Fügen et al., 2003a). Language independent acoustic models also proved to be very useful for the recognition of language mixed pronunciations of proper names, as in name-dialing applications and non-native speech recognition, which will be described in more detail in Chapter 9. Language independent acoustic models can save memory and CPU resources, which is extremely valuable for small devices such as PDAs and cell phones (Badran et al., 2004). Last but not least, multilingual components are attractive to keep speech processing systems compact and maintainable.

Finally, the development of language adaptive acoustic models in combination with highly efficient adaptation schemes allow a rapid adaptation of acoustic models to new target languages, even if only a very limited amount of data of the target language is available. Recently, language independent and language adaptive acoustic models have been applied in the area of multilingual speech synthesis. This will be further described in Chapter 7.

We introduced the development cycle of compiling data in multiple languages, combining the acoustic model to language independent and language adaptive model sets, and then adapting them to new target languages. However, current experiments in language independent acoustic model combination reveal that sharing data across languages results in more general models that are less specific for the languages that are included in the training set. In most cases so far, sharing results in less sharp models that give worse results when compared to language specific models. However, model combination can be very beneficial if the data of the target language is very limited. It can be concluded that it requires careful consideration as to when and which knowledge is to be shared when languages are combined. When new languages need to be bootstrapped with small amounts of training data, language independent and language adaptive models prove to be favorable over language dependent models of other languages.

Although speech processing systems have been developed in a large variety of languages, only a small number of languages have so far been addressed compared to the vast amount of languages in the world. Each language exhibits individual difficulties at one level or another. In fact, some languages are relatively "easy" at one level but extremely "hard" at another, such as Japanese, which has a rather simple system of sounds that are easy to recognize but has a rather complicated higher-level semantic structure. Therefore, the question is, How can we optimally prepare ourselves for the hundreds of languages to be handled in the future? Have we already encountered all major language dependent properties? If not, can we characterize a subset of the world's languages that covers all peculiarities? How can such a subset be selected, and would the subset be small enough to be collected beforehand? How can we technologically support the collection of such languages even if many of them are currently not of great economic value? How do we reach native speakers of (low-density) languages that are about to disappear?

Another experience is that it is surprisingly time and cost intensive to build speech recognizers for a new language with a reasonable performance although the technology for training and evaluating speech recognizers is widely available. One reason has to do with the peculiarities that come with a new language, but another one is the lack of language experts—especially for low-density languages. It is extremely difficult to find native speakers who have enough insight into their own language and the technology background to be entrusted with the development of speech processing systems. Without such a skilled language expert, speech recognition developers face the time-consuming situation of either familiarizing themselves with the language or educating an unskilled language expert with the necessary technology. How can we bridge this gap between language and technology expertise? What kind of tools do we need to enable "native" language experts to build a speech processing system? How can we accelerate this process of learning and teaching?

One method to accelerate the development cycle has been proposed in this chapter, namely the use of language adaptive acoustic models to rapidly bootstrap acoustic models in new languages with very limited data resources. In order to define language independent or adaptive acoustic models, we assumed that the production of human speech is similar across all languages. Is this a valid assumption? Do we know enough about *all* languages of the world to make this claim? We furthermore applied the idea of sharing phoneme models across languages. Is this appropriate? A phoneme is the smallest sound unit that makes a difference in words of *one* language. As a consequence, a phoneme is always a language specific unit. Is it appropriate to share them across languages? Who knows enough about two or more languages to decide about the similarities between phonemes? Last, but not least, sharing has been applied to phonemes, implicating that phonemes are appropriate units to model speech sounds in *any* language. As already pointed out, many scientists are in strong favor of alternative sound representations. The poor performance of phoneme-based systems in adverse conditions supports the trend to turn to alternative units. We discussed articulatory features as alternative units and showed multilingual and crosslingual experimental results. Are articulatory features the appropriate solution when it comes to rapid adaptation to new languages?

Multilingual speech processing provides a great opportunity to revisit lingering challenges. First, current speech technology is challenged by the peculiarities of many languages, which increases the probability to detect inappropriate modeling assumptions. Second, recognition of multiple languages, especially their simultaneous recognition, can be viewed as an extreme instance of model mismatch and can therefore serve as a testbed for model adaptation and other robustness techniques. Third, due to the need for technology-educated language experts, we are forced to think about speech and language technology education in general. Last but not least, the high demand of speech processing systems in new languages encourages the development of tools and methods that automate the building process.

Chapter 5

Multilingual Dictionaries

Martine Adda-Decker and Lori Lamel

5.1 Introduction

Substantial progress in speech technologies over the past decade has led to a variety of successful demonstration systems and commercial products. In an international context where potential users speak different languages, speech-based systems have to be able to handle multiple languages as well as code-switching and non-native accents. Multilingual environments are very common with a wide spectrum of potential applications. To increase the usability of speech systems, the challenges of multilinguality and non-native speech must be addressed efficiently. Porting a given system to another language usually requires significant linguistic resources as well as language-specific knowledge in order to obtain viable recognition performance. Speech recognizers are often quite sensitive to non-native speech, with notable performance loss when compared to native speech. The two main research directions taken to address this problem have been training acoustic models on non-native speech to implicitly model the accents, and adapting pronunciation dictionaries to take into account some known characteristics for a given accent.

Research in multilingual speech recognition has been supported by the European Commission when dealing with multiple languages (there are now 20 official languages of the European Union, not counting regional languages) and, more recently, by the Defense Advanced Research Project Agency (DARPA) for a relatively limited number of languages (Mariani and Lamel, 1998; Armstrong et al., 1998; Chase, 1998; Mariani and Paroubek, 1998; Culhane, 1996; Pallett et al., 1998). Over recent years, there has been growing interest in reducing the costs (in terms of effort and money) to bootstrap the development of technologies for previously unaddressed languages.

The vast majority of the approximately 6,900 languages in the world do not have an acknowledged written form. Only 5–10% of all languages use one of about 25 writing systems (see Chapter 2 for a classification of writing systems).

To date, speech processing has primarily addressed languages for which there is a commonly accepted written form, with the exception of recent studies on dialectal forms of Arabic (Vergyri and Kirchhoff, 2004) and minority languages such as Mapudungun within the Avenue project.[1] This is largely due to the need to represent the language in a written (normalized symbolic) form for further downstream processing. Particularly for automatic speech recognition, the core functionality of a system is the automatic generation of a written representation of speech. However, for other tasks, such as speech-to-speech translation, the written form of a language can be considered less crucial, and ongoing research in this field will show to what extent and under what conditions it will be possible to bypass a written form of the language.

In this chapter, only languages for which written resources are available are considered. For relatively close dialects of standard written languages (for example, some Arabic dialects), automatic transcription may be able to bootstrap off the standard form. For other spoken languages, automatic processing tools may offer help to linguists working to define phonemic and morphological systems, aiding progress toward definition of a writing system.

From the speech recognition point of view, there is generally the need for at least a minimal knowledge of the linguistic characteristics of the language of interest and the means to obtain the necessary linguistic resources.

[1] http://www.cs.cmu.edu/~avenue

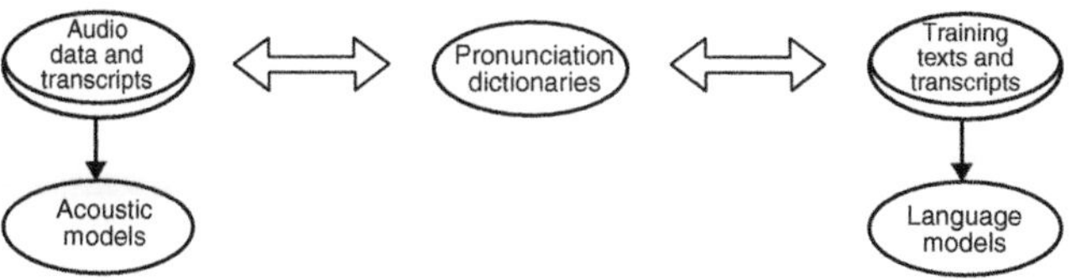

Figure 5.1: Language-dependent resources for transcription systems.

As shown in Figure 5.1, there are typically three primary language resources required for system development: texts for training language models; audio data for training acoustic models; and a pronunciation dictionary. While there is a tendency to treat these related activities as separate research areas (acoustic modeling, pronunciation modeling, and language modeling), there are close links between all three. The transcriptions of the audio data and the language model training texts are typically used in defining the recognition vocabulary; the pronunciation dictionary is the link between the acoustic and language models.

The main focus of this chapter is on multilingual dictionaries for use in automatic speech recognition. There are some common aspects with issues discussed in several other chapters—in particular Chapter 7, concerning multilingual speech synthesis; Chapters 4 and 6, on multilingual acoustic and language modeling; and Chapter 9, on non-native speech.

5.2 Multilingual Dictionaries

What is meant by multilingual dictionaries? Can we build a super-dictionary as the union of monolingual dictionaries? What is the intersection of two monolingual dictionaries of two different languages? When looking first at monolingual dictionaries, one can find, in variable proportions, entries from other languages. For example, it is known that the vast majority of words in French are derived from Latin. However, approximately 13% of the French vocabulary is imported from other languages (Walter, 1997). Imported items may be subject to some graphemic assimilation transformations, as shown by the Italian and Germanic examples in

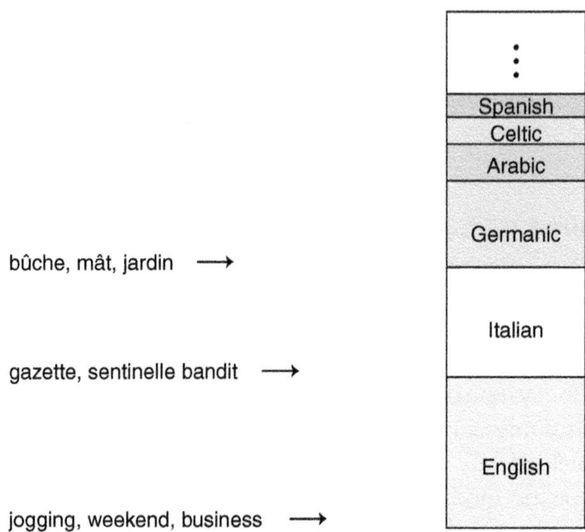

Figure 5.2: About 13% of French's entries are imported from other languages, mainly English, Italian, and Germanic (after Walter, 1997).

Figure 5.2. More recently adopted English words mainly keep their original orthography, even though it is quite different from the French system. Any living language continually imports items from languages in contact.

Similar observations of multilingual permissiveness in monolingual dictionaries hold for other languages (e.g., Spanish or French entries in English dictionaries, French and English entries in German dictionaries). The adoption of foreign words can even imply a change in the writing system (e.g., Japanese). An interesting problem is raised by proper names for all languages. The same person or the same location may be designated by many different surface forms. For instance, in English news texts the following spellings of Muammar Kadhafi are found: *Kadafi, Kadaffi, Kaddafi, Khadafi, Khaddafi, Khadafy, Khaddafy, Khaddaffy, Qaddafi, Qadhafi, Qadaffi, Qadafi, Qhadafi.*

If languages are close cousins within a family of languages (e.g., Italian and Spanish in Romance languages), a number of identical words can be found. For example, the pairwise intersection of the *N* most frequent words in the Indo-European languages French, Spanish, and German results in an overlap of less than 1% for the most frequent 1,000 words in each language, but 10% for the most frequent 20,000 words. Although the vast majority of shared words are proper names and place names, some of the other words (ignoring accents) are *union, region, club, normal,* and *via.*

Concerning the current state of the art in large vocabulary speech recognition, multilingual pronunciation dictionaries are generally collections of monolingual dictionaries that are selectively applied, depending on the identity of the language hypothesized for the speech signal. However, there is ongoing research in speech recognition, speech synthesis, and speech-to-speech translation on how to couple dictionaries from different languages more tightly.

There are three main considerations when designing pronunciation dictionaries: the definition of words in the particular target language, selection of a finite set of words, and determining how each of these words is pronounced. Each of these aspects will typically require a variety of decisions that may be more or less language dependent and have a set of consequences that are interrelated with the two other considerations. The three main aspects of the global language independent design process are represented in Figure 5.3.

While in many languages the definition of a word may at first appear to be straightforward for written texts (e.g., a sequence of alphabetic characters separated by a space or some other specified marker), for other languages, this is not the case (e.g., Chinese, cf. Chapter 2). Automatic procedures have been successfully used to propose a word-like splitting of the continuous character flow. These are also relevant in the context of language modeling, as further described in Chapter 6.

Word definition for speech recognition needs to meet two contradictory requirements. On one hand, the number of distinct entries needs to be within reasonable limits, such that good coverage of the system's vocabulary is guaranteed while still enabling the reliable estimation of language model probabilities. This condition favors smaller units. On the other hand, there is a tendency to prefer longer items in order to provide context for pronunciation and acoustic modeling. For these antagonistic criteria, a trade-off is sought, which depends on the amount of available training texts, the limit for the vocabulary size, and the speaking style of the data to be handled. To overcome fixed-size vocabulary limitations, there has been growing interest in open-vocabulary speech recognition, for example, dynamic adaptation of the recognizer vocabulary (Allauzen and Gauvain, 2005b, 2003), which can help reduce errors caused by out-of-vocabulary (OOV) words.

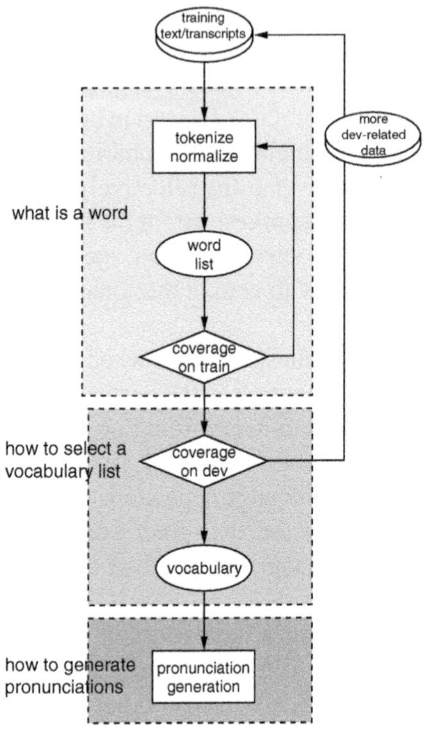

Figure 5.3: Language independent processing steps for pronunciation dictionary generation.

Pronunciation generation can be carried out automatically for languages with a close-to-phonemic writing system (e.g., Italian, Spanish); for others, preexisting pronunciation dictionaries or manual dictionary development are required (e.g., English). The writing system of languages such as Arabic, which specifies only consonants and long vowels, underspecifies the pronunciations, resulting in a high degree of ambiguity. Recent research has addressed using graphemes directly for speech recognition without using an explicit phonemic or phonetic representation of word pronunciations (Billa et al., 2002; Kanthak and Ney, 2003; Killer et al., 2003; Stüker and Schultz, 2004). An important criterion, whatever the adopted method of pronunciation generation, is consistency: if different lexical entries share (partly) the same pronunciation, the phonemic transcription should be (partly) identical.

Some of these considerations are discussed in more detail below.

5.3 What Is a Word?

The question of what is a word seems to be trivial when we are considering languages with a well-established orthography. In general, each of these languages implies a clear sociocultural status, with an education system promoting and relying on a writing standard, and a high production of written language resources. For many spoken languages, however, there are no established writing conventions. For example, in Amharic, any written form that reflects what is said is acceptable (Yacob, 2004). The same applies to Luxembourgish before its writing reform some 20 years ago. For such languages, the orthographic variability and its manifestation at the acoustic level are major challenges for automatic speech recognition. The question of how to define a word for spoken languages with no established orthography is primarily a concern of linguistic research. However, automatic processing may contribute to future progress in addressing such problems.

The following discussion focuses on how to define a word for automatic speech processing in languages with widely adopted writing conventions. Even in this situation, nontrivial questions arise for adopting appropriate lexical units.

A lexicon's word list generally consists of a simple list of lexical items observed in running text.[2] A straightforward definition of a lexical item as a graphemic string between two blanks is too simplistic to be applied without generating a large number of spurious items. Depending on the form of the training texts, different word lists can be derived.

The sample lexicon shown in Figure 5.4 may be obtained from texts like:

> Mrs. Green is a member of the Greens Garden Club.
> Bob Green's car is green.

[2]It can also be a list of root forms (stems) augmented by derivation, declension, and composition rules. This approach is more powerful in terms of language coverage, which is a desirable quality for recognizer development but more difficult to integrate in present state-of-the-art recognizer technology.

Normalization 1		*Normalization 2*	
graphemic form	*phonemic form*	*graphemic form*	*phonemic form*
green	gri:n	green	gri:n
Green	gri:n	greens	gri:nz
Green's	gri:nz	's	z
greens	gri:nz		
Greens	gri:nz		

Figure 5.4: Sample word lists obtained using different text normalizations, with standard base-form pronunciations.

> The Greens all like eating greens.
> Green's her favorite color.

Various text forms can be generated by applying different normalization steps to the text corpora (Adda et al., 1997), which may then result in different word lists and pronunciation lexica.

While for many years a case-insensitive text form has been used for large vocabulary conversational speed recognition (LVCSR) in American English (in part due to the availability of common language models in this form) (Paul and Baker, 1992), there has been a move toward maintaining (or re-adding) case to avoid loss of syntactic and semantic information, which can be important for further downstream processing—in particular, named-entity extraction and indexing.

If during text processing case is ignored and the apostrophe (or single quote) is considered as a word boundary mark, the sample lexicon is reduced to the forms shown in the right part of Figure 5.4. In English, the apostrophe has a limited impact on lexical variety. It is mainly used to build the genitive form of nouns, but it can also be found in contracted forms, such as *won't, you'd, he'll, she's,* and *we've,* or in proper names (*D'Angelo, O'Keefe*). In general, English lexicons are represented without considering apostrophe as a boundary. For example, in a lexicon containing 100,000 entries, only 4% of the words contain an apostrophe. In contrast, in French, the *apostrophe* is very frequent, occurring in word sequences, such as *l'ami, j'aime,* and *c'est.* If all forms containing the apostrophe are included as separate lexical-entries, there is a huge expansion in the lexicon size. Therefore, different text normalizations have to be considered depending on the language's characteristics.

5.3.1 Text Normalization

A common motivation for normalization in all language is to reduce the lexical variability so as to increase the coverage for a fixed-size task vocabulary. In addition, more robust language models can be estimated; however, normalization may entail a loss in syntactic or semantic resolution. Whereas generic normalization steps can be identified, their implementation is to a large extent language-specific. In the following, the most important normalization steps implemented for processing languages such as English, French, German, Spanish, and Arabic are highlighted, and case studies are presented for French and German.

The definition of a word in a language is carried out iteratively, as was illustrated in Figure 5.3. After relatively generic text normalization steps (formatting punctuation markers, numbers), the appropriateness of the resulting words can be measured as the lexical coverage for a fixed-size vocabulary. Depending on these measures, more or less language-specific normalization can be identified and added to the processing chain. Taking, for example, the 65,000 most frequent words in the available processed training data yields a lexical coverage close to 100% for English, but only about 95% for German. This means that with the same type of normalization procedures, German has a much higher lexical variety. The sources of this variety must be identified to efficiently address this problem.

For large-vocabulary conversational speech recognition applications, some of the most readily available sources of training texts are from electronic versions of newspapers.[3] Much of the speech recognition research for American English has been supported by DARPA and has been based on text materials that were processed to remove case distinction and compound words (Paul and Baker, 1992). Thus, no lexical distinction is made between, for instance, *Gates, gates or Green, green.* In the French *Le Monde* corpus, capital letters are kept distinct for proper names, resulting in different lexical entries for *Pierre, pierre* or *Roman, roman,* for example (Adda et al., 1997). In German, all substantives are written with a capitalized first letter, and most words can be substantivized, thus generating a large lexical variety and homophones. Even so, the overall impact of

[3]While not the subject of this discussion, the text data contain errors of different types. Some are due to typographical errors, such as misspellings (MILLLION, OFFICALS) or missing spaces (LITTLEKNOWN); others may arise from prior text processing.

this kind of variability remains small. The out-of-vocabulary rate reduction when going from a case-sensitive to a case-insensitive word list in German is only about 0.2% (from 4.9% to 4.7%) with a 65,000 word lexicon.

Different types of text normalization may be explored, depending on the characteristics of the language under study. In order to illustrate problems in word definition, case studies are given for the French and German languages in which a greater variety of graphemic forms are observed than for English. For French, an extensive study on different types of normalization was reported in Adda et al. (1997), using a training text set of about 40 million words from the *Le Monde* newspaper (years 1987–1988).[4] For German, the effect of word compounding on the vocabulary has been studied, and a general corpus-based decomposition algorithm is described.

Case Study I: Effect of Normalization Steps on French Vocabulary

French lexical variety stems mainly from gender and number agreement (nouns, adjectives) and from verb conjugation. A given root form produces a large number of derived forms, resulting in both low lexical coverage and poor language model training. The French language also makes frequent use of diacritic symbols, which are particularly prone to spelling, encoding, and formatting errors. Some of the normalization steps can be considered part of the process of establishing a baseline dictionary. These include the coding of accents and other diacritic signs (in ISO-Latin 1); separation of the text into articles, paragraphs, and sentences; preprocessing of digits (10 000 → 10000) and units (kg/cm^3), as well as the correction of typical newspaper formatting and punctuation errors; and processing of unambiguous punctuation marks. These are carried out to produce a baseline text form. Other kinds of normalization generally carried out include the following:

N_0: processing of ambiguous punctuation marks
(hyphen -, apostrophe ') not including compounds
N_1: processing of capitalized sentence starts

[4]Evaluating n different types of text normalization entails producing (at least temporarily) n times the training text volume. For this reason, the study has been carried out on a limited subset (40 million words) of the complete training text material available at the time (200 million words).

N_2: digit processing (110 → *cent dix*)
N_3: acronym processing (*ABCD* → *A. B. C. D.*)
N_4: compounding punctuation (*arc-en-ciel* →*arc en ciel*)
N_5: remove case distinction (*Paris* → *paris*)
N_6: remove diacritics (*énervé* → *enerve*)

N_0, N_2, N_3, and N_4 can be considered as "decompounding" rules, which change tokenization (and thus the number of words in the corpus). N_1, N_5, and N_6 keep word boundaries unchanged but reduce intraword graphemic variability.

The elementary operations $N_0 \ldots N_6$ can be combined to produce different versions of normalized texts. Eight such combinations based on the normalization operations $N_0 \ldots N_6$ are shown in Table 5.1. Only the baseline normalizations are used to produce the reference text V_0. The N_0 and N_1 normalizations make use of two large French dictionaries: BDLEX (Pérennou, 1988) and DELAF (Silberztein, 1993) to produce V_1 and V_2 texts. A more detailed description of the normalizations can be found in Adda et al. (1997).

While any normalization results in a reduction of information, the amount of information loss varies for the different types of normalizations. It is straightforward to recover a V_0 text (or an equivalent form) from a V_5 text using some simple heuristics. It is nearly impossible to recover the original V_0 forms from the V_6 and V_7 texts without additional knowledge sources. Furthermore, the V_7 texts seem poorly suited for speech recognition, since a high level of lexical ambiguity is introduced.

Table 5.1 For each version V_i ($i = 0, \ldots, 7$) of normalized text, the elementary normalization steps N_j ($j = 0, \ldots, 6$) are indicated by 1 in the corresponding column.

	N_0	N_1	N_2	N_3	N_4	N_5	N_6	Comment
V_0	0	0	0	0	0	0	0	baseline normalizations
V_1	1	0	0	0	0	0	0	V_0 + ambiguous punctuations
V_2	1	1	0	0	0	0	0	V_1 + capitalized sentence starts
V_3	1	1	1	0	0	0	0	V_2 + digits
V_4	1	1	1	1	0	0	0	V_3 + acronyms
V_5	1	1	1	1	1	0	0	V_4 + decompounding
V_6	1	1	1	1	1	1	0	V_5 + case-insensitive
V_7	1	1	1	1	1	1	1	V_6 + no diacritics

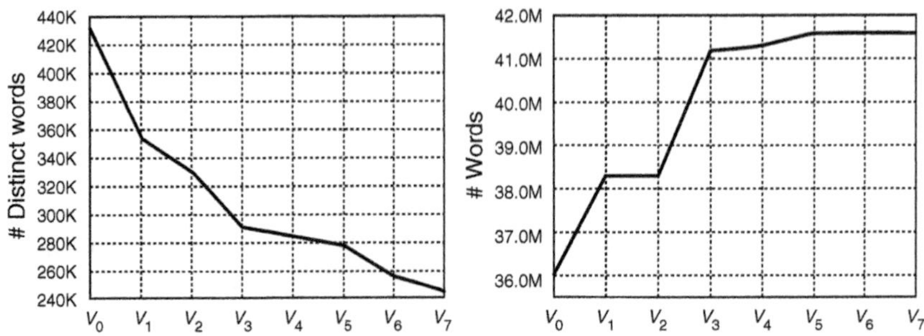

Figure 5.5: Number of distinct (left) and total number (right) of words in the training data for different normalization combination V_i.

Using these different normalization combinations, the number of distinct words and the lexical coverages for each text version can be compared with the training data to help evaluate the relative importance of each step. In the plot on the left-hand side of Figure 5.5, three regions can be distinguished: (1) the region between V_0 and V_3, in which the curve indicates a strong decrease in lexical variety, dropping from 435,000 to 290,000 word forms, (2) the middle region between V_3 and V_5 in which the curve is relatively flat, and (3) the region on the right of V_5, in which the curve gently decreases toward 245,000 word forms. The most important normalization steps are N_0 and N_2, which are case-independent "decompounding" rules. These account for 65% of the gain achieved by the best version V_7. The impact of the decompounding rules on the total number of words in a given text is shown in Figure 5.5 (right); an increase is observed for text versions V_1, V_3, and V_5, where the difference with the previous version is an additional normalization of type N_0, N_2, and N_4, respectively. Figure 5.6 shows the corresponding OOV rates (complementary measure of lexical coverage) of the training data using 64,000 entry lexica (containing the most frequent 64,000 words in the corresponding normalized training data). The OOV rate curve is seen to parallel the #-distinct-word curve of Figure 5.5. A large reduction in OOV rate is obtained for the V_1, V_2, and V_3 text versions, which correspond to the processing of ambiguous punctuation marks, sentence-initial capitalization, and digits. Subsequent normalizations improve coverage, but to a lesser extent.

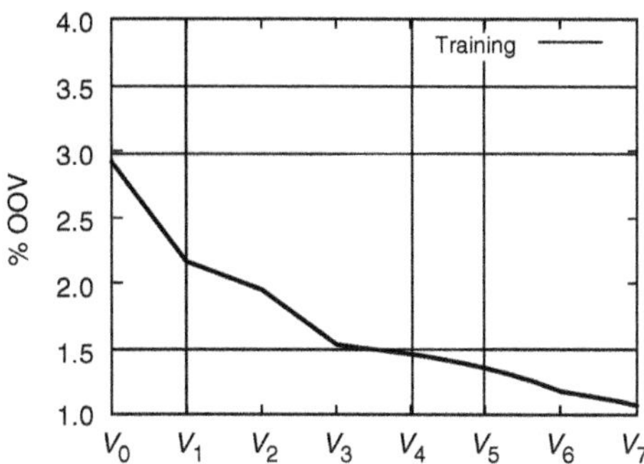

Figure 5.6: OOV rates for different normalization versions V_i on the training data using 64,000 word lists.

This shows the importance of processing punctuation marks and numbers prior to word-list selection. Generally speaking, the influence of ambiguous punctuation-mark processing can be considered as language-specific, whereas number processing is probably important for all languages. If a case-sensitive text output is a desirable feature of the recognizer, capitalized sentence-start processing also has a significant impact on lexical coverage; the other normalization steps turn out to be less important. The V_5 text form achieves a good compromise between standard correct French system output and lexical coverage (Adda et al., 1997; Gauvain et al., 2005). The final choice of a given normalization version has to be chosen as a compromise between the best possible graphemic form and the highest lexical coverage. This compromise is largely application driven.

Case Study II: Effect of Compounding on German Vocabulary

German lexical variety is mainly due to declensions and word compounding. In order to gain a deeper insight into the relative importance of both mechanisms several word-length measures can be compared. Compounding can have a multiplicative effect on word length, whereas declensions typically add just two or three characters.

Figure 5.7 shows the lexical distribution of word lists obtained for German, English, and French in a text corpus of 300 million words per

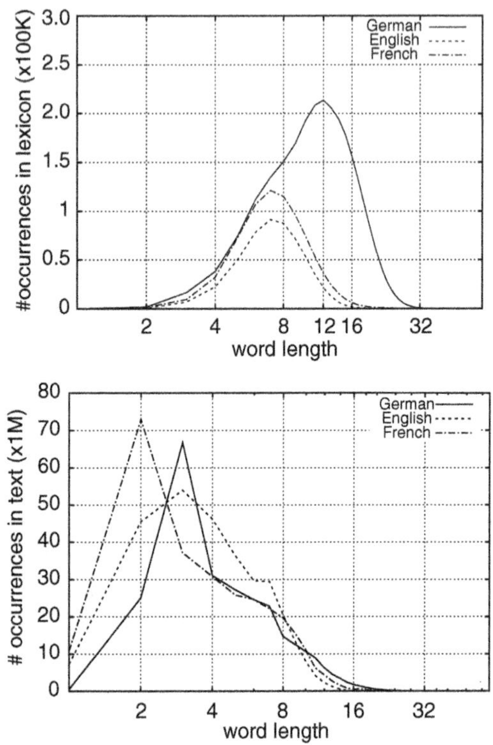

Figure 5.7: Number of words as a function of length (in characters) for German, English, and French from 300 million words running texts in each language. Number of **distinct entries** in the full **lexicon** (top). Number of **occurrences** in the **corpus** (bottom).

language. The top curves show the number of lexical entries as a function of word length in the three languages. The English and French curves are quite similar and have a maximum number of entries with a word length around 7. The French curve is higher than the English one, which can be explained by a larger number of distinct surface forms for French verbs. For example, the verb *faire* (to do) has about 40 distinct forms, whereas one of the most productive verbs in English is *to be,* with 8 distinct forms (*be*, *am*, *are*, *is*, *was*, *were*, *been*, and *being*). In German, verb conjugation as well as noun and adjective declension add significant variety to the word lists. For instance, more than 10 distinct inflected forms can be found for the adjective *schnell* (fast) among the 65,000 most frequent entries, including the comparative and superlative forms (faster, fastest). The general slope of the German curve is quite different, with an inflection point around length 8. This characteristic can be related to **word compounding** and suggests that for German, compounding generates a large number of additional lexical entries. The curves in the bottom half of the figure show the distribution on the text corpus for each language by word length. It can be seen that French has, in general, the shortest words, with a sharp peak at 2 characters compared to 3 for German and a broader distribution (2–4 characters) for English.

In order to study more closely the importance of compounding as a function of part of speech, the German word list was separated by case, since the words starting with a capital letter are mainly nouns (or proper names). This division clearly demonstrates that compounding involves mainly nouns, and that while nouns account for an overwhelming percentage of the lexicon, their occurrence in the text corpus is much more limited (Adda-Decker, 2003).

As an example, of the 65,000 most frequent words in the corpus, about 100 distinct entries start with the noun *Stadt* (town), for example: *Stadtamt, Stadtarchaeologen, Stadtautobahn, Stadtbahn, Stadtbaurat.* In the total word list, there are more than 4,000 compounds beginning with *Stadt.*

Automatic Language-Independent Decomposition Algorithm

When developing a speech recognizer for a previously unseen language, it is necessary to assess the lexical variety of the available texts or transcripts: How many different units can be extracted from a given amount of data? How long are these units? If a high lexical variety is measured with a large proportion of long units, there are several reasons to consider reducing the variability. Smaller units will provide better lexical coverage for a given sized word list, easier development of pronunciations, more efficient spelling normalization, and more reliable N-gram estimates for language modeling.

German is a well-known example of a language that makes intense use of compounding to create new lexical units—a characteristic shared with other Germanic languages, such as Dutch and Luxembourgish; for the

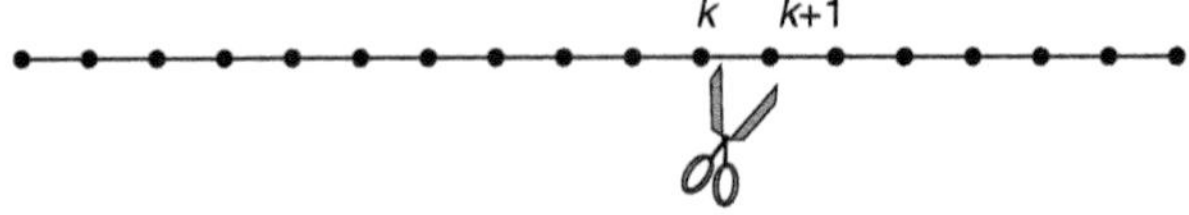

Figure 5.8: Can a word be decompounded after letter k.

latter, there is significantly less written material available, so decomposition can be an important processing step. In 1955, Zellig Harris described an algorithm to locate morph boundaries in phonemic strings (Harris, 1955) based on a general characteristic of spoken language: the number of distinct phonemes that are possible successors of the preceding string of phonemes reduces rapidly with the length of that string unless a morph boundary is crossed. This feature is easily transposable to written language: the number of potential distinct letters that are possible successors of a given word start reduces rapidly with the length of the word start.

The written language decomposition problem as illustrated in Figure 5.8 can be stated as: given a word of length K, is there a morpheme boundary between letters k and $k + 1$? A straightforward solution is to check whether the decompounded items exist in a language's baseline vocabulary. Table 5.2 gives some example words with multiple possible decompositions. When the boundary is ambiguous, more information is required to make a decision. This information can be easily extracted from the corpus

Table 5.2 Example words with ambiguous decompositions.

compound	⇒	*decompounded*
Fluchtorten	⇒	Flucht-Orten (right)
Fluchtorten	⇒	Fluch-Torten (wrong)
Musikerleben	⇒	Musik-Erleben (right)
Musikerleben	⇒	Musiker-Leben (right)
Regionalligatorjäger	⇒	Regional- Liga-Tor-Jäger (right)
Regionalligatorjäger	⇒	Region-Alligator-Jäger (wrong)
Gastanker	⇒	Gas-Tanker (right)
Gastanker	⇒	Gast-Anker (wrong)
weiterdealt	⇒	weiter-dealt (right)
weiterdealt	⇒	weit-erde-alt (wrong)

Table 5.3 Given a word start *Wbeg*(k) of length k, the number of character successors #*Sc*(k) generally tends toward zero with k. A sudden increase of #*Sc*(k) indicates a boundary due to compounding. #*Wend*(k) indicates the number of words in the vocabulary sharing the same word start.

k	*Wbeg*(k)	#*Wend*(k)	#*Sc*(k)	*Examples*
1	K	147731	62	Klasse, Kopf, Kritik, Kind, Köln, Kurs
2	Ka	29068	41	Kampf, Kanzler, Kairo, Kauf, Kappe
3	Kap	2131	25	Kapuze, Kapriolen, Kapitän
4	Kapi	1281	14	Kapielski, Kapillaren, Kapitel
5	Kapit	1218	8	Kapitän, Kapitel, Kapitulation, Kapitol
6	Kapita	974	4	Kapital, Kapitain, Kapitan
7	Kapital	968	27	Kapitalismus, Kapitals, K-erhöhung

by organizing the word list in grapheme-node-based lexical trees. For a given grapheme node at depth k, the higher its branching factor (number of successor nodes), the more reliable is its boundary hypothesis.

As can be seen in the last entry of Table 5.3, which gives the successor information for the word start *Kapital*, at the location of a lexical (or morphemic) boundary, the number of successors significantly increases. This general behavior is schematically depicted in Figure 5.9.

Using this type of analysis, the boundary location for the following two examples can be resolved. The second example is ambiguous. If a word start has more than 10 distinct successor letters, the successor number is displayed.

Pfirsichtorten	$\text{P}_{58}\text{f}_{25}\text{i}_{14}\text{rsich}_{17}\text{torten}$	Pfirsich-Torten
Fluchtorten	$\text{F}_{60}\text{l}_{22}\text{u}_{25}\text{ch}_{15}\text{t}_{27}\text{orten}$	Flucht-Orten

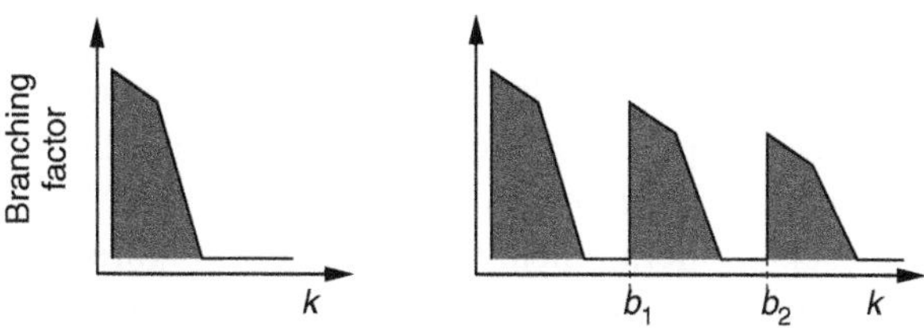

Figure 5.9: Goëlette profile for decomposition: branching factor as a function of length k for a simple word (left) and a three-word based compound (right).

Table 5.4 Examples of decomposition rules, including composita with imported English and French items; the number of occurrences of the decomposed items is given in parentheses.

3-word composita			
Theateraufbruchstimmung	→	Theater - Aufbruch•stimmung	(29205 - 788)
Schmerzensgeldanspruchs	→	Schmerzens•geld - Anspruchs	(1001 - 336)
Schlechtwettereinbruchs	→	Schlecht•wetter - Einbruchs	(65 - 253)
German-English composita			
Programmhighlights	→	Programm - High•lights	(38977 - 653)
Lightgetränk	→	Light - Getränk	(638 - 562)
Imagezuwachs	→	Image - Zuwachs	(7279 - 3890)
English-English composita			
Streetlights	→	Street - Lights	(6522 - 97)
Lightdesigns	→	Light - Designs	(638 - 232)
Breakthrough	→	Break - Through	(811 - 59)
German-French composita			
Weltraumrendezvous	→	Welt•raum - Rendez•vous	(1815 - 622)
Avantgardezeitungsprojekt	→	Avant•garde - Zeitungs•projekt	(2559 - 61)
Luxusboutiquen	→	Luxus - Boutiquen	(2564 - 410)

For the first word, the compound boundary can be unambiguously placed after the word *Pfirsich* with the locally highest branching factor. Similarly, in the second example, there is a large increase in branching factor after the letter *t*, indicating that the boundary should be placed after the word *Flucht*.

Table 5.4 gives examples of decompounded items for some typical German 3-word compounds and for compounds mixing German, French, and English items. The results of a one-step decomposition are shown; that is, only one boundary is located. Remaining boundaries are indicated with a •. The decompounding algorithm can be applied iteratively. After each iteration, the word lists, lexical trees, and successor-node information are updated. Decomposition rules already extracted for shorter items can be applied to partially decompounded items. The resulting decomposition can be represented hierarchically, as shown in Figure 5.10. It provides a semantic structuring, which may be useful for certain applications, such as translation and indexing.

Table 5.5 highlights the importance of decomposition as a normalization step for compounding languages. The larger the vocabulary size, the higher

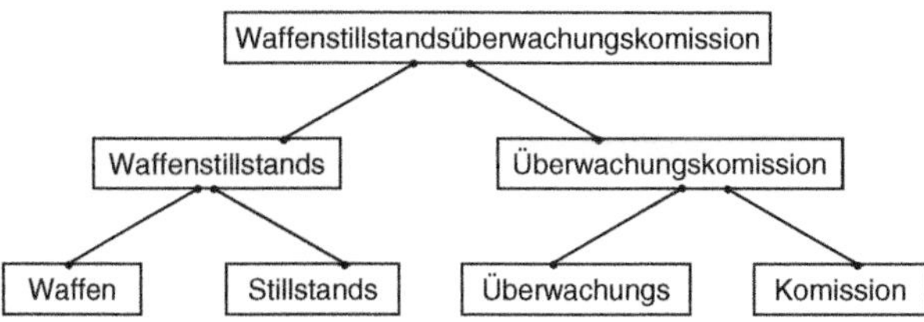

Figure 5.10: Hierarchical representation for a complex decomposition.

Table 5.5 Lexical coverage and complementary OOV rates measured for different-size vocabularies on a 300-million word German text corpus. Measures are given with and without decomposition. The last two columns indicate the absolute and relative gains in OOV reduction rates.

Vocab.	$\%cov_{orig}$	%OOV	$\%cov_{decomp}$	%OOV	Δabs	Δrel
65k	94.8	5.2	96.0	4.0	1.2	23
100k	96.1	3.9	97.2	2.8	1.1	28
200k	97.6	2.4	98.5	1.5	0.9	37
300k	98.3	1.7	99.0	1.0	0.7	41
600k	99.0	1.0	99.5	0.5	0.5	50

the relative OOV reduction rate. The OOV rate of a 300,000 vocabulary on 300 million words of training data is about 1% with the decompounded text version, whereas the original text OOV rate is close to 2%.

The decomposition algorithm presented here is language independent and only requires large corpora. It can thus be straightforwardly adapted to any language to minimize the impact of compounds on lexical coverage. Beyond its utility for tuning lexical coverage in a given language, the knowledge of lexeme (and morpheme) boundaries may be important for pronunciation generation. This point is addressed later in the chapter.

5.4 Vocabulary Selection

Recognizer vocabularies—that is, word lists for LVCSR—are generally defined as the *N* most frequent words in training texts. This guarantees optimal lexical coverage on the training data. As the resulting word list depends heavily on the content of the training material, it is not necessarily optimal under testing conditions. For large vocabulary speech recognition, a major requirement for the word list is high lexical coverage during testing. In order to achieve this, the training text materials should be closely related (in time and topics) to the test data. In the following, the dynamic properties of living languages are discussed, and some measures highlighting the importance of epoch adequacy between training and test data from similar sources are presented. This is followed by a discussion of spoken language specific problems, the differences between spoken and written language, and how word-list development can accommodate these. Finally, prospective developments for multilingual dictionaries are presented.

5.4.1 Vocabulary Changes Over Time

Research on automatic transcription of broadcast news speech has highlighted the importance of word lists keeping pace with language usage across time. Diachronic word list and language model adaptation is a research area of its own (Allauzen and Gauvain, 2005a; Federico and Bertoldi, 2004; Khudanpur and Kim, 2004; Chen et al., 2003, 2004). The usage of a word can decay or increase with time, and completely new items may appear. An existing word can, at a given moment, be boosted by an important personality (*abracadabrantesque*, used by French President Chirac), by new techniques (*toile*, "*net*"), or by its usage in another language (the English words *road map* has boosted the usage of the translation *feuille de route* in France, which has become very popular in political speeches but also in everyday conversations). New items (neologisms and proper names) may also be introduced. In the last ten years, new items such as *européiste, solutionner*, *cédérom*, *Internet*, and *cyber-café* have appeared, which originated either in morphological combinations of existing items or as a result of new technological developments. However, in a system's word lists, most new items correspond to proper names.

5.4.2 Training Data Selection

It is common practice to use a set of development data in order to select a word list representing the expected test conditions. In practice, the selection of words is done so as to minimize the system's OOV rate by including the most useful words. In this context, *useful* refers to (1) being an *expected input to the recognizer*, and (2) being *trainable for* **LMs** *given the training text corpora.* In order to meet the latter condition, one option is to choose the N most frequent words in the training corpora. This criterion does not, however, guarantee the usefulness of the lexicon, as stated by the first requirement. Selection or weighting of the training data can be a step in this direction.

For transcription of general news, problems of lexical coverage can appear if the training corpora are either too small or too remote in time from the test data. To illustrate this problem, French is again used as an example, although similar behavior could be expected for any other processed language. In order to measure the lexical coverage under similar training and test conditions a development set (dev_{96}) of about 20,000 words was extracted from the *Le Monde* newspaper from the month of May 1996. The impact of training corpus size and epoch on lexical coverage was measured by defining two additional training corpora: $T_{87\text{–}95}$ and $T_{91\text{–}95}$. The training text sets compared are:

$T_{87\text{–}88}$: 40 million words from 1987–1988
$T_{94\text{–}95}$: 40 million words from 1994–1995
$T_{87\text{–}95}$: 185 million words from 1987–1995
$T_{91\text{–}95}$: 105 million words from 1991–1995 of more recent data

Figure 5.11 compares OOV rates using 64,000 word lists (containing the most frequent words) obtained on the $T_{87\text{–}88}$ and $T_{94\text{–}95}$ training sets to the OOV rates on the dev_{96} data. For the word list derived from the $T_{87\text{–}88}$ training texts, the OOV rates for the dev_{96} set are significantly higher than those for the training data for all text versions. In contrast, for the word list derived from the $T_{94\text{–}95}$ training texts, the OOV rates on the training and dev_{96} sets are quite similar. This comparison measures the impact of training data epoch using a constant amount of training material, and illustrates the need for up-to-date data. As mentioned before, an important proportion of the word list consists of proper names related to current events, which are strongly time and topic dependent.

Figure 5.12 shows that the use of larger and more recent training texts ($T_{87\text{–}95}$ or $T_{94\text{–}95}$) significantly reduces OOV rates on test data. The OOV

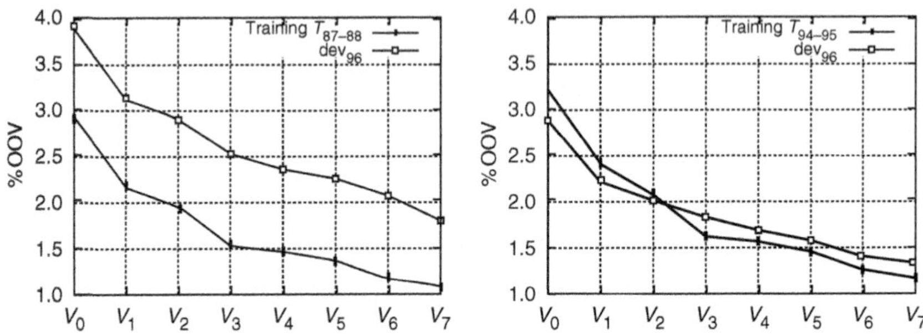

Figure 5.11: OOV rates on training and dev_{96} data for different normalization versions V_i and 64,000 most frequently words from 40 million training data highlighting the importance of training epoch. **Left:** T_{87-88} training data (years 1987–1988). **Right:** T_{94-95} training data (years 1994–1995).

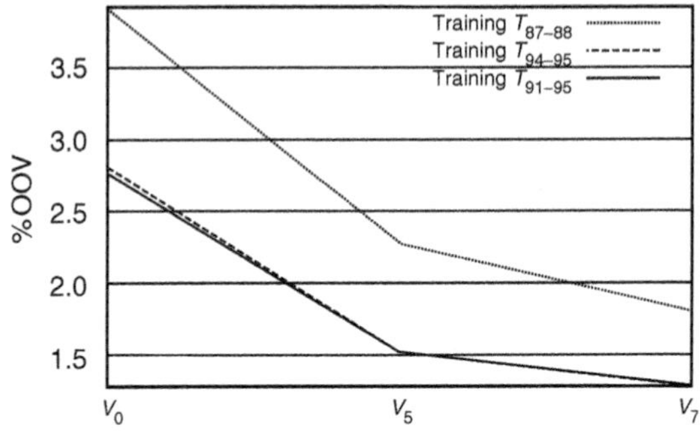

Figure 5.12: OOV rates for normalization versions V_0, V_5, and V_7 on dev_{96} text data, using 64,000 word lists derived from different training text sets.

rates for the T_{94-95}, T_{87-95}, and T_{91-95} curves are close to 1.5% for V_i, versions ($i > 0$), with the absolute difference from the T_{87-88} curve of close to 1%. The time proximity between the training and test data is more important than the use of additional older data to minimize the OOV rate. Appropriate selection of training material for a given test condition is also seen to be more important for reducing the OOV rate than some of the elementary normalization steps (compounding, punctuation, case sensitivity).

Optimized training-data selection can be carried out by weighting recent training texts more than the older text material. This optimization can even eradicate the effects of some minor normalizations (Adda et al., 1997).

5.4.3 Spoken Language-Specific Vocabulary Items

When processing spontaneous speech, there are additional considerations that should be taken into account. Since it is generally easier to have access to written texts than to have transcriptions of spoken language, such texts constitute the vast majority of language model training material. However, there are important differences between written and spoken language, including the vocabulary items, their frequencies, and reduced forms found essentially only in speech (Boula de Mareüil et al., 2005). For example, first- and second-person (singular and plural) pronouns and related verb forms are much more frequent in speech transcripts than in written texts. Thus, among the top words in speech transcripts are words such as *I, we, you, am,* and *are* but also hesitations and discourse markers. However, numbers and acronyms have a relatively low usage in speech transcripts as compared to news text sources. Speech transcripts are therefore particularly important for adapting language models for spontaneous speech.

There are also a number of abbreviated words or clipped words in spoken language that are rarely found in standard written texts. The use of clipped words seems to be quite language dependent, and is frequent in vernacular French. Some examples of general clipped words in French are *appart* (*appartement*), *aprèm* (*après-midi*), *cata* (*catastrophe*), and *compèt* (*compétition*) (Huot, 2001). Similar items, though less frequent, can be found in spoken English, e.g., function words such as *cuz* (*because*) but also some content words, such as *vet* (*veterinarian*), *corp* (*corporation*), and *deli* (*delicatessen*).

Some of the processing specific to spoken language concerns frequent word sequences, corresponding to function words, letters of acronyms, or words composing a date or a complex number. Important temporal restructuring can be observed here (Adda-Decker et al., 2005); in particular, short words may change significantly or even disappear. Hence it is common practice to represent a limited number of acronyms as distinct lexical entries (as opposed to a sequence of individual letters) and to represent some frequent word sequences subject to reduction as compound words (Gauvain et al., 1996, 1997, 2002; Finke and Waibel, 1997; Ma et al., 1998; Stolcke et al., 2000). There are different strategies for selecting these items, ranging from simply including the most frequent *N*-word sequences, to including all *N*-word sequences containing a certain set of words, to complete manual specification based on linguistic knowledge or observed reductions.

Speech transcripts, if produced thoroughly, also contain disfluencies, such as hesitation items and word fragments (incompletely uttered word starts). Depending on the application, it can be interesting to map all hesitations or filler forms such as *uh, hmm, hum*, and *uhm* to a single unique form under the hypothesis that only the fact that there is a hesitation is important, and that the particular manifestation is a personal choice and therefore unimportant. In other situations, this information may be of interest since it can be indicative of the speaker or of the language being spoken or even of the native language of a speaker using a different language (Candea et al., 2005). Such forms may also serve as back channels during communication and in some languages (e.g., English) indicate agreement (*uh-huh*) or disagreement (*uh-uh*). Word fragments are generally ignored in word lists as well as singletons (words occuring only once), which are likely to be errors.

These examples illustrate some major differences between spoken and written language, and are meant to underline the crucial importance of using speech transcripts as training data.

5.4.4 Multilingual Considerations

For all languages, recognition vocabularies are generally composed of:

- function words (hundreds)
- general content words (thousands)
- technical content words (thousands)
- proper names (millions)
- foreign proper names (millions)

These word classes are listed by decreasing frequency of occurrence and an appropriate number of types in each class is shown in parentheses.

These different types of words raise different problems for modeling and particularly for pronunciation modeling, which is addressed in the following section. Whereas function words and general content words are strongly language-specific, technical words and proper names tend to be shared more easily (after accounting for some writing convention adaptations across languages). In order to give an idea of the number and types of lexical entries shared among languages, the number of common entries among the top N words in recognizer word lists were compared in pairs, for the French, Spanish, German, and English languages. Results are shown in Figure 5.13. If for the top 50,000 words 10,000 words are shared, this represents 20% of the word list. This proportion is almost achieved for the English-French and the English-Spanish pairs. As can be expected with a higher top N limit, the shared word percentage increases since the proportion of technical items and proper names becomes larger. Of course, shared proportions depend on the language pairs and the type of corpus. For the same type of speech corpora, English and French share more words than German and Spanish (see Figure 5.13, top). When full word forms are compared, the German language shares the lowest number of entries with the other languages. A 50,000 word list is not large enough here to include many technical words or proper names, as declension, conjugation, and—more importantly—word compounding produce many distinct general language entries. Figure 5.13 (bottom) compares a Luxembourgish word list, extracted from parliamentary debates, to French, German, and English. The curves are relatively similar to the top part of the figure except for French, which is known to be largely used in official speech. However, the curves indicate that the proportion of shared proper names is smaller: the Luxembourgish corpus's proper names mostly refer to national personalities, whereas for the other languages, proper names taken from broadcast news data include more international names.

Some frequent words with common orthography in French and English are *but, or, son, me, mine, met, as, fond, sale, sort, note, type, charge, moment, service*, and *occasion*. Shared entries may be identical only in their surface forms or may share some of their meanings. Shared entries with some common meanings are *me, charge, moment, type, service,* and *occasion*, but others have entirely different meanings. The word *sale* in French means "dirty," the equivalent of the English *sale* being *soldes*; the French word *son* means "his," the English to French translation of *son* being *fils*.

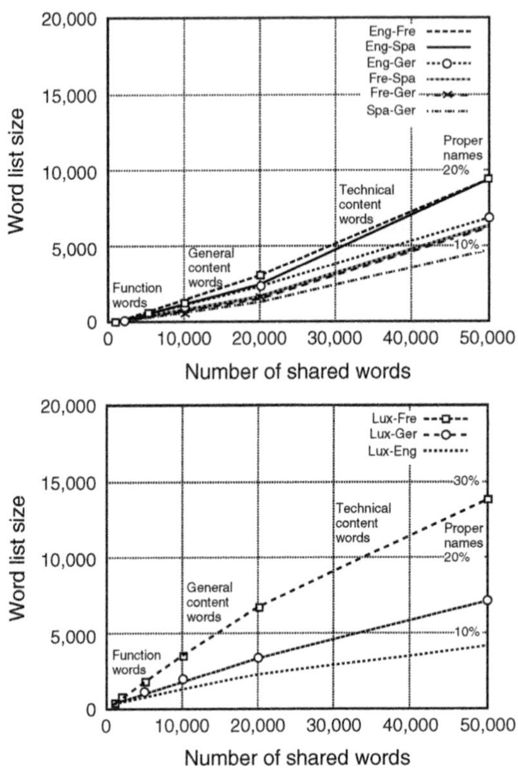

Figure 5.13: Word list comparisons between pairs of languages. The number of shared words is shown as a function of word list size (including for each language its *N* most frequent items). **Top:** language pairs are among English, French, Spanish, and German. **Bottom:** language pairs are Luxembourgish versus French, English, and German.

To date, multilingual dictionaries have been investigated for a few applications in which the language of the user may not be known in advance (Shozakai, 1999; Micca et al., 1999; Übler et al., 1998). Typically these applications are very task-specific, which entails relatively small vocabularies. Languages can be processed separately by different language-specific systems in parallel, or by one single "polyglot" system applying multilingual acoustic models as further discussed in Chapter 4.

It is likely that future automatic speech and text processing algorithms will be based on huge multilingual word lists of millions of entries (for languages from a given family that share similar writing conventions). Depending on the type of text normalizations (e.g., removal of accents and diacritics), a significant part of the vocabulary will then be shared among languages. Language-specific lexical entries can typically be limited to some tens of thousands of items. These aspects, though not yet addressed—in the context of open-domain speech recognition—offer new perspectives for multilingual automatic processing.

5.5 How to Generate Pronunciations

In order to correctly recognize an utterance of a given word, the corresponding acoustic word models must take into account the observed variations in the acoustic signal. Acoustic feature extraction and acoustic modeling techniques (see Chapter 4) provide powerful means either to reduce the variability or to take it properly into account. ASR is not, however, a bottom-up process, and the contribution of language models is very important in ranking acoustically similar candidates (see Chapter 6). The more discrimination among words is provided by the language model, the less discrimination needs to be provided by the acoustic word models.

Word pronunciations specified in the recognition dictionary provide the link between sequences of acoustic units (phones) and words as represented in the language model. Whereas spoken and written sources are relatively easy to collect for major languages, pronunciations are generally not directly available. Generating word pronunciations for ASR requires a modicum of human expertise and thus cannot be carried out fully automatically. However, if the primary purpose is to transform the acoustic signal into a word string without additional annotations, the question arises as to whether acoustic models can be directly linked to graphemic units, rather than using phone-based acoustic models. Given the close relationship in many languages between the graphemic and phonemic form, there has been growing interest in bypassing the explicit step of pronunciation generation in favor of using graphemes directly for speech recognition. However, the relation between graphemes and phonemic forms may be at least locally ambiguous. French and English are examples of languages with a high proportion of ambiguous grapheme-phoneme relations. For example, the English grapheme sequence *ough* can be pronounced as /ʌf/, /o/, or /u/ depending on the carrier word (*rough, thorough, through*). Word context can help in resolving this ambiguity to a certain extent. Conversely, the English phoneme /f/ can be written as either *f, ff, ph,* or *gh.* French writing conventions include letters that carry information about words' etymological origins and that are mute with respect to pronunciations. Mute consonants in French word endings are very common. The sound /o/ can be written as *o, au, eau, ô, oh, aux, ault, eaux.* The word *est* ("is" or "east") can be pronounced as /e/, /ɛ/, or /ɛst/ or even /ɛstə/; the letter sequence *ent* can either be mute or be pronounced as IPA schwa symbol /ə/, /ã/ or /ɛ̃/. The corresponding grapheme-based acoustic models need to implicitly include all these variants and share parts of them among different graphemic units. Grapheme-based acoustic modeling has been successfully addressed for different languages by several teams, including Bisani and Ney (2003), Billa et al. (2002), Killer et al. (2003), Kanthak and Ney (2003), and Schultz (2004). It has the clear advantage of being straightforward and fully automatic and is of particular importance for rapid porting of an existing recognition system to a new language. However, ambiguous letter sequences necessarily produce ambiguous acoustic models, which is a drawback if the lexicon and the language model information are not able to solve the ambiguity. As seen in previous sections, a language is a living entity composed of a relatively small kernel of language-specific items (function and general content words). A huge number of items with low occurrences in the language are composed of grapheme sequences that escape language-specific regularities (thus the importance of the Onomastica Project discussed later in the chapter). For a more detailed discussion of grapheme-based acoustic modeling in a multilingual setting and a comparison across languages, the reader is referred to Chapter 4, Section 4.3.

For languages with a close correspondence between writing and pronunciation conventions, ASR can be carried out without pronunciation generation. However, explicitly specified pronunciations allow spoken language to be modeled more accurately. A pronunciation-based approach includes the potential for reducing the ambiguity of a given language's writing system. Beyond its importance for speech recognition, a pronunciation-based approach contributes a finer tuning of oral dialogue components; to the development of educational and medical services (L2 acquisition, orthophony) (Seneff et al., 2004; Flege, 1995); and to research in linguistics (phonetics, phonology, dialectology, sociolinguistics) (Gendrot and Adda-Decker, 2005; Durand and Laks, 2002).

Most state-of-the-art ASR systems use phone-based representations for acoustic modeling.[5] The strength of phone-based approaches is that acoustic word models can be built for any word, even if it has never been observed in the acoustic training data. The weak point is that phone-based pronunciations are fixed a priori, meaning they are not optimally integrated in the acoustic-model training process. Other units have been explored to model some well-known contextual factors that affect the acoustic realization of phones, ranging from demiphones to demisyllables, syllables, and automatically selected subword units (Holter and Svendsen, 1997; Jones et al., 1997; Marino et al., 1997; Pfau et al., 1997; Tsopanoglou and Fakotakis, 1997; Bacchiani and Ostendorf, 1998; Kiecza et al., 1999). Studies have addressed factors such as syllable position, lexical stress, and coarticulatory influences of the neighboring phones (Schwartz et al., 1984; Chow et al., 1986; Lee, 1988; Shafran and Ostendorf, 2003; Lamel and Gauvain, 2005).

Once a recognition vocabulary has been selected, it is necessary to generate a pronunciation for each entry. This process can be decomposed into several independent steps, as shown in Figure 5.14. The first step consists of producing **canonical pronunciations** for each lexical entry. This can be done by (1) relying on master dictionaries for the language under consideration, by (2) an automatic letter-to-sound module if the language has a relatively unambiguous writing system with respect to pronunciations, or by (3) manual specification of pronunciations by a human expert. In practice, a combination of different methods is chosen.

Next, pronunciation **variants** are added, both the canonical pronunciations and their variants being specified as phone sequences. Different types of variants may be introduced depending on the precision of the phone set. The choice of the phone set is also important for acoustic modeling as discussed in Section 5.5.3. When adding variants, one has to consider the types of speech that will be processed in order to add relevant pronunciation variants for genre and style. Is the speech formal in style (e.g., broadcast

[5]The following discussion focuses on phone-based pronunciation dictionaries. The notion of phone and phone inventory is described in the section addressing pronunciation variants.

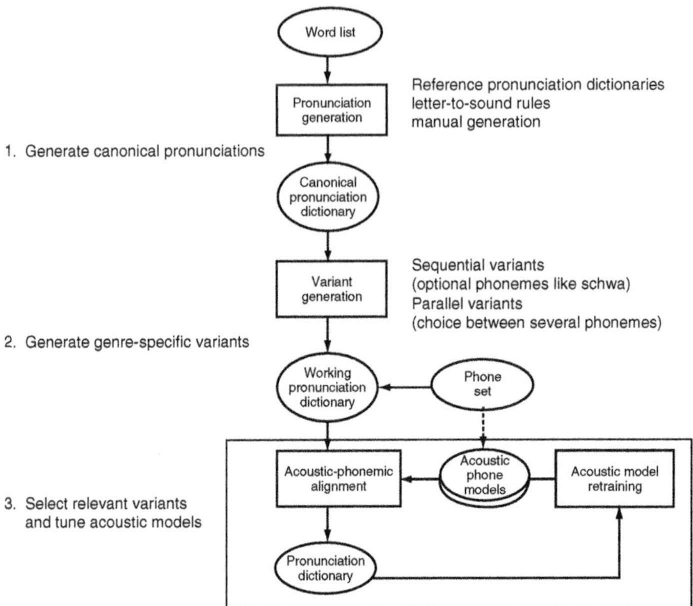

Figure 5.14: Pronunciation dictionary development for ASR system.

speech)? If so, the pronunciations will tend to remain close to canonical forms, and few variants are required. Is the speech vernacular? In this case, phonetic changes can be observed both within words and across word boundaries (Labov, 1966), which implies the need for a higher proportion of pronunciation variants, and in some cases, even a change of lexical unit definition. ASR researchers working on both styles of speech in different languages have become aware of the significant differences between both speaking styles at all modeling levels: word lists, pronunciations, and acoustic and language models.

After generating a basic set of pronunciations, an **acoustic corpus-based validation process** can be carried out, which aims at selecting the variants that are useful given the acoustic models and the type of speech under consideration. Different variants can be assigned probabilities, e.g., based on their occurrence in a training corpus. However, as word occurrences follow a Zipf distribution, which seems to be language universal, only a few words will have reliable estimates of pronunciation variants.

The problem is then to generalize pronunciation probabilities among similar words.

In the following discussion, the three steps of canonical pronunciation generation are discussed further for phone-based pronunciation dictionaries, as well as the addition of variants and corpus-based validation. While in practice these three steps are not necessarily separated, they provide a language independent methodological guideline for pronunciation dictionary development.

5.5.1 Canonical Pronunciations

To generate initial canonical pronunciations, one of the following approaches is generally used:

- **Completely manual:** The developer (often an expert in linguistics or phonetics) types in the phone sequence for each lexical entry. This approach is only viable for relatively small vocabulary tasks and poses the problem of pronunciation consistency.
- **Manually supervised:** Given an existing pronunciation dictionary, rules are used to infer pronunciations of new entries. This requires a reasonably sized starting dictionary and is mainly useful to provide pronunciations for inflected forms and compound words.
- **Grapheme-to-phoneme rules:** These are usually developed for speech synthesis and work well for many languages. Special care needs to be taken to ensure that the text normalization is consistent with the pronunciation rules.
- **Manually supervised grapheme-to-phoneme rules:** Manual supervision is particularly important for languages with ambiguous written symbol sequences. For any language, proper names—particularly those of foreign origin—may not be properly spelled or may require multiple pronunciations.

In practice, it is common to use a combination of the above approaches, in which an existing pronunciation dictionary is used to provide pronunciations for known words, and new entries are added in a semiautomatic manner with possible pronunciations provided by rule. The resulting entries can be simply scanned into or to be displayed by a text editor,

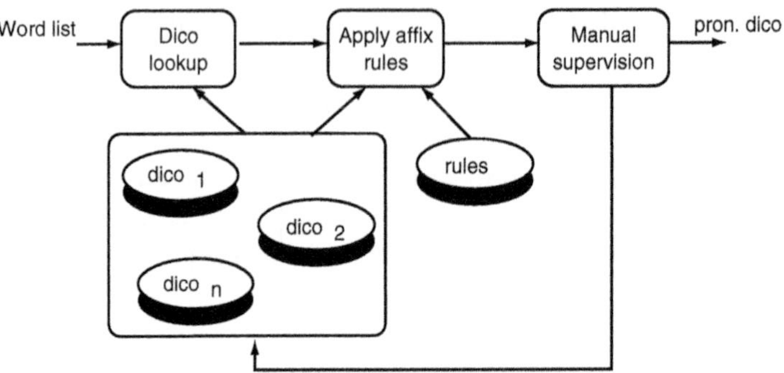

Figure 5.15: Pronunciation generation tool.

or can be presented to a human via a specially adapted tool. Such a tool was developed for American English (Lamel and Adda, 1996), in which grapheme-to-phoneme conversion is often ambiguous, and straightforward rule application can produce erroneous phone transcription even for function and common words. For example, a rule to generate a pronunciation for the word *used* will derive a correct pronunciation from the root *use* but an incorrect one from the word *us*. An example tool to propose pronunciations for American English is shown in Figure 5.15, which applies rules as illustrated in Table 5.6. The rules try to remove either prefixes (P) or suffixes (S) from the word, and specify ordered actions (strip, strip+add), which are applied to the words (letter), and context dependent actions to modify the resulting pronunciations.[6] For example, if the word *banned* is unknown, the letter sequence *ed* is removed and the *n* undoubled. If the word *ban* is located in one of the source dictionaries, the phone /d/ is added to the returned pronunciation. If multiple rules match, all possible pronunciations are returned along with their source.

This type of tool is useful for deriving inflected forms of words already in the pronunciation dictionary. But once a reasonably sized master dictionary is available, new words tend to be proper names and acronyms, which fall out of the scope of a morphologically based pronunciation tool. If multiple monolingual dictionaries or a multilingual dictionary

[6]The algorithm was inspired by a set of rules written by David Shipman when he was at MIT in the 1980s.

Table 5.6 Some example rules to strip and add affixes used by a pronunciation generation tool. Affix types are P (prefix) and S (suffix).

Affix type	Rule type	Remove affix	Add affix	Add phonemes	Context A/V/UV/C	Example word
P	strip	anti	-	/æn{t}[Iɑʲ]/	any	
P	strip	pre	-	/pri/	any	*preconceived*
S	strip+add	ier	y	/i/	any	*dirtier*
S	strip+add	iness	y	/nIs/	any	*happiness*
S	strip	ness	-	/nIs/	any	*carelessness*
S	strip	ally	-	/[l̩l]i/	any	*critically*
S	strip+	ed	-	/əd/	t,d	*admitted*
	undouble			/d/	V	*banned*
S	strip+add	ed	e	/əd/	/t,d/	*acted, ceded*
				/d/	V	*praised*
				/t/	UV	*faced*

is available, this tool can be used to propose spellings for proper names of different origins, with appropriate mapping of the graphemic form and phone spelling (Schultz and Waibel, 1998a).

In languages with a close or at least regular correspondence between the written and spoken forms (such as French, German, Italian, Portuguese, and Spanish), an initial set of base-form pronunciations can be generated using grapheme-to-phoneme rules. The rules can be derived manually or can be developed in a data driven manner. (See Chapter 7 for references on letter-to-sound conversion.) Such letter-to-sound conversion systems typically make use of several hundred rules and a list of exceptions that may contain thousands of items. Spanish and Portuguese can be processed using about 100 rules, whereas German and French require about 300 and 500 rules, respectively. Some example rules and exceptions for French, German, and English are shown in Figure 5.16. The problem of homographic heterophones (words spelled alike but pronounced differently) is more or less important depending on the language and will be addressed in the following section. For example, the Arabic language is particularly challenging since in Modern Standard Arabic, written texts are produced without vowel diacritic marks, there can be many vowelized forms (each which can have multiple possible pronunciations)

French	example	"l" Letter : /l/j/ / *mute*/sound ambiguity	
ctx letters ctx	sound	regular examples	Some exceptions
vi ll	l	ville, Albertville, bougainvilliers	pavillon, Chevilly
mi ll	l	mille, millier, millésime, millilitre	millet, Millon
Ci ll	j	fille, billet	pillule, bacille
^ V ll	l	elle, aller, illustre	
[aeu] il $	j	détail, pareil, seuil	
1	l	le, loi, alors, filet, total, vil,	gentil, fusil, outil *mute letter*

German	example	"s" letter: /z/s/ʃ/	sound ambiguity
ctx letters **ctx**	sound	regular examples	some exceptions
ss	s	Masse, Fluss, gefasst	Abgassteuer *morpheme bound*
sch	ʃ	schön	Volkscharakter *morpheme bound*
sh	ʃ	shoot, shirt *English import*	geisteshell *morpheme bound*
s**[pt]**	ʃ	spassen, streiten	alterspassende *morpheme bound*
s**C**	s	skrupellos, slawish, snobistisch *import words*	
s**V**	z	sieben, Faser	service *(imports)* preisaggressiv *morpheme bound*

English	example	"oo" letter: /ə/ʊ/ /ɔ/ʌ/	sound ambiguity
ctx letter **ctx**	sound	regular examples	some exceptions
oo **r**	ɔ	door, floor	coordinator, hooray, zoology
l oo **d**	ʌ	flood, blood	
oo	[əʊ]	room, football, balloon, childhood	

Figure 5.16: Example of letter-to-sound rules standard French, German, and English, and related exception. Rule precedence corresponds to listed order; **ctx** specifies letter contexts; **C** is a constant; **V** is a vowel; and ^ and **$** signify the word start and end.

associated with each lexical entry (Messaoudi et al., 2004). Each entry can be thought of as a word class containing all observed (or even all possible) vowelized forms of the word. However, if a vowelized written form is available, it is straightforward enough to derive reasonable canonical pronunciations.

In any case, even when grapheme-to-phoneme conversion is a viable solution for a given language, the huge class of words belonging to proper names (either domestic or imported) needs to be treated separately, as standard pronunciation rules often do not apply. Here, dictionaries from other languages can serve as initial knowledge sources.

Some Available Pronunciation Dictionary Resources

The Linguistic Data Consortium (see Chapter 3) lists pronunciation dictionaries for American English, German, Japanese, Mandarin Chinese, Spanish, Egyptian Colloquial Arabic, and Korean. The dictionaries range in size from 25,000 to over 300,000 words. Pronunciations for words in these dictionaries were either derived by letter-to-sound conversion with some manual supervision or derived from other resources. For example, the German dictionary is based on the Celex (http://www.ru.nl/celex/) lexical resources for which large dictionaries are available for the Dutch, English, and German languages. Also important to mention are the pronunciation dictionaries developed for the Eurom and SpeechDat family of corpora distributed by ELDA (see Chapter 3) and the Carnegie Mellon University Pronouncing Dictionary (CMUdict) (CMU, 1998) for North American English.

The difficulty in generating reasonable full-form pronunciations for proper names is a well-known problem: the potential list of proper names is unbounded, and pronunciation rules are certainly less reliable here than for general language words. The generation of proper name pronunciations for speech technology has been explicitly addressed by the Onomastica Project and at Bellcore (Spiegel, 1993) with the development of the Orator speech synthesizer for American English names. The Onomastica Project (Onomastica Consortium, 1995) funded by the European commission under the LRE program, developed pronunciation dictionaries of proper names and toponyms in 11 languages (Danish, Dutch, English, French, German, Greek, Italian, Norwegian, Portuguese, Spanish, and Swedish). The Orator system uses rules to generate name pronunciations with a classification of the ethnic origin of the name, complemented by an exceptions dictionary (Spiegel, 1993; Spiegel and Macchi, 1990). The availability of pronunciation dictionaries in many languages contributes to fostering multilingual speech technologies for numerous applications.

5.5.2 Pronunciation Variants

A given lexical entry may be assigned multiple pronunciations for different reasons: ambiguous writing conventions; phonological variants, which may be due to coarticulation, speaking style, dialect, or accent; loan words; and proper names. Adding appropriate variants seems particularly useful if significant differences can be observed in the acoustic realizations of a given word, and if these differences are unlikely to be represented properly by the acoustic models.

Variants need to be added for homographic heterophones, the proportion of which is to a large extent language dependent. Text normalizations may also contribute to this type of ambiguity. As mentioned earlier, Modern Standard Arabic written texts are produced without vowel diacritic marks, which means that each consonantal root form can be associated with a large number of vowelized forms (Messaoudi et al., 2004). Each consonantal root can be considered a word class, representing all possible vowelized forms of the root. As an example, the transliterated word *ktAb* (*book*) corresponds to four vowelized written forms: *kitaAb, kitaAba, kitaAbi*, and *kutaAbi*. The French language produces mainly morphosyntactic homographs (*président* /prezidã/ [noun] or /prezid/ [verb], *désertions* /dezɛʁsjõ/ [noun] or /dezɛʁtjõ/ [verb]) but also some homographs with unrelated lemmas (*as* /a/ [(*you*) *have*] or /as/ [*ace*], *est* /ɛ/ [(*he*) *is*] or /ɛst/ [*east*]). For Arabic the explicit writing of short vowels eliminates the need for variants, whereas in French, the explicit morphosyntactic information would be required.

Among very common phonological variants is the optional schwa vowel in many languages (French, German, Dutch). Other common phonological variants are due to assimilation. Some example alternate pronunciations for American English and French are given in Figure 5.17 using IPA symbols. For each word, the base-form transcription is used to generate a pronunciation graph to which word-internal phonological rules are optionally applied to account for some of the phonological variations observed in fluent speech. The pronunciation for *counting* allows the /t/ to be optional, as a result of a word-internal phonological rule. The second word, *interest*, may be produced with 2 or 3 syllables, depending on the speaker; in the latter case, the /t/ may be omitted and the [n] realized as a nasal flap. For the next word, *excuse*, the different pronunciations reflect different parts of speech. The suffix *-ization* can be pronounced

counting	kɑʷntɨŋ kɑʷnɨŋ	
interest	IntrIst Intɚ·Ist Inɚ·Ist	
excuse	Ekskjuz Ekskjus	
amortization	əmɔrtəzeSxn æmɔrtəzeSxn əmɔrtɑʲzeSxn æmɔrtɑʲzeSxn	
company	kʌmpəni kʌmpni	
coupon	kjupɑn kupɑn	
republique	repyblik repyblikə	word-final optional schwa
les	le lez lɛ lɛz	liaison phoneme /z/, vowel harmony
prendre	prɑ̃drə prɑ̃dr prɑ̃d	word-final consonant cluster reduction
dix	dis dis{ə} di diz	variants on numbers
DM	dœtSmark deɛm	abbreviations
Morgan	mɔrgɑ̃ mɔrgɑn	multilingual proper name

Figure 5.17: Examples of alternate valid pronunciations for American English and French.

with a diphthong (/ɑʲ/) or a schwa (/ə/). Another well-known variant is the palatalization of the /k/ in a /u/ context, such as in the word *coupon* (/ku/ versus /kju/). In the spectrogram on the left of Figure 5.18, the word was pronounced /kjupɑn/, whereas on the right, the pronunciation was /kupɑn/. If the correct pronunciation is not predicted, during acoustic model training the speech frames will be aligned to the standard pronunciation unless the recognizer can handle pronunciation variants and applies a flexible alignment. If the recognizer can only handle single pronunciations, the variant /kju/ will be implicitly modeled by the model sequence /ku/ and that all pronunciations /kju/ can be decoded as /ku/, which is not always desirable (e.g., *Cooper* versus *cue*).

The French examples illustrate major variant phenomena: word-final optional schwa; vowel harmony; consonant cluster reduction; liaison consonants, which may be optionally produced before a vowel; ambiguous written forms (abbreviations and proper names).

For Spanish, rules can be used to generate multiple pronunciations of the grapheme *acci*—as in *accidental* and *acciones*—allowing for a realization as /akz/ or /az/, and for the grapheme *cion*, to be realized as an /s/ or a /z/. Similarly, rules can be used to propose the deletion or insertion of schwas in certain contexts, vowel reduction to schwa, voicing assimilation, or stop deletion, just to mention a few common phenomena in many languages.

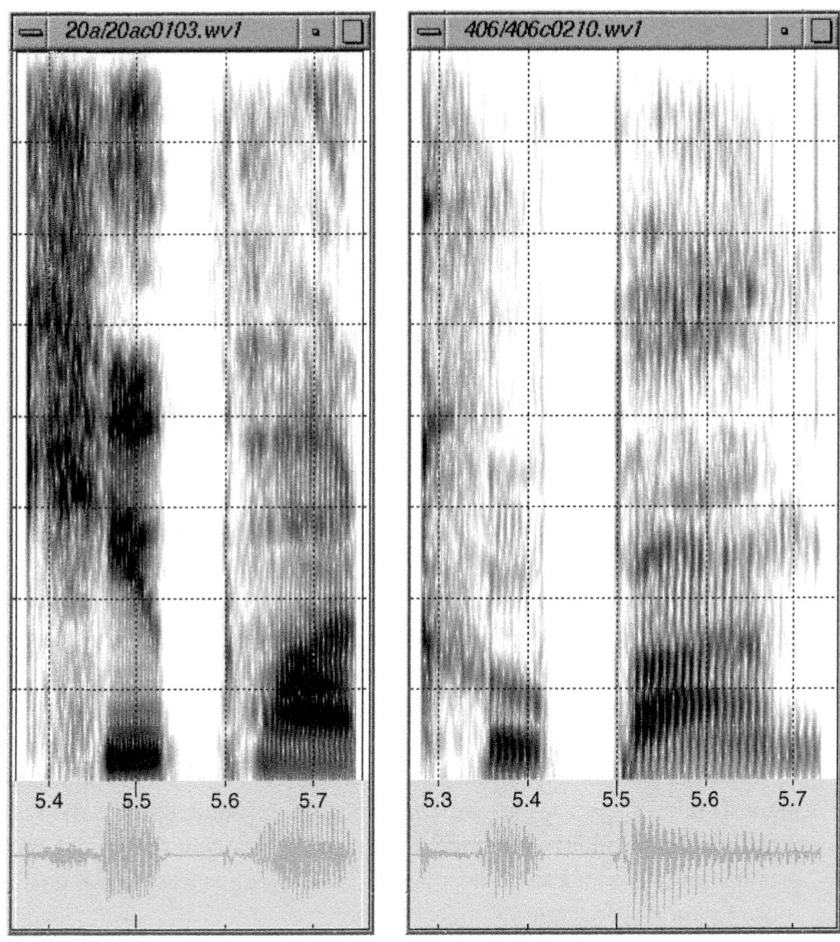

Figure 5.18: Two example spectrograms of the word *coupon:* (left) /kjupɑn/ and (right) /kupɑn/. The grid is 100 ms by 1 kHz.

Including variants into the pronunciation dictionary becomes particularly important when severe temporal mismatches are likely to occur between the full-form pronunciation and the produced utterance. This is more frequent in casual speech than in formal speech. The part of the vocabulary that is shared by vernacular and formal speech (function words, common verbs, nouns, and idiomatic expressions) is more prone to phonological variants than are technical items or proper names, for instance. This is a general observation that can be made for all languages for which we have developed LVCSR systems (English, French, German, Arabic, Spanish, Mandarin Chinese, and Portuguese). An explanation can be proposed on an information theoretic level: since these words are very frequent, their information content is low. They are very often highly predictable from their context. Similar observations hold for morphological units (subwords) corresponding to recurrent morphemic items, such as declension specifications, prefixes, and affixes. On an articulatory level, simplified articulations (pronunciations) can be favored as a result of the repetitive production of these words. On a perceptual level, one can also hypothesize an accelerated activation, which is due more to context than to objective acoustic observations.

Dates and numbers are subject to pronunciation simplifications since they are frequent and contain a fair amount of redundant information. This is particularly true when the contextual information is sufficient for understanding. For example, for the number 88 (*quatre-vingt-huit* in French), the /v/ is often essentially deleted. Similar observations can be made for numbers in German (99, nominally *neun-und-neunzig*, is frequently pronounced as *neun'-n'-neunzig*) and English (150, nominally *one hundred and fifty*, where the word *and* can be heavily reduced or even disappear).

Simplified pronunciations can also be observed across word boundaries for function-word sequences. For example, the German word sequence *haben wir* (*do we have*), with a full-form pronunciation /habən viɐ/, can be reduced in vernacular speech to approximate pronunciations such as /ham vɐ/ or even /hamɐ/. In French, the sequence *c'est quelque chose* (*it's something*), which has a canonical pronunciation (/sɛkɛlkəʃoz/), can be severely reduced, keeping only six phonemes [sekʃoz].

As already mentioned, proper names are particularly difficult to handle, since their pronunciation can be quite variable, depending on the speaker's general knowledge, the origin of the name, and influence of other languages. For example, Worcester—a city in Massachusetts—should be pronounced /wustɝ/, but those not familiar with the name often mispronounce it as /wɔʧɛstɝ/. Similarly, the proper names *Houston* (the street in New York is pronounced /hɑʷstən/ and the city in Texas is /hyustən/), *Pirrone*, and *SCSI* may be pronounced differently depending on the speaker's experience.

Experiments with automatically transcribing different styles of speech (public speech from broadcast news, conversational speech over the telephone) have highlighted the important differences in pronunciations between formal and casual speech, particularly concerning its temporal structure. Pronunciation modeling will contribute to a better knowledge of these spontaneous speech-specific phenomena.

5.5.3 Phone Sets and Acoustic Modeling

Typically a pronunciation dictionary will use a specific phone set. Different dictionaries for the same language may have slightly different numbers of units. For example, commonly used phone sets range from about 25 for Spanish to about 50 for English, French, Arabic, and German to about 80–100 for Mandarin when tone is explicitly modeled. For speech modeling, pronunciations are expressed using a phonemic alphabet, which is then shared by the acoustic models. This alphabet can allow for more or fewer distinctions with more or less detailed IPA (International Phonetic Alphabet) symbols. For automatic speech processing, it is important to consider at what level speech variation should be modeled. As pronunciation generation (base-forms and variants) involves human expertise, it is desirable to limit its relative importance in the overall system. This implies that it is preferable to model variation implicitly within the acoustic models rather than explicitly in the pronunciation dictionary. Different parameters must be taken into account when choosing a phone set:

- For the purpose of acoustic modeling, the granularity of the phone set, phone frequencies, and temporal modeling capacity must be considered.
- For pronunciation dictionary development, the granularity of the phone set, the consistency of the resulting pronunciations, and the level of human effort required must be taken into account.
- Finally, for multilingual applications, the portability of the phone set to different languages should be a criterion.

Frequency of occurrence of phones is an important criterion. In order to obtain reliable estimates of the properties of a sound, especially of a sound in different phonemic contexts, it is vital to have enough observations. This is the reason why xenophones (i.e., phones from different languages) are generally not added to phone lists in a monolingual setup. Both during training and recognition, each phone needs to be aligned with an acoustic segment. In phone-based hidden Markov model (HMM) recognizers (see Chapter 4), it is common to use a 3-state left-to-right model with a minimum duration (typically 3 frames, corresponding to 30 ms), as illustrated in Figure 5.19. Beyond frequency of occurrence, this minimum duration constraint can also have an impact on the definition of the system's phone inventory. Complex phonemes such as affricates (ʧ, ʤ, ts) and diphtongs (ɑʲ, ɑʷ, ɔʲ, ʲu) can be represented by either one or two phone symbols, which implies modeling with one or two HMMs, as shown in Figure 5.20. On the one hand, a possible advantage of using a single unit is that the minimum duration is half that required for a sequence of two phones, and may be more appropriate for fast speaking rates or casual speech. On the other hand, a representation using two phones may provide more robust training if the two component phones also occur individually and diphthongs only occur infrequently in the training data.

Most contextual variation is implicitly taken into account by training multiple models for a given phone, depending on its left and right phone contexts (Schwartz et al., 1984; Chow et al., 1986; Lee, 1988). The selection of the contexts to model usually entails a trade-off between resolution

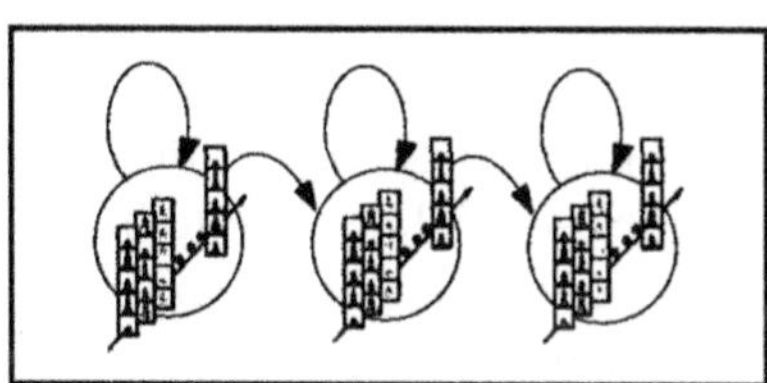

Figure 5.19: An acoustic phone like segment is temporally modeled as a sequence of three states, each state being acoustically modeled by a weighted sum of Gaussian densities (see Chapter 4 for more details on acoustic modeling).

Figure 5.20: Impact on acoustic/temporal modeling depending on the choice of one or two symbols for affricates or diphthongs.

and robustness, and is highly dependent on the available training data. Different approaches have been investigated, from modeling all possible context-dependent units to using decision trees to select the contexts (see Chapter 4), to basing the selection on the observed frequency of occurrence in the training data. In all cases, smoothing or back-off techniques are used to model infrequent or unobserved contextual units. Numerous ways of tying HMM parameter have been investigated (Young et al., 1994; Gauvain and Lamel, 1996).

If the phone symbol set makes fine distinctions (such as between different stop allophones—unreleased, released, aspirated, unaspirated, sonorant-like) many variants will be needed to account for the different pronunciation variations. This raises the problem of completeness and consistency of the pronunciation dictionary, and increases the amount of human effort in pronunciation-dictionary generation. If the basic phone set remains close to a phonemic representation, pronunciation variants are necessary only for major deviations from the canonical form or words for which there are frequent alternative pronunciation variants that are not allophonic differences.

When porting a recognizer from one language to another, standard practice is to use acoustic models from already modeled languages as initial seed models. In doing so, a mapping must be made between the phones in the target language and those in the other language(s). Generally there is a preference to use more generic context-independent models to reduce the influence of the original language. Sometimes it is interesting to use a particular context-dependent model in order to better approximate a phone in the target language that does not exist in any of the other languages. There also exist language-independent and cross–language acoustic modeling techniques to port recognition systems from one language to another without language-specific acoustic data. However, these data remain valuable for acoustic model adaptation. See Chapter 4 for more details.

5.5.4 Corpus-Based Validation

For many years, the use of pronunciation variants was considered risky, since too many variants could potentially increase the number of homophones; therefore, they were only sparsely introduced into pronunciation dictionaries. The availability of very large transcribed speech corpora enables exploration of a new approach to introduce variants by using rules to overgenerate pronunciations in a preliminary working dictionary and validating their selection on a large amount of data (see Figure 5.14) (Adda-Decker et al., 2005). The corpus-validation step aims at a coupled tuning of acoustic and pronunciation models in order to minimize speech recognition errors.

Even if multiple pronunciations can be hypothesized for a given lexical entry, they are not equally useful. Whereas multiple pronunciation lexicons are often (at least partially) created manually, several approaches have been investigated to automatically learn and generate word pronunciations and to associate probabilities with the alternative pronunciations (Cohen, 1989; Riley and Ljolje, 1996; Cremelie and Martens, 1998). The estimation of pronunciation probabilities commonly relies on pronunciation variants in large corpora. As an example, Table 5.7 gives the pronunciation counts for different variants of four inflected forms of the word *interest* in 150 hours of broadcast news data and 300 hours of conversational telephone speech. It can be seen that there is a larger proportion of reduced pronunciations (fewer phones, nasal flap) in conversational telephone speech (CTS) than in broadcast news (BN). For a given style of speech, longer entries (*interesting*) tend to have more reduced variants than shorter entries (*interest*). As word occurrences follow Zipf's law, pronunciation probabilities can be reasonably estimated for several thousand lexical entries if several hundred hours of transcribed speech data are available. The problem

Table 5.7 Pronunciation counts for inflected forms of the word *interest* in 150 hours of broadcast news (BN) data and 300 hours of conversational telephone speech (CTS).

Word	*Pronunciation*	*BN (150h)*	*CTS (300h)*
interest	ɪntrɪst	238	488
	ɪntɚɪst	3	33
	ɪnɚɪst	0	11
interested	ɪntrɪstəd	126	386
	ɪntɚɪstəd	3	80
	ɪnɚɪstəd	18	146
interesting	ɪntrɪstɨŋ	193	1399
	ɪntɚɪstɨŋ	8	314
	ɪnɚɪstɨŋ	21	463

is then to estimate probabilities for pronunciation variants of words that are not sufficiently observed in the training corpus, which is the case for most pronunciations in any large dictionary. In practice, for frequent words (mainly function words but also some content words and idiomatic items), unobserved variants are removed or given a minimal count. By default, all pronunciation variants of infrequent words are considered as equiprobable. One major outstanding challenge is a realistic generalization of the observed pronunciation probabilities via phonological rules and variant types (used to generate the variants). Information about stressed and unstressed syllables in polysyllabic words is certainly a factor to take into account for probability generalization.

While modeling of pronunciation variants and estimation of pronunciation probabilities has attracted much research attention over the years, recent work reported in Hain (2005) proposes using only a single pronunciation for LVCSR. Recognition tests demonstrate that there is no loss in performance, and in certain cases, performance can be improved. This is an interesting case of a corpus-based selection process pushed to its limit. First, a large set of possible pronunciation variants is generated using varying degrees of human supervision. Then, the single most representative pronunciation is selected from all variants for a given word using appropriate acoustic training data. Experience shows that the most representative variant is likely to change with speaking style of the training data, at least for the most frequent items. This implies that for a given language, different pronunciation dictionaries are used depending on the speaking situation. Future multilingual pronunciation dictionaries will certainly need such a validation step on multilingual acoustic data.

5.6 Discussion

For most automatic speech recognition systems, multilingual pronunciation dictionaries are still collections of monolingual dictionaries. However, as was observed for the different languages examined in this chapter, the proportion of imported words—that is, words shared with other languages—increases with vocabulary size. For example, for word lists containing the most frequent 50,000 words in broadcast news texts, 10–20% of the lexical entries are shared between language pairs. Proportions are particularly high for languages of smaller linguistic communities, which tend to more easily incorporate words from other languages. Almost 30% of the lexical entries are shared between French and Luxembourgish. French is often used in addition to Luxembourgish in official situations. This means that monolingual word lists naturally evolve toward multilingual ones with increasing vocabulary size. If common word lists are developed for processing different languages, pronunciations and acoustic models need to be adapted appropriately. The size of the vocabularies used in state-of-the-art speech recognition systems has been growing, and it is likely that system word lists will contain over 500,000 words in the near future.

This chapter has addressed the various steps in lexical development, including the normalization, choice of word items, the selection of a word list, and pronunciation generation. Tokenization and normalization were first addressed in the context of written sources, which often form the basis of language modeling material and are required to ensure viable lexical coverage and word N-gram estimates, particularly concerning the treatment of punctuation, numbers, abbreviations, acronyms, and capitalization. Lexical coverage measures on training data give a good indication of whether the normalizations are properly addressed, For compounding and agglutinative languages (e.g., German, Dutch, Turkish, and Finnish), decomposition techniques are required to optimize lexical coverage. A simple corpus-based and mostly language-independent decomposition algorithm was presented, based on Zellig Harris's algorithm for finding morphemes from phonemic strings (developed half a century ago).

To select efficient word lists for a given speech recognition application, it is important to tune the system's word list using appropriate development data with respect to epoch (in those cases where the data is time-sensitive, such as broadcast news text) and topics (see Chapter 6 on language modeling for more details). For automatic speech recognition, research has shown the importance of using speech transcripts in addition to written sources. Speech-specific items, such as disfluencies, discourse markers, respirations, and fragments do not exist or are only weakly present in written sources. Therefore, their observation probabilities can only be reasonably estimated from speech transcripts. Here, the problem of weighting different types of sources for vocabulary list and language model definition is an important issue. Moreover, to achieve reasonable accuracy in acoustic modeling of reduced word sequences, *multiwords* (function word sequences, idiomatic expressions, frequent acronyms) may be necessary for casual speech.

Concerning word pronunciations, general guidelines can be given for choosing a phone symbol set that produces a reasonable compromise between pronunciation and estimation accuracy: in a monolingual setup, rare symbols and xenophones need to be eliminated. A multilingual setup may at least partially solve this problem. If fine distinctions are allowed by the phone symbol set, more human effort may be required to ensure dictionary completeness and consistency. Even if acoustic phone models can potentially model a lot of implicit acoustic variation, pronunciation variants should be added for ambiguous written forms (homograph heterophones) and for the most important phonological variants. For all languages, adding variants is very important for the N most frequent words (with N lower than 1,000). Frequent function words often have pronunciation variants with different temporal structures, which do not necessarily generalize to less frequent items. As a result, implicit modeling within triphones can be harmful for the global acoustic modeling accuracy. In our view, explicitly modeling pronunciations contributes to a finer tuning of multilingual models, which can in turn be useful in the development of educational and medical services (L2 acquisition, orthophony) and for linguistic research (phonetics, phonology, dialectology, sociolinguistics). With the recent availability of very large spoken corpora, corpus-based explorations may develop into an important research direction.

Future work will include automatic processing based on huge multilingual word lists of millions of entries (for languages from a given family that share similar writing conventions). Depending on the type of text normalizations (e.g., removing accents and diacritics), a significant part of the vocabulary can then be shared among languages such that language-specific lexical entries can typically be limited to some tens of thousands of items. These aspects, though not yet addressed—at least in the context of open-domain transcription—offer new perspectives to multilingual automatic processing.

Chapter 6

Multilingual Language Modeling

Sanjeev P. Khudanpur

6.1 Statistical Language Modeling

A language model is a probability assignment over all possible word sequences in a natural (human) language. Its goal, loosely stated, is to assign relatively large probability to meaningful, grammatical, or merely frequent word sequences compared to rare, ungrammatical, or nonsensical ones. For example, the latter might be assigned zero probability.

For concrete examples of what these goals ought to be and how a language model facilitates achieving them, consider the classical communication channel model of automatic speech recognition (Jelinek, 1997), as depicted in Figure 6.1.

A speaker is presumed to draw a word sequence $\mathbf{W} = \langle w_1, w_2, \ldots, w_k \rangle$ from some source distribution $P(\mathbf{W})$, and utter it using her articulatory apparatus. The listener, be it another human or a machine, is presumed to perform some processing of the received acoustic signal, producing a sequence of observations $\mathbf{A}$, and then "decode" the spoken words based on $\mathbf{A}$. The stochastic realization of $\mathbf{W}$ as a particular sequence of acoustic

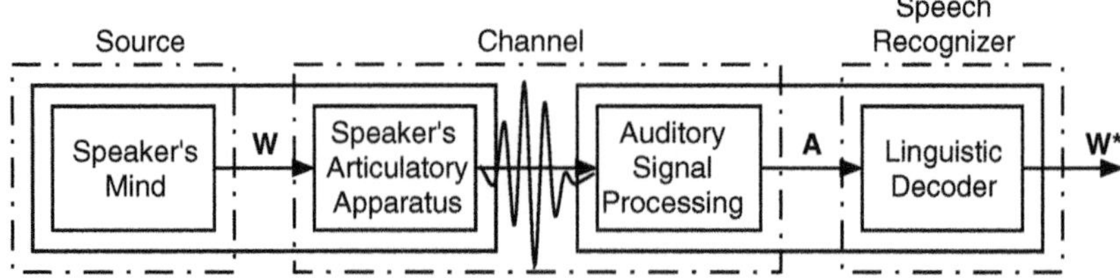

Figure 6.1: The classical communication channel model of automatic speech recognition.

observations **A** is described by the *acoustic* model $P(\mathbf{A}|\mathbf{W})$: as discussed in Chapter 4, and the probabilistic model for the generation of words is the language model $P(\mathbf{W})$. In this formulation, therefore, the language model provides *prior* probabilities for word sequences, which the decoder uses to search for the most likely word sequence given **A** as

$$\begin{aligned}\mathbf{W}^* = \arg\max_{\mathbf{W}} P(\mathbf{W}|\mathbf{A}) &= \arg\max_{\mathbf{W}} \frac{P(\mathbf{A},\mathbf{W})}{P(\mathbf{A})} \\ &= \arg\max_{\mathbf{W}} P(\mathbf{A}|\mathbf{W})P(\mathbf{W}), \qquad (6.1)\end{aligned}$$

where **W** ranges over all possible word sequences in the language, and $P(\mathbf{A})$ in the denominator may be ignored when seeking the maximizer $\mathbf{W}^*$.

Furthermore, the language model is also used in various heuristics to speed up the *search* for the maximizer $\mathbf{W}^*$ over the large set of candidate **W**s. For this reason, one often seeks the left-to-right decomposition of the language model via Bayes rule, namely

$$P(\mathbf{W}) = \prod_{n=1}^{k} P(w_n|w_{n-1},\ldots,w_1). \qquad (6.2)$$

Some of the most popular language models therefore have the form of (6.2), and a large body of literature exists that discusses parameterization of the conditional probability of (6.2), the estimation of the parameters from data, and the introduction of hidden variables to capture the linguistic structure underlying the observed word sequence **W**. For instance, the simple N-step

Markov approximation

$$P(\mathbf{W}) \approx \prod_{n=1}^{k} P(w_n | w_{n-1}, \ldots, w_{n-N}) \tag{6.3}$$

has been found to be extremely useful even for small values of N, typically one to three. Other methods include parsing the sentence prefix $\langle w_1, \ldots, w_{n-1} \rangle$ to obtain a parse-tree Φ, and then extracting linguistically motivated predictors of w_n, such as head-words of exposed syntactic constituents (Chelba and Jelinek, 2000; Charniak, 2001; Roark, 2001).

$$P(\mathbf{W}) \approx \prod_{n=1}^{k} P(w_n | w_{n-1}, \ldots, w_{n-N}, \Phi(w_1, \ldots, w_{n-1})) \tag{6.4}$$

This is illustrated by the prediction of the word *consider* from the word *analysts* in the utterance

> Financial analysts *in the former British colony* consider the contact essential for . . . ,

where the interceding prepositional phrase (italicized) is subsumed in the partial parse-tree by the noun it modifies—that is, Φ extracts the most immediate exposed head-word:

$$\Phi\,(\text{Financial analysts in the former British colony}) = \langle \text{analysts} \rangle. \tag{6.5}$$

Still others have attempted to capture coarse semantics through an equivalence classification of the extended context $\langle w_1, \ldots, w_{n-1} \rangle$ into topic-like categories (Iyer and Ostendorf, 1999; Clarkson and Robinson, 1997; Bellegarda, 2000). These models also have the form of (6.4) and (6.5), with Φ taking values in a set of topics.

The goodness of a language model is ultimately measured by how well it facilitates application-specific goals, such as the transcription accuracy for a speech recognition system, the fluency and adequacy of the text output for a machine translation system, or the rate of successful transactions for a spoken dialog system. But end-to-end performance of a system depends on complex interactions between the language model and other components, such as the acoustic model in (6.1), and is therefore not always instructive in developing good language modeling principles that may generalize to other applications. A popular application-independent measure of the goodness of a language model is the average probability it assigns to naturally occurring text that was held out during model estimation. The idea is that given two language models, the probability of all possible word sequences in the language is unity under both, and the model that assigns higher probability to a representative set of meaningful and grammatical sentences must necessarily assign lower probability to ungrammatical and nonsensical ones, and hence improve system performance. Specifically, given held-out text $\mathbf{W} = \langle w_1, \ldots, w_n \rangle$, the **perplexity** of a language model $P(\cdot)$ is

$$\text{perplexity}(P; \mathbf{W}) = \exp\left(-\frac{1}{n}\log P(\mathbf{W})\right). \tag{6.6}$$

Perplexity may be interpreted as the number of *equally likely* word choices faced, on average, when using $P(\cdot)$ to predict w_n given $\langle w_1, \ldots, w_{n-1} \rangle$. The perplexity of a perfect predictor, namely a model that assigns probability one to the held-out text, is unity, while the perplexity of a model that assigns uniform probability to all words in a (finite) vocabulary in every context in (6.2) is equal to the size of the vocabulary. In the limit of large n, it may be argued from an information-theoretic perspective that the language model that minimizes perplexity is also the model that is closest in a statistical sense to the *true* probability distribution of the language. For this reason, language models are often designed to minimize perplexity.

Alternatively, a language model may be required to play a discriminative role rather than the role of a prior (generative) probability. In particular, one may design an error-correcting language model (e.g., Roark et al., 2004) or optimize some other figure of merit, such as accuracy of content-bearing words. In the context of other applications, such as information-seeking spoken dialog (e.g., Glass et al., 1995) or call routing (e.g., Riccardi et al., 1997), the language models may be weighted finite state transducers that directly extract shallow semantics from the observed acoustics $\mathbf{A}$ without explicitly seeking the most likely word transcription $\mathbf{W}^*$ as an intermediate step. We will discuss these briefly in Section 6.5, but the bulk of the following discussion is dedicated to the basic language model of (6.2). For an excellent discussion of the (almost current) state of the art in language modeling, see Rosenfeld (2000).

We have adopted a markedly statistical notion of language modeling. Most of the discussion that follows will take the view that the issues in language modeling involve deciding what constitutes an utterance or a word, how the words in an utterance are statistically dependent, and how to estimate the parameters of this statistical dependence from data. However, note that the definition of (6.2) is general enough that in principle, it does not rule out handcrafted grammars or other nonstatistical notions of a language model. Even a context-free grammar may be cast in the form of (6.2)—for instance, by using the so-called left-corner transformation (Jelinek and Lafferty, 1991). Viewing everything from a statistical perspective simply reflects a personal bias of the author.

6.1.1 Language Models for New Languages and Multilingual Applications

Considerations of multilingual processing arise in two qualitatively different settings, leading to two related but different avenues of research.

The first setting is the development of speech and language technology in a new language. Historically, language technology research has focused on a few major languages in the world—such as English, French, German, Spanish, and Japanese—broadly reflecting costs associated with the collection of large data resources, and the potential economic payoff. Languages added more recently to this list include Mandarin Chinese and Modern Standard Arabic. The main questions to be answered in this setting are (1) whether techniques and insights developed for or from the major languages will continue to be effective, if not optimal, for other languages; and (2) whether resources available in the major languages can somehow be leveraged for rapid development of systems in other languages.

A quintessential example of this setting is the MALACH project for providing access to more than 100,000 hours of recorded oral histories of survivors of the Jewish holocaust (Byrne et al., 2004). The recordings are in more than 40 different languages, most of which do not have any existing speech recognition technology; the first of the two questions above must be addressed in order to develop such technology. Furthermore, about 10,000 hours of the English recordings have been extensively (manually) annotated for providing automated access using textual queries. For instance, a hierarchical thesaurus of more than 25,000 indexing terms has been created, and lists of place and person names pertinent to the period of interest have been compiled. The second question above, namely whether having such considerable linguistic resources in English can help improve performance in the other languages by exploiting the synergy in the domain, also arises naturally.

The second setting that has multilingual implications is a truly multilingual system, in which a single recognition engine must accept spoken input and produce spoken output in multiple languages. A third pertinent question in this setting is whether a joint design for all the languages is more effective than a separate design for each individual language or language pair.

A typical example of such an application is two-way speech-to-speech translation to facilitate communication between, say, a French-speaking doctor and a Kiswahili-speaking patient in a refugee camp or in an international disaster relief effort.

In the following, we will attempt to summarize solutions proposed in the literature to address the three questions listed above. Before addressing the multilingual or crosslingual questions, however, we will briefly summarize some problems that arise in language modeling in general, and their standard solutions for English. This will hopefully aid the reader in making the crosslingual comparisons in the sequel.

6.2 Model Estimation for New Domains and Speaking Styles

It should be clear upon a little reflection that even estimating the simple model of (6.3)—called a *bigram* for $N = 1$, a *trigram* for $N = 2$, and an N-gram in general—requires enormous amounts of text in electronic form. For instance, with a modest vocabulary of 20,000 words, the bigram model has nearly 400 million free parameters, and the trigram, nearly eight trillion (10^{12}). Such large vocabularies are, of course, necessary to obtain adequate coverage in many domains. For instance, a 20,000 word vocabulary covers about 98% of word tokens in the original two million word Switchboard corpus of transcribed English telephone conversations (Godfrey et al., 1992).

Even with large text corpora for estimating the models, sufficient counts are not available to estimate a conditional model in most contexts due to the nature of human language. For instance, there are nearly 800,000 distinct word triples in the Switchboard corpus mentioned above, and nearly half of them appear only once. In other words, one has a single sample of w_n, given a particular $\langle w_{n-1}, w_{n-2}\rangle$, to estimate the model of (6.3). Relative frequency estimates, therefore, are grossly inadequate.

This data-sparseness problem has received much attention in the last two decades. Several techniques have been developed to estimate N-gram models from limited amounts of data, and we describe below only one such technique for the sake of completeness. For the trigram model, for instance, one may use

$$
\begin{aligned}
&P(w_n = c \mid w_{n-1} = b,\ w_{n-2} = a) \\
&\quad = \begin{cases} \dfrac{\mathrm{count(a,b,c)}-\delta}{\mathrm{count(a,b)}} & \text{if } \mathrm{count(a,b,c)} > \tau, \\ \beta(a,b)\, P(w_n = c \mid w_{n-1} = b) & \text{if } \mathrm{count(a,b,c)} \leq \tau, \end{cases}
\end{aligned}
\tag{6.7}
$$

where the threshold τ is usually set to 0; count($\cdot$) is the number of times a word-triple or -pair is seen in the "training" corpus; $\delta < 1$ is a small, empirically determined *discount* applied to the relative frequency estimate of $P(w_n|w_{n-1}, w_{n-2})$; and $\beta(\cdot)$, the *back-off weight*, is set so that the conditional probabilities sum to unity over possible values of w_n. The bigram model $P(w_n|w_{n-1})$ may be recursively estimated from discounted counts in the same way, or based on other considerations.

The discounting of the relative frequency estimate is colloquially called *smoothing* and the process of assigning unequal probabilities to different unseen words in the same context, based on a lower-order model, is called *back off*. (For a comparison of several smoothing and back-off techniques proposed in the literature see the empirical study by Chen and Goodman [1998].)

It turns out that such smoothed estimates of N-gram models, for $N = 2$ and 3, are almost as effective as the best-known alternatives that exploit syntax and semantics. This, together with the ease of estimating them from data, has led to their widespread use in almost every application of human language technology.

6.2.1 Language Model Adaptation

Due to their purely statistical nature, however, N-gram models suffer from several shortcomings that one would find annoying, even unacceptable, in a language model. Most notable among them is the lack of portability across domains and speaking styles. For a test-set of Switchboard telephone conversations, a trigram model estimated from nearly 140 million words of transcripts from radio and television shows is outperformed by a trigram model estimated from as little as the 2 million words of the Switchboard transcriptions. Lest one suspect that this is an aberration due to the fact that the radio and television transcripts are primarily from the news domain while Switchboard conversations are on a variety of other topics, Rosenfeld (1994) points out a similar discrepancy within newswire texts themselves: trigram models estimated from the Associated Press newswire are remarkably worse at modeling *Wall Street Journal* text than a model estimated from the same amount of Dow-Jones news wire.

However, obtaining in-domain text in every domain of interest is an expensive proposition, particularly if, unlike the broadcast news domain, closely related text such as newswire does not naturally exist, and the speech needs to be manually transcribed.

Attempts at bridging the domain gap in language modeling include those by Iyer et al. (1997), Clarkson and Robinson (1997), and Seymore and Rosenfeld (1997). They usually involve estimation of separate out-of-domain, language models $P_{\text{out}}(w_n|w_{n-1}, w_{n-2})$ from one or more resource-rich domains, and a language model $P_{\text{in}}(w_n|w_{n-1}, w_{n-2})$ from whatever in-domain text is available, according to (6.7) above, followed by interpolation. For trigram models, for instance, this takes the form

$$P(w_n|w_{n-1}, w_{n-2}) = \lambda P_{\text{in}}(w_n|w_{n-1}, w_{n-2}) + (1 - \lambda) P_{\text{out}}(w_n|w_{n-1}, w_{n-2}), \quad (6.8)$$

where the interpolation weight λ is empirically estimated from some held-out in-domain text. The interpolation weight may depend on the count of the conditioning event $\langle w_{n-1}, w_{n-2} \rangle$ in the in-domain text, the N-gram count in the out-of-domain text, etc. Alternatives to model interpolation include interpolation of the counts in (6.7), which may be interpreted as a Bayesian estimate of the in-domain language model from a small amount of data, with the out-of-domain model providing the hyperparameters of the prior. Empirical evidence suggests that model interpolation (6.8) is usually more accurate.

More sophisticated methods of domain-adaptation based on minimum Kullback-Leibler divergence have also been developed (Rao et al., 1995; Dharanipragada et al., 1996). In a manner similar to the estimation of maximum entropy models (Rosenfeld, 1994), a linear family P_{in} of admissible models is specified based on limited in-domain data. Specification of the family is usually in the form of constraints on the probability that an admissible model must assign to N-grams seen frequently within the domain, leaving the probabilities of most (unseen) N-grams un(der)specified. From this family, the model that is closest to the out-of-domain model $P_{\text{out}}(w_n|w_{n-1}, w_{n-2})$ is chosen—that is,

$$P^* = \arg\min_{P \in P_{\text{in}}} D(P(c|b,a)||P_{\text{out}}(c|b,a) \mid P_{\text{in}}(a,b)), \tag{6.9}$$

where $D(A||B \mid C)$ denotes the divergence between the two conditional probabilities A and B averaged using the (marginal) probability C of the conditioning events.

6.3 Crosslingual Comparisons: A Language Modeling Perspective

The preceding description of a language model is in such generic terms that one may be led to conclude that techniques developed for English apply without much specialization to almost any language. It is indeed true in practice that estimating the N-gram language model of (6.3) simply requires collection of sufficient amounts of text in the language of interest, followed by a sequence of language-independent parameter-estimation procedures. Yet things are not as straightforward as they may appear at first sight.

- *Words*: In languages such as English and French, it is fairly natural to define a word as a sequence of characters separated by whitespace. But it may not be as natural or suitable in other languages.

  Morphological Richness: This definition of a word as a basic unit of language modeling may perhaps survive highly inflective languages such as Arabic, Czech, and Serbo-Croatian, where a single root form or lemma has dozens of derived forms with closely related meanings. Czech, words, for instance, use morphological inflections to mark for tense, number, gender, case, aspect, mood, and so on. Using fully inflected word forms as the units of language modeling may adversely impact the robustness of the statistical estimates if one treats the morphological forms as distinct and unrelated.

  Compounding and Agglutination: This definition of a word begins to show its limitation in languages that permit productive use of compounding, such as German and highly agglutinative languages like Turkish (Oflazer, 1994). An oft-cited Turkish "word" is osman.lı.laş.tır.ama.yabil.ecek.ler.imiz.den.miş.siniz, which means "(behaving) as if you were of those whom we might consider not converting into an Ottoman." Turkish words usually do not have any explicit indication of morph boundaries; in this example, the dots between morphs have only been placed by the author to enhance readability. It is clear that some manner of morph-synthatic or morpho-phonemic decomposition is need for such languages.

  Free Word Order: Finally, many languages, particularly those with rich morphology, permit free word order to varying degrees, as illustrated in Table 6.1. Therefore, while the N-gram models have been fairly successful with languages such as English and French, where the relative order of constituents within phrases and phrases within clauses is fairly consistent, it is not clear that the same would hold if the word order were free and were used variably to make other distinctions, such as the topic and the focus of a sentence—that is, the distinction between old and new information.

- *Segmentation*: In languages such as Mandarin Chinese, sentences are written using a sequence of characters (logographs) without any whitespace between words, and while the language admits a word as a bona fide linguistic unit, and most readers agree on the identities (boundaries) of most words in any given sentence, even manual

Table 6.1 Czech is a free-word-order language, as illustrated by this set of sentences from Kuboň and Plátek (1994).

Czech Word	Označený	soubor	se	nepodařilo	úspěšně	otevřít
Part of Speech	adjective	noun	pronoun	verb	adverb	inf. verb
English Gloss	Marked	file	itself	failed	successfully	to open.
Czech sentence (1)	Označený soubor se nepodařilo úspěšně otevřít.					
Czech sentence (2)	Nepodařilo se úspěšně otevřít označený soubor.					
Czech sentence (3)	Označený soubor se úspěšně otevřít nepodařilo.					
Czech sentence (4)	Označený soubor se nepodařilo otevřít úspěšně.					
Czech sentence (5)	Úspěšně otevřít označený soubor se nepodařilo.					
Czech sentence (6)	Nepodařilo se označený soubor úspěšně otevřít.					
Czech sentence (7)	Nepodařilo se označený soubor otevřít úspěšně.					
English translation	The marked file failed to be opened successfully.					

segmentation of text into words is not possible with 100% interlabeler agreement. Thus, it begs the question whether a word is an optimal (or even a valid) unit for modeling such languages.

- *Spoken versus Written Forms*: There are several languages whose formal or standard (written) form is significantly different from the spoken or colloquial form. Colloquial Arabic is the better known example of this problem, but it is prevalent in many other languages, including Chinese and Czech (Kirchhoff et al., 2002a). It is sometimes not possible in these languages to faithfully transcribe what was spoken using the orthographic conventions of the written language. At other times, the phonetic nature of the orthography permits transcription, but only at the cost of highly inconsistent representations of different instances of the same spoken word. The extreme form of this problem, of course, is the transcription of a spoken language that has no written form at all.

Such differences between language families, and other issues that may be specific to particular languages, call for a typology of the world's languages from the point of view of statistical language modeling. A useful comparison by Cutler from the viewpoint of acoustic processing provides a good starting point for interested readers (Cutler, 1997).

In the following subsections, we will discuss some of these issues in greater detail.

6.3.1 Language Modeling for Morphologically Rich Languages

Morphology refers to the word-forming elements and processes of a language that transform an underlying word or lemma to other words that are related in meaning but that play different syntactic or semantic roles. The English words *ran*, *runner*, and *runaway* are morphological forms of *run*. Among the languages of the world, English is perhaps morphologically less productive than many. Spanish, Arabic, Czech, Russian, and Serbo-Croatian are progressively richer, while Mandarin Chinese is considerably less productive than English.

Morphological processes transform underlying word forms in several different ways, and most languages exhibit instances of multiple morphological processes.

Inflective transformations usually mark an underlying form for qualities like number (singular, dual, plural); gender (masculine, feminine, neuter); or tense (present, past). Table 6.2 illustrates morphological variations of the underlying Czech noun *žena* (woman). Serbo-Croatian and Russian also exhibit similarly high degrees of inflection, while Spanish and Italian do so to a lesser degree. The English words *runs* and *ran* are inflected forms of *run* that mark for number and tense, respectively. Note from the English examples in Table 6.2 that the grammatical case encapsulated by the inflected forms in Czech

Table 6.2 Illustration of inflected forms in Czech for the underlying noun *žena* (woman).

Singular	Plural	Case	English example
žena	ženy	nominative	Women were angry.
ženy	žen	genitive	The women's fury knew no bounds.
ženě	ženám	dative	You provided inspiration to women.
ženu	ženy	accusative	They underestimated women.
ženě	ženách	locative	Your message ignited the passion in women.
ženou	ženami	instrumental	You won the election with women.
ženo	ženy	vocative	Well done, young woman!

Table 6.3 Illustration of derived forms in Arabic for the roots ktb (write) and drs (study) from Kirchhoff et al. (2002a).

Root: **ktb**		Root:**drs**	
Arabic	Gloss	Arabic	Gloss
ka**t**a**b**a	he wrote	**d**a**r**a**s**a	he studied
ki**t**aa**b**	book	**d**a**r****s**	lesson
ku**t**u**b**	books	**d**u**r**uu**s**	lesson
'a**k****t**u**b**u	I write	'a**d****r**u**s**u	I study
ma**k****t**a**b**	desk or office	ma**d****r**a**s**a	school
ma**k****t**a**b**a	library	mu**d**a**r**ri**s**	teacher
kaa**t**i**b**	writer	**d**a**r**ra**s**a	he taught

is captured in English by particular prepositions and by a particular sentential word order.

Derivational transformations result in morphological forms that are still semantically related to the underlying form but serve even more grammatical functions than the inflected form. Table 6.3 illustrates some of the derived forms for two Arabic roots. Other Semitic languages, such as Hebrew, exhibit similar richness, while languages such as English do so to a lesser extent. Note, however, that while the root forms in some languages (like English) are words in their own right, the same may not always be true in other languages. Furthermore, derivations could simply add prefixes and suffixes to the root word or, in more complex cases, add infixes, change the vowels, or transform the word in even more significant ways.

Agglutinative transformations and *compounding* refer to the process of joining multiple underlying word forms to construct new words. German is well known for its prolific use of noun compounds, while the Turkic languages exhibit a high degree of agglutination. A hallmark of compounding is that but for minor modifications to obtain syllabic and phonotactic well-formedness, the constituent morphemes are usually well preserved through the transformation and are easily identified, as illustrated in Table 6.4 for Inuktitut, an Amerindian language. The words *runaway* and *runabout* may be considered examples of compounding in English, as may the word *blackboard*.

Table 6.4 Illustration of an agglutinated form in Inuktitut for the root word *tusaa* (hear).

Inuktitut	tusaatsiaqjunnaqnngittualuujunga					
Morphemes	tusaa	tsiaq	junnaq	nngit	tualuu	junga
Gloss	to hear	well	be able to	not	very much	(1st person singular)
English	I'm unable to hear very well					

Morphological richness, relative to languages like English, leads to two issues that directly impact statistical language modeling: a more rapid growth of the vocabulary and a freer word order. Of the two, vocabulary growth has received considerable attention in the literature and will be discussed further. The impact of free word order is not as well studied.

The Curse of (a Very Large) Vocabulary

If one were to retain a word as the unit of language modeling, then a language model has to accommodate much larger vocabularies in morphologically rich languages. As a result, the vocabulary needed to cover a comparable corpus of text grows much more rapidly in such languages compared to languages like English, which are not as morphologically productive (Barnett et al., 1995; Lamel et al., 1995; Byrne et al., 1999). Table 6.5 shows the number of distinct words seen in different corpora of comparable sizes, and the rate at which out-of-vocabulary (OOV) words are encountered in some held-out text from the same corpus, for a fairly large vocabulary of approximately 60,000 words.

Using Morphs Derived from a Linear Decomposition of Words as Units of Language Modeling

It has been noted that the rate at which new morphemes are seen in these languages is often comparable to that in English (Geutner, 1995). This naturally leads to the idea that perhaps one should construct a language model not on the full word forms but on the morphemes. This is particularly straightforward to formulate for languages in which the morphology primarily adds prefixes and suffixes: one simply decomposes a word

Table 6.5 Typical vocabulary growth and rate of out-of-vocabulary words for various languages.

Language	Corpus Name	Corpus Size (in millions of words)	Vocabulary (thousands of words)	OOV@ 60K
English	*Wall Street Journal*	19	105	~1%
Czech	*Financial News* (Byrne et al., 1999)	16	415	~8%
Serbo-Croatian	Internet (Geutner et al., 1998)	11	350	~9%
Modern Standard Arabic	*Al Nahar News*	19	690	~11%
Turkish	Newspapers (Schultz and Waibel, 2001)	16	500	~14%

sequence $\mathbf{W} = \langle w_1, \ldots, w_n \rangle$ further into a sequence of morphemes $\langle m_1, \ldots, m_{n'} \rangle$, and estimates N-gram language models on the morpheme alphabet. It, of course, follows that the effective *span* of a trigram model of morphemes will be shorter than that of a word trigram model, which will, in principle, lead to less accurate predictions. The morpheme trigram model may also assign positive probability to illegal or impossible morphological compositions. On the positive side, however, the smaller morpheme alphabet will lead to more robust estimates, not to mention the coverage of hitherto unseen words, whose constituent morphemes have been seen before. Similar arguments may be made for language models other than N-grams. With trade-offs on either side, it therefore becomes an empirical question as to whether this straightforward morphological decomposition is better or worse than a simple word-based language model.

Using morphemes for language modeling is, of course, predicated on the availability of morphological analyzers, such as those for Czech and Modern Standard Arabic, whose creation, in turn, requires detailed linguistic knowledge about the morphology of the language (Hajič, 1999; Buckwalter, 2002). Note, however, that if the goal of the morphological analysis is to decompose words into morphemes to control vocabulary growth, the linguistic validity of the decomposition is less important than the resulting ease of model estimation and reduction in, say, the perplexity (6.6) of the language model. Thus, data-driven techniques, such as those investigated for Russian, may be used when robust morphological analyzers are not available (Whittaker, 2000).

By and large, N-gram models based on linear morphological decomposition seem to improve perplexity, but do not lead to significant improvements in recognition accuracy over word N-gram models (Geutner et al., 1998; Çarki et al., 2000; Byrne et al., 2001).

A secondary effect of linear decomposition of words into morphs is that the resulting morphs could be used as dictionary units and even as acoustic model units. This could result in a significant reduction of the dictionary size and therefore a decoding speed-up. Assuming suitable cross-word context, this strategy could fairly accurately generalize to unseen morphological variants. However, the length and nature of morphs might increase the confusability among the dictionary units. Furthermore, their usefulness as acoustic model units depends on the ratio of morph types to the amount of training data.

Factored Language Models

An elegant view of the morphological decomposition of a word as a feature bundle, or a bundle of *factors*, not a linear string of morphemes (Kirchhoff et al., 2003). For one, it applies equally to languages with a morphology consisting primarily of affixation and those such as Arabic, in which the morphology results in in-fixing, vowel changes, and so on. For another, it allows a more direct modeling of more complex relations between words—for example, if a verb and subject need to agree in gender, then a factored model is aware of exactly which factors of the words in the verb and subject positions designate gender. In the morpheme N-gram model, specific factors are not explicitly identified.

For a simple illustration of the factored language model, consider a language in which each word w_n is decomposed into a root r_n and an ending e_n. Then the word w_n could be predicted jointly from the preceding roots and endings

$$P(w_n|w_{n-1}, w_{n-2}) = P(w_n|e_{n-1}, e_{n-2}, r_{n-1}, r_{n-2}) \tag{6.10}$$

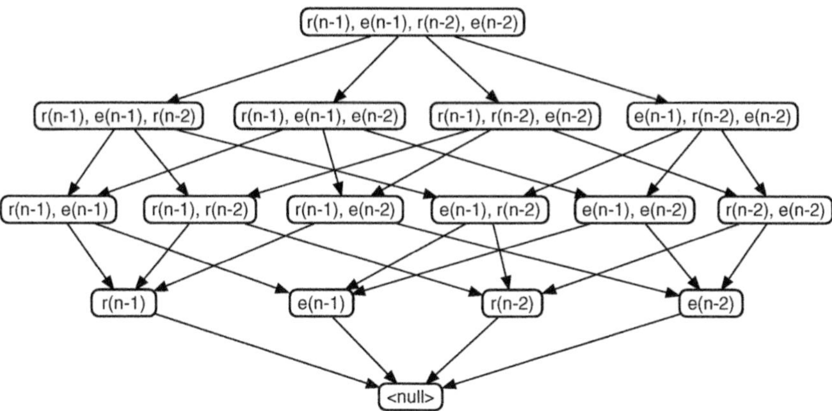

Figure 6.2: Illustration of multiple back-off paths implicitly followed in a factored language model to evaluate the probability $P(w_n|r_{n-1}, e_{n-1}, r_{n-2}, e_{n-2})$ of (6.10).

and the probability for this model would have to be estimated by some suitable smoothing and back-off scheme. The back-off, in particular, would truncate the context from right-to-left in the order listed in (6.10) above or in some other predetermined order. The factored language model permits a richer back-off scheme instead, as illustrated in Figure 6.2.

Specifically, since different back-off paths may be optimal for different *configurations* or values of the four-tuple $\langle r_{n-1}, e_{n-1}, r_{n-2}, e_{n-2}\rangle$ or the five-tuple $\langle r_n, r_{n-1}, e_{n-1}, r_{n-2}, e_{n-2}\rangle$, multiple back-off paths are implicitly or simultaneously followed, and the lower-order probability estimates from these alternative paths are combined. This combined estimate replaces the back-off to a predetermined lower-order model in (6.7).

A number of evidence combination techniques may be tried to derive the optimal modified lower-order estimate, including different smoothing techniques along alternative back-off paths, and different ways of computing the lower-order back-off weight β. An adequate discussion of these issues may be found in Vergyri et al. (2004).

Hypothesis Driven Lexical Adaptation

It has also been observed that when recognition errors are caused by a morphological variant being OOV, it is often substituted by an incorrect variant of the same root form. Furthermore, Ircing and Psutka (2001) note that even though the Czech test set they worked with had an OOV rate of 13% for a recognition vocabulary of nearly 60,000 words, less than 0.5% of the tokens in the test set were *truly unseen* among the 650,000 distinct word forms encountered in their language model training text (2001). This suggests an alternative two-pass strategy called **hypothesis-driven lexical adaptation** (Geutner et al., 1998). The core idea is to perform a first recognition pass using a moderately large vocabulary, use the recognition output to adjust the vocabulary and language model to include morphological variants of all the likely hypotheses, and perform a second recognition pass. Some heuristics have been developed for deciding which word forms to include in the second pass, including drawing additional words from the training text and drawing on a large resource, such as the Internet, which significantly reduces the effective OOV rate.

Significant improvements in recognition accuracy have been reported using hypothesis-driven lexical adaptation in Czech and Serbo-Croatian (Ircing and Psutka, 2001; Geutner et al., 1998).

Note that the lexical items that are added to the vocabulary in the second pass are usually rare words, which is why they were excluded from the first-pass vocabulary. Adding them to the vocabulary (and reestimating the relevant language model probabilities) usually appears to result in significantly *higher* perplexity of the reference transcriptions of the test set compared to the first-pass language model. This is due, however, to a common error in the interpretation of perplexity. The newly added words that are present in the reference transcription were either excluded from the perplexity calculations of the first-pass language model, or mapped to a generic unknown-word token <*unk*>. In either case, their contribution to the perplexity of the first-pass model is not comparable to that in the second-pass model, where they are bona fide words in the vocabulary.

Free Word Order Isn't

Morphologically rich languages tend to permit greater freedom in word order, since the grammatical functions are marked on the words themselves. In other words, if *John* was marked as the subject and *cake,* the direct object of the predicate *eat*, then "John eats cake" and "cake eats John" would have the same meaning. It seems natural to contemplate whether richer morphology leads to greater difficulty in language modeling—particularly with N-gram language models—in the so-called free-word-order languages.

If preliminary investigations in Czech are any indication, this is not a major issue or at most a second-order effect (Byrne et al., 1999, 2000, 2001). Once the consequences of having a large vocabulary are factored out, perplexity and recognition error rates in Czech broadcast news seem to follow trends in English broadcast news for bigram language models—for instance, in the relative gain from increased amounts of training text.

This is primarily because even in free-word-order languages, while the sentential ordering of clauses and phrases may be determined by considerations of information structure, with old information (topic) placed toward the beginning of the sentence, and new information (focus) placed toward the end, the words within phrases still tend to follow a canonical order; and since N-grams have very short spans, they capture local order fairly well without being affected by global reordering. In other words, N-gram statistics in naturally occurring texts are largely insensitive to the effect of the language having a free word order, particularly for small N.

Relative gains in going from bigrams to higher-order N-grams, however, are lower in Czech than in English, and similar results have been reported for Arabic. This is because in morphologically rich languages, a word encapsulates parts of the phrase structure modeled in English by prepositions and other function words. Hence, an Arabic bigram may model an *entire phrase*, and the predictive power of an N-gram diminishes quickly beyond phrases. This phenomenon, rather than the free word order, motivates the investigation of grammatically structured language models for morphologically rich languages.

Free word order is of greater consequence for language models that exploit syntactic structure. It is argued, for instance, that statistical parsing of Czech is more difficult than parsing of English (Hajič et al., 1998). Indeed, a crucial step in parsing morphologically rich languages is the automatic morphological analysis, since a significant portion of the syntactic structure of the sentence is contained in it. Therefore, the three key components in language modeling for morphological rich languages are (1) accurate morphological analysis, (2) part of speech tagging, and (3) shallow parsing.

6.3.2 Languages Lacking Explicit Word Segmentation

Languages such as Chinese are written using an inventory of a few thousand logographic characters. Each character corresponds phonologically to a tone-marked syllable and may be viewed alternatively as being composed of a sequence of phonemes or a pair of demisyllables: an optional initial and a mandatory final. Words are composed of one or more characters, with a majority of words being two characters long.

Chinese is written without any intervening whitespace between words. From an automatic processing perspective, this poses a peculiar problem, since a sequence of characters may be *segmented* in more than one way to produce a sequence of valid words. The readability for humans, however, is not severely impacted by this apparent ambiguity, perhaps because the incorrect segmentations usually lead to meaningless sentences. This is not to suggest that human readers will agree entirely on the segmentation of any particular sentence. The disagreements, when they occur, are largely due to different segmentation conventions, although a few are due to genuine ambiguity or freedom in the sentence.

Automatic segmentation of Chinese text is needed for several applications. The precision in information retrieval may suffer if a query (key) word is assumed to be present in a document based on character matching, but segmentation of the document into words does not support such a match. Machine translation may also require segmentation of text into words, so that translation lexicons, domain ontologies, and other lexical resources may be brought to bear on the problem. Similarly, named entity extraction, sense disambiguation, and parsing are traditionally performed using words as the unit of analysis. Finally, speech synthesis requires word segmentation, for unit selection and prosody planning.

State-of-the-art methods in automatic segmentation of Chinese text use a combination of hand crafted lexicons and statistical techniques that exploit word frequencies, word collocation information, and sentence-wide optimization of the placement of segment boundaries.

Segmentation accuracy is measured in terms of the harmonic mean of *precision*, the fraction of the hypothesized boundaries that are correct, and *recall*, the fraction of the correct boundaries that are hypothesized. The best methods to date achieve accuracies of about 95%. The most difficult cases tend to be words not present in the lexicon—particularly proper names, that a human reader would nonetheless segment correctly. A recent series of benchmark tests in Chinese word segmentation, along with an excellent intellectual exchange of methodological details, has been taking place via the SIGHAN Workshops in conjunction with the meetings of the Association for Computational Linguistics (Sproat and Emerson, 2003).

It is not obvious that language modeling for automatic speech recognition would necessarily require word segmentation, since it is easy to use the characters themselves as the units of language and to estimate character N-gram models directly from the unsegmented text. However, more sophisticated language models that exploit syntactic structure or model topic-conditional variations in word usage require segmenting the language model training text into words. Further, since words, not characters, are the semantic units of the language, even N-gram models are likely to be more accurate if based on words rather than characters, even after the N-gram orders of the two are made comparable—for instance, by letting the character-based model look twice as far back as the word-based model.

Word based N-gram models are usually estimated from Chinese text by employing automatic methods to segment the training corpus. While it may be possible to view the segmentation as a hidden variable and use expectation maximization techniques to estimate the language model, this is not usually done.

The native output of a Mandarin Chinese speech recognition system that uses a word-based language model will be a sequence of words. It is customary to "unsegment" the output text before computing the *character accuracy* of the recognition system. This is done primarily because the reference transcriptions are usually created with Mandarin Chinese orthographic conventions without word segmentation. A secondary reason—or perhaps consequence—is that two different recognition systems may segment the same text in different ways, making it difficult to compare their word level outputs; their character level outputs are more directly comparable.

An indirect benefit of character-based evaluation is that as long as text segmentation is performed consistently on the training and test data, segmentation errors are masked in the evaluation. This may be one of the reasons that word segmentation is not considered a burning issue in Chinese language modeling at this time. Better segmentation, of course, may lead to more accurate language models.

Language model perplexity of Chinese text is also conventionally reported per character; the total log-probability of **W** in (6.6) is divided by the number of characters in the sentence even if the probability was computed by first segmenting the sentence into words and applying a word-based model. This too makes it possible to compare different language models regardless of the word segmentation they use.

6.3.3 Languages with Very Divergent Written and Spoken Forms

Human language is dynamic and evolving, and the written form of the language often lags behind its colloquial or informal use. For this reason alone, constructing models for spoken language would be a difficult problem. It was observed in Section 6.2 that if one needs to build a speech recognition system for conversational speech, statistical models estimated from written texts—while better than many handcrafted models—are fairly poor.

A more drastic disconnect is seen when the written form happens to be one particular dialect of the language or, worse, merely a modern day standardization of many closely related languages, none of which are actually spoken the way the standard form is written. Chinese and Arabic present two such examples.

- **Arabic:** Two broad categories of Arabic are spoken, respectively, in the western parts of North Africa and in the Middle East. Within each broad category, several regional varieties may be further distinguished, such as Egyptian, Levantine, and Gulf Arabic within the Middle Eastern category. The differences between these spoken varieties are comparable to differences between the Romance languages.

  More importantly, however, none of these spoken varieties of Arabic are used in any formal communication or for archival purposes, such as news broadcasts, government transactions, or record keeping. What is used is a modern-day standardization of classical Arabic originating in Egypt, called Modern Standard Arabic (MSA). Educated individuals speak MSA, particularly in formal settings and out of necessity when communicating with an Arabic speaker from a different region.

Small amounts of written dialectal Arabic may be gleaned from novels and chat rooms, but the quantity is not nearly enough—and the orthography not consistent enough—to make a significant impact on language modeling. Two corpora of transcribed conversational Arabic have been created at considerable cost for research purposes: the CallHome corpus of Egyptian colloquial Arabic (ECA) and the Fisher corpus of Levantine conversational Arabic (LCA) (Canavan et al., 1997; National Institutes of Standards and Technology, 2004). Both consist of telephone conversations, the former on a myriad of personal matters between family members, and the latter on preassigned topics between randomly chosen volunteers. Transcription of the colloquial speech in MSA orthography is augmented in the ECA corpus to include short vowels via diacritics. Short vowels are not transcribed in the LCA corpus, but a two-step transcription protocol has been used to attain greater interlabeler consistency in the orthographic transcription.

Yet using MSA text to estimate a language model for conversational Levantine Arabic is almost as challenging as using Italian text to estimate a language model for French. Several researchers have acknowledged, formally or in private communication, that previously known language model adaptation techniques have failed in exploiting MSA text to improve language models for ECA estimated from the CallHome corpus, and the same story is repeated for LCA from the Fisher corpus (Kirchhoff et al., 2002a).

- **Chinese:** At least six different languages, each perhaps more accurately described as a language group rather than as a single language, are spoken in China: Mandarin (Putonghua), Yue (Cantonese), Min (Fujianese and Taiwanese), Wu (Shanghainese), Xiang (Sichuan), and Gan (Zhejiang). Linguistically, these are distinct languages. However, due to cultural and geopolitical reasons, they are called Chinese *dialects*, with a particular (sub)dialect of Mandarin, named Putonghua, designated as standard Mandarin. Standard Mandarin is taught in schools all over Mainland China, and native speakers of different dialects speak to each other in standard Mandarin. However, they speak it with a strong and distinguishable non-native accent.

  From the point of view of language modeling, non-native speakers of standard Mandarin do not present as severe a problem as speakers of colloquial Arabic, since they are attempting to speak Putonghua. To a first approximation, this may be treated as a problem of pronunciation modeling or development of dialect-dependent pronunciation lexicons for Putonghua. Effects such as different lexical choice may be dealt with once the pronunciation issue is better understood.

Language modeling for the nonstandard dialects does, however, suffer from a problem comparable to that of colloquial Arabic: training texts are not easily found for the other dialects, since all current day official communication and record keeping in China uses standard Mandarin. Cantonese and Taiwanese are, of course, somewhat of an exception, again due to geopolitical and economic factors.

Language modeling, in situations in which the spoken and written forms diverge significantly, relies almost exclusively on manual transcriptions of conversational speech. This poses two problems: (1) transcriptions are expensive to obtain, and (2) interlabeler consistency in transcription is not very high, since the human transcribers are not schooled in writing the colloquial or spoken language, and others have to "invent" orthographic forms as they hear them.

Much of the recent research in speech recognition for conversational Arabic has been conducted under the auspices of the DARPA EARS research program and, prior to that, via the Egyptian CallHome corpus of telephone speech (National Institutes of Standards and Technology, 2004; Canavan et al., 1997). Other than simple word- and word-class-based N-gram language models and the factored language modeling effort described above, there is little research of note (to our knowledge) in addressing languages with very different spoken and written forms. As mentioned earlier, no major improvement has been demonstrated over and above language models estimated from the limited amounts of available transcribed speech.

Research in speech recognition of Mandarin spoken by non-native speakers was undertaken recently with the collection of a corpus of Wu-accented Mandarin, and preliminary results in acoustic modeling indicate that considerable hurdles remain in the acoustic model adaptation for non-native accents (Sproat et al., 2004). While not much language modeling research has been done on this corpus, existing adaptation techniques may suffice to adjust standard Mandarin language models to Wu-accented Mandarin speech.

6.4 Crosslinguistic Bootstrapping for Language Modeling

It is clear that availability of in-domain text in the language of interest is a crucial resource for language model estimation. It is equally clear that such text is not easy to come by—particularly if it requires manual transcription—be it because the speech is spontaneous in nature, is from a non-native dialect, or simply differs significantly from the written language. It is true in many applications, however, that in-domain text may be available in another language in which the application has already been developed.

- In the previously mentioned MALACH project, for instance, large amounts of English speech, and a modest amount of Czech and Russian, have been manually transcribed, while the remaining 40 or so languages are unlikely to be transcribed to any significant degree. All interviews, however, pertain to the same general period in time, describe similar experiences, refer to similar geographical entities and organizations, and are collected in a similar discourse style.
- Applications involving open-source intelligence, such as the DARPA TIDES project, are concerned with transcribing news broadcasts in multiple languages (Graff et al., 2000), as illustrated in Table 6.6. In addition to the common journalistic style, news stories in various languages on any given day pertain to the same set of world events, even if the viewpoints and perspectives of the broadcasters are different.
- Spoken dialog systems that can support multiple languages naturally share the language independent back-end task that is being accomplished via the interactive dialog, and provide an opportunity to develop language understanding components that can be shared across languages (Cohen et al., 1997; Glass et al., 1995).

Table 6.6 The TDT-4 corpus covers news in three languages (Strassel and Glenn, 2003).

	Arabic	English	Mandarin
Newswire	An-Nahar Al-Hayat Agence France Press	*New York Times* Associated Press Wire	Zaobao Xinhua
Radio	VOA Arabic	PRI The World VOA English	VOA Mandarin CNR
Television	Nile TV	*CNN Headline News* *ABC World News Tonight* *NBC Nightly News* *MSNBC News* with Brian Williams	CCTV CTS CBS-Taiwan

It is natural to investigate whether, in such situations, in-domain resources in other languages may be utilized to improve language modeling in a *resource-deficient* language.

6.4.1 Adaptation of Language Models Using Crosslingual Side-Information

The idea of using in-domain data in another language as *side information* to improve language models in a resource-deficient language has been explored by using Mandarin Chinese as a straw man for the resource-deficient language, and English as the resource-rich language in the broadcast news domain (Khudanpur and Kim, 2004; Kim and Khudanpur, 2004). Since, in reality, considerable Chinese text is available for language modeling, the efficacy of the techniques with varying amounts of in-domain Chinese text has been explored.

The first core idea is that if machine translation (MT) capability between English and Chinese is available, even if it is not optimized for the domain of interest, then in-domain English text may be automatically translated into Chinese and used for improving the Chinese language models

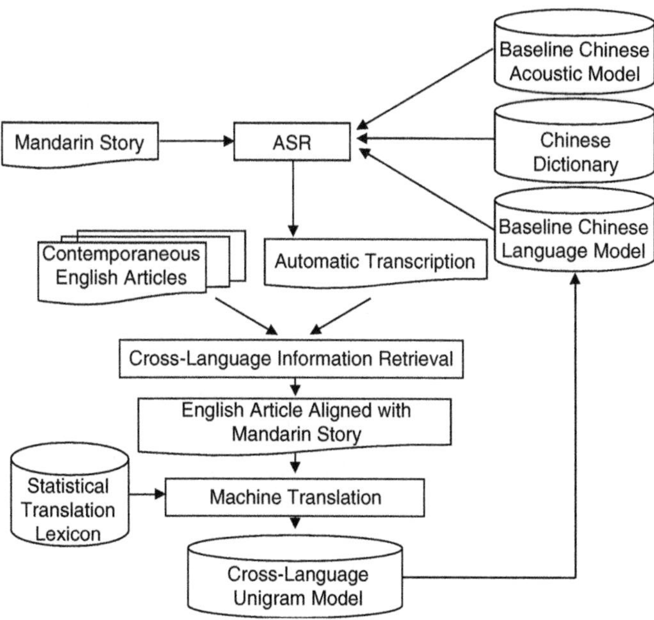

Figure 6.3: Dynamic adaptation of language models using contemporaneous data from another language.

(Khudanpur and Kim, 2004). The following basic scheme is illustrated in Figure 6.3.

1. A Mandarin Chinese story is transcribed using a baseline language model
2. The resulting recognition output for the entire story is used as a query to a crosslingual information retrieval (CLIR) system to obtain English documents, presumably on the same topic
3. The English documents are translated into Chinese
4. N-gram statistics are extracted from the translated text, and a language model is estimated
5. The resulting language model is interpolated with the baseline model as described in (6.8)
6. A second pass of recognition is performed with the adapted language model

Specifically, a unigram model is extracted from the translated Chinese text in step 4 and interpolated with a trigram language model in step 5, as

$$P(w_n|w_{n-1}, w_{n-2}) = \lambda P_{\text{Translated}}(w_n) \\ + (1 - \lambda)P_{\text{Baseline}}(w_n|w_{n-1}, w_{n-2}). \qquad (6.11)$$

Two important factors contribute to the reductions in both perplexity and recognition error rates as reported in Khudanpur and Kim (2004).

Story-Specific Language Modeling: A different subset of the English text is selected for translation, and a different interpolation weight λ is used in (6.11) for each Mandarin Chinese story being transcribed. The need for story-specific selection of the English text is easily understood, and it has been convincingly demonstrated that due to the varying degree to which matching English documents are found for different Mandarin Chinese stories and the varying accuracy in the translation of these documents, predetermined adaptation (interpolation) parameters are not effective. Fortunately, parameters chosen to minimize the perplexity of the first-pass recognition output turn out to be almost as good as parameters that would have minimized the perplexity of the reference transcriptions, had they been available.

Contemporaneous English Text: The English newswire covers the date of the Mandarin Chinese story, while the Chinese language modeling text is relatively older. While it is intuitively clear that this should matter, additional indirect evidence that the contemporaneous nature of the English text is a factor comes from the result that the crosslingual adaptation provides greater gains than monolingual topic adaptation of the language model on the (static) training text.

The existence of MT capability of reasonable quality, however, would be doubtful if the language of interest were truly resource-deficient; this situation has also been addressed (Kim and Khudanpur, 2004). The key observation here is that MT capability is assumed in two places: for CLIR in step 2, and for translating the retrieved English documents into Chinese in step 3; and in both places, alternatives may be substituted. Two of these alternatives are investigated below.

- **Crosslingual Lexical Triggers:** If pairs of documents in the two languages on the same story are available, then statistical measures such as *mutual information* may be used to automatically extract English-Chinese word pairs whose co-occurrence is more than mere chance. This extraction process is similar to obtaining monolingual lexical triggers, and the extracted pairs are analogously named crosslingual lexical triggers (Rosenfeld, 1994). The list of crosslingual lexical trigger pairs, once created, acts as a stochastic translation dictionary in steps 2 and 3. The (joint) probability that an English-Chinese word pair (e, c) is a translation pair is made proportional to their mutual information $I(e; c)$:

$$P_{\text{Trigger}}(e, c) = \frac{I(e; c)}{\sum_{(e',\, c')} I(e'; c')}, \qquad (6.12)$$

  with $I(e; c) = 0$ whenever (e, c) is not a trigger pair. Empirical evidence is presented to demonstrate that this is as effective (in the CLIR and subsequent N-gram language model estimation) as having a high-quality translation lexicon estimated from sentence-aligned text or even a large, manually crafted translation dictionary.
- **Crosslingual Latent Semantic Analysis:** Latent semantic analysis (LSA) has been used in monolingual and crosslingual information retrieval and requires as a resource a document-aligned corpus of bilingual text (Deerwester et al., 1990; Landauer and Littman, 1995; Berry et al., 1995). Therefore it is a natural alternative for the CLIR of step 2 above. Interestingly, it also turns out to be effective for translation in step 3.

  Assume that a document-aligned Chinese-English bilingual corpus $\{(d_j^C, d_j^E), j = 1, \ldots, N\}$ is provided. The first step in crosslingual LSA is to represent the corpus as a word-document *co-occurrence frequency matrix* W, in which each row represents a word in one of the two languages, and each column represent a document pair. If the size of the Chinese plus English vocabularies $|\mathcal{C} \cup \mathcal{E}|$ is M, then W is an $M \times N$ matrix. Singular-value decomposition is performed, and W is approximated by its largest R singular values and the corresponding singular vectors, for some $R \ll \min\{M, N\}$.

$$W \approx U \times S \times V^T, \qquad (6.13)$$

where columns of U and V are orthogonal left- and right-singular vectors of W, and S is a diagonal matrix whose entries $\sigma_1 \geq \sigma_2 \ldots \geq \sigma_R$ are the corresponding singular values (Berry et al., 1995).

The j-th column W_{*j} of W, or the document-pair (d_j^C, d_j^E), is represented in (6.13) by a linear combination of the columns of $U \times S$, the *weights* for the linear combination being provided by the j-th column of V^T. Equivalently the *projection*

$$V_{*j}^T \approx S^{-1} \times U^T \times W_{*j} \tag{6.14}$$

of the j-th column of W onto the *basis* formed by the column vectors of $U \times S$ provides an R-dimensional representation for (d_j^C, d_j^E). By providing a common R-dimensional representation for all documents via (6.14), LSA makes it possible to compare English documents in the contemporaneous side corpus in Figure 6.3 with the first-pass Mandarin transcription, and to therefore obtain English text on the same topic as the Mandarin story being transcribed.

Similarly, projecting the i-th row of W, which represents the distribution of the i-th word, onto the basis formed by rows of $S \times V^T$ provides an R-dimensional representation of words.

$$U_{i*} \approx W_{i*} \times V \times S^{-1} \tag{6.15}$$

Now, if the i-th row corresponds to a Chinese word, and the i'-th row corresponds to an English word, then their projections U_{i*} and $U_{i'*}$ in a common R-dimensional space make it possible to compare English and Chinese words.

By transforming this *semantic similarity* between words into word collocation probabilities, one may construct a language model based on LSA in the monolingual setting (Coccaro and Jurafsky, 1998; Bellegarda, 2000; Deng and Khudanpur, 2003), or a stochastic translation lexicon in the crosslingual setting, according to

$$P_{\text{LSA}}(e, c) = \frac{\text{sim}(e, c)^{\gamma}}{\sum_{(e', c')} \text{sim}(e', c')^{\gamma}}, \qquad \forall\, c \in \mathcal{C},\ \forall\, e \in \mathcal{E}, \tag{6.16}$$

where sim(e, c) measures the similarity between e and c, typically via the cosine of the angle between their R-dimensional representations, and $\gamma \gg 1$, as suggested by Coccaro and Jurafsky (1998).

Once English documents d_i^E are obtained via CLIR and stochastic translation probabilities have been estimated per (6.12) or (6.16), the story-specific crosslingual adaptation of (6.11) is carried out, with the probability of a Chinese word c calculated from the relative frequencies $\hat{P}(e)$ of English words in d_i^E as

$$P_{\text{Translated}}(c) = \sum_{e \in \mathcal{E}} \hat{P}(e) P_{\text{T}}(c|e), \tag{6.17}$$

where the translation probabilities $P_{\text{T}}(c|e)$ may be obtained from either (6.12) or (6.16).

Improvements in recognition accuracy comparable to the case when high-quality statistical MT is available have been obtained by using cross-lingual triggers and LSA.

6.5 Language Models for Truly Multilingual Speech Recognition

Some applications, such as spoken dialog systems, may be required to automatically handle spoken inputs in multiple languages. An illustration of such a system is a tourist information kiosk with international visitors, or an automated government information call center with an obligation to serve all members of the community.

- In the simpler and perhaps most frequent of the possible cases, each input utterance to such a system, or perhaps each dialog or conversation, will be in one of a given set of languages. The language model in such applications needs to assign a probability distribution over the *union of word sequences* from all the languages of interest.
- A more challenging situation is presented by a multilingual speaker, who may not only switch languages from utterance to utterance but mix languages even within an utterance. The language model needs

to assign a probability distribution over word sequences drawn from the union of the vocabularies of all languages of interest in order to correctly transcribe such speech.

The former, of course, is merely a special case of the latter from the perspective of probability theory. In practice, the former is a more important and much easier special case, with different engineering considerations and solutions, than the latter.

Note that one may simultaneously recognize an utterance using multiple language-specific systems without knowing which language it was uttered in, and choose one of the outputs depending on which recognizer assigns the highest confidence to its output. However, if the languages share acoustic models, shared computation of acoustic likelihoods may be possible or necessary, creating a need for a multilingual language model back-end.

Utterance Level Multilinguality

The most prevalent practice in handling multilingual input of one utterance at a time is to use finite-state grammars such as N-grams to model each language, and to place the N-gram models of multiple languages in *parallel* (Cohen et al., 1997; Glass et al., 1995; Weng et al., 1997a; Wang et al., 2002). Recall that an N-gram model of a language is, in the formal sense of finite state automata, equivalent to a weighted regular language. The prevalent practice, therefore, may be described as follows.

1. Define a distinct (nonoverlapping) vocabulary Σ_i for each language $i = 1, \ldots, M$, of interest.
2. From training text for each language, estimate a weighted finite state acceptor G_i that accepts Σ_i^*.
3. Construct the multilingual language model by taking the finite state *union*: $\bigcup_{i=1}^{M} G_i$.

Since regular languages are closed under union, the result is still an N-gram language model. Powerful algorithms for manipulating finite state machines may be brought to bear on the N-gram language models, for instance, to determinize and minimize the model and create small recognition networks suitable for embedded systems. (Readers unfamiliar with these techniques are referred to tutorial papers on the application of finite state transducers in speech recognition by Mohri et al. (2000).)

A noteworthy characteristic of this approach, particularly for spoken dialog systems, is that the finite state acceptor describing each individual language is merely the front end—more formally, the projection on one side—of a transducer that performs domain-dependent interpretation of the spoken inputs, and its projection on the other side is used by a *language-independent* text-understanding module to accomplish an interactive task. In a flight information system, for instance, the transducer may perform some syntactic and semantic parsing of the text to extract the city of origin or destination pertaining to the information request. The common language-independent back-end processes suggest that considerable design overhead may be shared across the languages of interest. It is indeed possible, at least in highly constrained domains, to construct an *interlingua* for the language understanding component. (For a more detailed discussion of these ideas, see a recent survey by Waibel et al. (2000).)

Word Level Multilinguality and Code-Switching

It is very common for multilingual speakers in informal settings to engage in code-switching—the act of using words from different languages in the same sentence. Sometimes this is done because while one of the languages happens to be in use at the moment, the other is more suitable for expressing a particular concept. This phenomenon is observed frequently in highly multilingual or multicultural societies such as India, where (educated) people often speak to each other in English but frequently use well-known Hindi expressions. At other times, the speaker simply accesses the lexical item in the other language faster, or does not know the word in the primary language of the utterance. Transcription of such speech requires a model that permits code-switching at every word.

Several techniques have been investigated for modeling such multilingual speech (Cohen et el., 1997; Glass et al., 1995; Weng et al., 1997a; Wang et al., 2002; Fung et al., 1997; Fügen et al., 2003b).

- One alternative has been to merge the language model training text in the different languages, perhaps *tagging* or marking words in each language to distinguish those that happen to have the same orthographic form in both languages, and then estimating a single N-gram language model from the merged corpus.

- Another alternative is to merge the vocabulary of the multiple languages, but to estimate an N-gram model for each language separately and interpolate the models. In this scheme, words of one language appear with zero counts in the training corpus of the other languages and therefore get assigned a small probability from the so-called *back-off* state. Each language model therefore assigns nonzero probability to all words, permitting code-switching at every word.

Recognition performance from these mixed-language models tends to be worse than the utterance level merging of language models described above, with data merging being considerably worse than word-level model interpolation. Typically, the errors come from frequent words in an incorrect language replacing infrequent but similar sounding words in the correct language. This trend is exaggerated in the data-merging approach if the amount of training texts in the different languages are not balanced.

Most of the investigations cited above do not explicitly target code-switching but merely contrast the parallel N-gram models described earlier with the code-switching language models. This is evident from the fact that while their test sets contain multilingual speech, the language is constant throughout an utterance. We are not aware of any study that addresses recognition performance relative to code-switching. This might be partly due to the lack of data.

6.6 Discussion and Concluding Remarks

Multilingual speech and language processing is rapidly becoming an important application area, if it has not done so already, due to increasing globalization. Several indicators of this trend may be seen: the multilingual nature of content on the Internet, multilingual versions of operating systems, multilingual ownership and management of major businesses, and the multilingual nature of spam received via e-mail, to name a few. Language models play a central role in most if not all language processing technologies, including speech recognition, machine translation, information retrieval, and several aspects of information extraction. Consequently, multilingual language modeling is becoming increasingly important, and some of the issues that will need to be addressed follow.

Standardized Character Encoding: An interesting trend was observed with the emergence of Hindi on the Internet, which may impact other languages that will surely follow. Hindi is written using the Devanagari script. Now, dozens of newspapers, as well as many government agencies, use proprietary character encodings of Devanagari to publish in Hindi, partly because creating content in the Unicode standard (see Section 2.4.2 within Chapter 2) is quite cumbersome, and partly because client-side support for rendering Devanagari Unicode fonts is not very widespread. As a result, while large corpora are presumably available for language modeling in Hindi, it resulted in a developers' nightmare when an effort was made to collect and process these resources in a DARPA-sponsored technology demonstration exercise (Oard, 2003; Strassel et al., 2003; He et al., 2003). A pressing need in enabling language technology for new languages is rapid promulgation of character encoding standards and dissemination of tools that content providers can use to create content in the standard encoding.

Language Modeling for Spoken Variants of Written Languages: The colloquial Arabic experience shows in no uncertain terms that the statistical methods so successful in English will run into data-sparseness problems with many languages in which the spoken form diverges significantly from the written form.

Considerable research needs to be undertaken to understand language model portability issues. After all, should not enormous quantities (100 million words) of MSA text help to significantly improve language models over a few hundreds of thousands of words of transcribed colloquial Arabic? Is it that colloquial Arabic is so strongly influenced over the centuries by other regional languages that each dialect needs to be treated as a case for crosslingual adaptation? Or is the major component of divergence the fact that MSA is used in formal settings, while topics touched on in informal and personal settings are very different, calling for domain and topic adaptation as well? What aspects of MSA are retained in a particular Arabic dialect, and how do the other aspects change?

Other emerging languages and applications are likely to pose similar questions, particularly as the population of users of speech recognition systems grows. Approaches such as the factored language model need to be developed within a language model adaptation framework to incorporate linguistic constraints from the written language that are by and large followed in the spoken language.

Exploring Bridge Languages: Just as Chinese exposed word segmentation issues and Arabic and Czech exposed issues of morphologically rich languages, other languages in the world are likely to expose other limitations of the current technology. With a community-wide effort, research should be initiated to cover key bridge languages in each major language group in the world. This will provide insights for rapid development of language models in the nearly 400 languages spoken by 100,000 or more people each, by grounding the development effort in experience with a closely related language. Language groups that come to mind are the Indo-European and Dravidian languages from South Asia; the group of languages spoken in the Indonesian isles; and the Turkic family, including languages from the former Soviet states in Central Asia.

Translation Aware Language Modeling: A speech recognition system, which contains a language model, often serves as the front-end of a translation system. The back-end of such a system is either another human language or a database access and manipulation language. In case of spoken interaction with databases, considerable progress has been made in paying special attention to parts of the source language sentence that matter for the application. There is only preliminary effort in analogously constructing language models specifically geared toward translation to another human language (Ney, 1999). Much more research needs to be done in this direction.

Going Away from Verbatim Transcription: An ambitious solution to the speech recognition problem is to aim not for a verbatim transcription but for a fluent written record of the speech that merely preserves the essential meaning of the utterance to the extent permitted by orthographic conventions. Going further along these lines, the language of the output need not be the language of the input; colloquial Arabic speech, for instance, may be transcribed to produce MSA text with the same meaning, or a language with no written form may be transcribed to produce English text. Perhaps a modest beginning in this direction is to transcribe conversational English speech to produce a fluent English paraphrase.

Language modeling in languages of emerging economic, social, political, or military importance will need to address these as well as other hitherto unexplored issues.

Chapter 7

Multilingual Speech Synthesis

Alan W. Black

Speech systems often need to be able to use speech output to communicate with users in an understandable and natural way. The most successful early work in speech synthesis often utilized **formant synthesis** techniques, which required significant low-level phonetic knowledge in the target language to develop voices. More recently, as computing resources have increased, there has been a move toward **concatenative synthesis**, in which subword units of recorded natural speech are joined to form new words and phrases. Automatic techniques for construction of new voices based on recorded samples of the target language are now becoming feasible.

Although there are many examples of high-quality commercial speech synthesis systems for the major languages, it is still a nontrivial issue to provide speech output support for the remainder of the languages of the world. This chapter is concerned with the issues involved in creating and using speech output in multiple languages, that is, multilingual speech synthesis, and describes some of the current technologies to build synthetic voices in new languages.

7.1 Background

There are several ways in which multilingual requirements affect voice building for text-to-speech (TTS) applications. First, we can design a completely new voice for a new language, building all of the individual parts of the voice from scratch. Second, we can adapt an existing voice in some related language to our target voice. As both of these techniques can be viewed as extremes on a continuum, it is reasonable to find some intermediate combination of both to achieve the goal of a voice in a new language. The particular way in which these techniques are mixed will depend on the particular language and the actual resources available; we will discuss these issues in detail below.

A third class of challenges in multilingual TTS arises when multiple languages are required from the same system. Ideally, in this case, the same voice should be used to deliver the data rather than switching between different voices. In a world where multilinguality is increasingly common, there is a strong demand for synthesizers with multilingual capabilities, which include not only delivering complete sentences in different languages whenever a full language shift is required, but also mixing different languages within the same sentence. For example, many English words have crept into German, resulting in the need for English phoneme support; for example, the name "Windows" has become common, and most Germans will pronounce the initial phoneme as English /w/ rather than the more traditional /v/. Another example is the even more mundane and nontrivial issue of car navigation systems to be used within the European Union: a decision is required on how to pronounce local names in the desired language. Should "Straße" be pronounced "Street" for the British driver in Germany, or "North" be pronounced "Nord" for the French driver in England? We will not directly address these issues here, rather concentrating on how to provide basic spoken output support in any language, although such issues of mixed-language synthesis are already becoming a research topic (Pfister and Romsdorfer, 2003).

7.2 Building Voices in New Languages

In this section, we outline the basic process of building a new voice in a new target language. It must be noted that very high-quality speech synthesis is still very much an art, and a set of instructions alone is insufficient. However, the procedure outlined below, in combination with careful execution, typically yields practical speech synthesis of acceptable quality.

The basic steps involved in building a synthesis in a new language are as follows:

define a phone set: The basic sound units of the target language need to be determined; often, it is helpful to use existing phone set definitions. The amount of dialectal and **xenophone** influence (perhaps from English or some other related language) should be taken into account. For example, a German synthesizer should probably include /w/ even though it is not a native German phoneme but it is fairly common in technology-related texts, and a synthesizer without it may sound uneducated.

define a lexicon: For each word in the language, a definition of its pronunciation within the given phone set is required. For many languages, this will be an explicit list of words and pronunciations. In some languages, the relationship between the written form and the spoken form may be close enough that simple grapheme-to-phoneme rules can be used. In Spanish, for instance, the grapheme-to-phoneme mapping is fairly straightforward, and there are only a few letters with multiple pronunciations. English, on the other hand, does not have such a close relationship between written and spoken form, and it is harder to write rules for the mapping. In either case, we also need to provide a mechanism for pronouncing unknown words. For this reason, some mechanism of letter-to-sound mapping will be required, even if it is only used occasionally.

design a database to record: Next, an easily readable, phonetically, and prosodically balanced set of sentences needs to be constructed. The exact requirements will depend on the particular waveform synthesis technique to be used (see below).

record the database: The selected set of prompts needs to be recorded with a well-spoken, acceptably accented, consistent speaker (a so-called **voice talent**). It is important to choose a good speaker and to use the highest-quality recording equipment available.

build the synthesizer: The next step involves pairing the recorded waveforms with the database prompts and determining the phone boundaries. This can be done automatically or by hand, or by some combination of both. Once this has been completed, the synthesizer itself can be built.

text normalization: Raw text has a surprisingly large amount of non-words in it. Numbers, symbols, and abbreviations all need to be converted to a spoken word form before they can be synthesized. In English, for example, a string of digits like "1998" will be pronounced differently depending on whether it is a year, a quantity, or a telephone number.

prosodic models: Speech is not just a string of phonemes; there is additional prosodic variation, such as duration and intonation, that helps understanding by indicating phrasing. In addition, there are aspects of prosodic and phonetic variation resulting from the word context—for example, vowel reduction or liaison in French. These can sometimes be modeled implicitly from the recorded data, sometimes simple rules suffice, but in other cases, explicit models must be created.

evaluate and tune: The resulting voices should be evaluated with native speakers of the target language.

language-specific issues: In our experience, there are almost always language-specific issues that are difficult to anticipate. Examples include word segmentation in languages like Chinese and Japanese, vowel prediction in Arabic, and morphology in Polish. Additional time and effort needs to be spent on resolving these and other problems.

A widely available text-to-speech tool kit is the **Festival Speech Synthesis System** (Black et al., 1998b); the **FestVox** distribution (Black and Lenzo, 2000a) includes additional details, scripts, and samples for building synthetic voices.

For waveform synthesis, **concatenative synthesis** is the easiest technique and produce high-quality output. Earlier **formant synthesis** techniques (Allen et al., 1987), most famously represented in the *DECTalk* system, require too much design expertise and therefore are prohibitively difficult to port to new languages, even though the quality may be very understandable.

In the field of concatenative synthesis, there are two fundamental techniques: **diphone synthesis** and **unit selection**. Diphone synthesis (Olive et al., 1993) follows the observation that phone boundaries are the most dynamic portions of the acoustic signal and thus the least appropriate places for joining units. Since the middle of a phone is more likely to be acoustically stable, it provides a better join point. Thus, the unit inventory for diphone synthesis consists of segments extending from the middle of one phone to the middle of the next. The number of units in a diphone synthesizer will be approximately the square of the number of phones ("approximately" because some diphone pairs might not actually exist in the language). Although diphone synthesis is currently considered a somewhat old-fashioned technique, it can still produce perfectly understandable synthesis.

A diphone inventory list can be built from carefully selected natural speech or by designing an exhaustive set of nonsense words. Nonsense words have the advantage of better guaranteeing coverage because the desired phones can be explicitly incorporated into the prompts. In designing inventories, is it common to consider carrier words for the various diphones so that the desired diphones appear in fluent speech and not just in isolation. We encourage our voice talents to "chant" their diphone list so that the words are always delivered with the same vocal effort, duration, and fundamental frequency (F_0). One easy way to construct a diphone prompt list is to devise carrier words for each vowel-consonant, vowel-vowel, consonant-vowel, consonant-consonant, silence-phone, and phone-silence. For example, for English, a diphone list may start as follows and they cycle through all possibilities.

t aa b aa b aa
t aa p aa p aa
t aa m aa m aa . . .

Unit selection speech synthesis is currently the most popular technique for waveform synthesis. It is based on the concatenation of appropriate subword units selected from a database of natural speech. Unit selection was first successfully implemented for Japanese (Sagisaka et al., 1992a) but has since been generalized to other languages (Black and Campbell, 1995). From a multilingual point of view, one clear advantage of unit selection is that it abstracts away from low-level phonemic aspects of the voice, dialect, and language, making it easier to apply to new languages for which detailed phonological knowledge is unavailable.

There are a number of different algorithms used to find the most appropriate units in a database. In general, all algorithms index the units and measure their appropriateness based on two parameters: the **target costs**—determined by phonetic context, position in word and phrase, and prosodic information—and the **join costs**, which indicate units that best fit together. One of the many algorithms used is the **clunits algorithm** (Black and Taylor, 1997), which is freely available in the Festival Speech Synthesis System. In this case, units labeled with the same type (e.g., phoneme) are acoustically clustered and indexed by symbolic features, including phonetic and prosodic context. This technique, like most other unit selection algorithms, is language independent.

In order to cluster similar units, it is necessary to define an **acoustic cost**, which should be correlated with human perception of acoustic similarity. Although an ideal measure is still an area of research, most systems use some form of distance based on Mel Frequency Cepstral Coefficients (MFCC), F_0, and power features. In the clunits algorithm, the actual distance measure is defined as

$$Adist(U,V) = \begin{cases} \text{if } |V| > |U| & Adist(V,U) \\ \text{else } \frac{D*|U|}{|V|} * \displaystyle\sum_{i=1}^{|U|}\sum_{j=1}^{n} \frac{W_j.(abs(F_{ij}(U) - F_{(i*|V|/|U|)j}(V)))}{SD_j * n * |U|} \end{cases},$$

where $|U|$ = number of frames in U; $F_{xy}(U)$ = parameter y of frame x of unit U; SD_j = standard deviation of parameter j; W_j = weight for parameter j; and D = duration penalty. This can be described as a weighted Mahalanobis frame distance, averaged over interpolated frames. Often, pitch-synchronous frames are used with default fixed frame advances in unvoiced regions. This technique does benefit from accurate pitch period identification.

Given these acoustic distances, we can then cluster the units and index them using features that are available at synthesis time. Using standard CART techniques (Breiman et al., 1984), we can build decision trees to find clusters of around 10 to 20 similar acoustic units. At synthesis time, we simply use the decision tree to find the best cluster and apply the Viterbi algorithm to find the best path through the candidates from the clusters, minimizing the join costs (also based on MFCC and F_0 coefficients).

However, there is another technique that is gaining acceptability; instead of selecting actual pieces of natural speech, we can find the "average" of the clusters. This **HMM-based generation synthesis** (Tokuda et al., 2000) has the advantage of providing a parametric representation of the speech, which allows further processing, such as transformations. The resulting quality is not as crisp as unit selection, but the results can be more consistent and more robust to errors in the database.

When using diphone, unit selection, or HMM-based generation synthesis, it is necessary to label the database with phoneme boundaries.

For smaller databases, like diphones, a simple Dynamic Time Warp (DTW) technique works well (Malfrere and Dutoit, 1997). If the prompts are first generated by a synthesizer, we will know where the phoneme boundaries are. The synthesizer prompts are time-aligned with the naturally recorded utterances, which allows us to map the boundary times from the synthesized prompts to the natural ones.

For larger databases, in which a significant amount of phonetically balanced data is available, we can build database-specific acoustic models and then force-align the prompts to the natural speech.

Both of these techniques can work well with good data, but the largest issue in automatic labeling of databases for speech synthesis is actually detecting errors in the databases. Often the voice talent will not follow the prompt script exactly, or the synthesizer's output will not be pronounced in exactly the same way as the voice talent pronounces it. It would be useful to be able to detect such errors, especially if this can be done automatically. Kominek et al. (2003) developed an error detection method based on duration: phones whose labeled duration significantly differs from their predicted duration are either hand-checked or excluded from the database.

7.3 Database Design

Any form of corpus-based synthesis, whether unit selection or HMM-based generation synthesis, is greatly affected by database content. It is therefore important to take great care in designing databases. Ideally, a database should contain all phonetic and prosodic variations within the target language. Since this is not typically feasible, some approximation must be found. Traditionally, databases either were designed by hand (Fisher et al., 1986) or consisted of a large number of sentences in the hope that sufficient coverage would be achieved. Since it is not easy for the voice talent to read thousands of sentences without error and in a consistent manner, it is better to design the content of the database to be concise, such that each utterance significantly adds to the richness of the database. A number of competing factors—phonetic coverage, prosodic coverage, and ease of speaking—need to be optimized to minimize errors between the voice talent's utterances and the designed prompts.

Designing each sentence by hand will tend to produce less natural sentences, but is a good way to achieve high-quality synthesis if the domain of the desired application is well defined (Black and Lenzo, 2000b). Selecting naturally occuring sentences, by contrast, requires a fair amount of thought and care, as one can never be exactly sure how a voice talent will render the text. It is therefore helpful to take the final application into account (e.g., information presentations, dialog systems, or weather forecasts), since voices will always sound better when they are delivering data closer to the style of the originally recorded database.

The requirement of ease of speaking limits the content of the prompt set. The longer each sentence, the more likely it is for an error to occur. Thus, sentence length is typically restricted to range between 5 and 15 words. Words within the chosen sentences should be easy to pronounce without any ambiguity. Although algorithms for producing pronunciations of unknown words tend to be fairly reliable, they are not perfect. Therefore, sentences that contain words not in the lexicon should be omitted. Next, homographic words—words with the same written form but different pronunciations, e.g., *bass* (as in "fish") and *bass* (as in "music")—should be omitted. Although text-to-speech algorithms may be able to correctly predict the pronunciation, human readers may choose a valid but different realization. Finally, the sentence database may be supplemented by a single word database, since the pronunciation of single words is very different from their pronunciation within fluent sentences.

It should be added that care should be taken when applying the "no-homograph" constraint. For example, English *a* can be pronounced as a reduced vowel (schwa) when used as an indefinite determiner, but is sometimes pronounced as a full /ae/ when it represents the letter. Excluding all sentences with the word *a* in them would be overzealous and result in a prompt list without the word *a*, which would not be typical of normal text.

It is also prudent not to include symbols, digits, and the like in the prompts, since the text normalization front end and the voice talent may disagree on their pronunciation. Subtle differences in, for example, the pronunciation of "$1.23," in which the synthesizer expands it to "one dollar and twenty-three cents" but the speaker says "a dollar twenty-three" will cause labeling and synthesis problems. It is easier to avoid these issues by having all prompts be standard words. Another constraint that can be easily applied is to limit sentences to those containing high-frequency words. This is especially useful when no lexicon has been constructed for the target language. Sentences with unusual words or even misspellings and grammatical mistakes will often provide more unusual phoneme strings, which is counterproductive.

For Latin-alphabet languages, it can also be useful to restrict the occurrence of nontarget-language words. For example, in constructing a Spanish prompt set, we introduced a large number of words from other European languages, as the original text corpus was news text. In particular, names of European soccer teams and players were included. Although these names offer interesting phonetic variation, there is no guarantee that a Spanish-speaking voice talent would know the pronunciation unless he or she were a soccer fan. Thus we excluded sentences with words that were already contained within our English lexicon, assuming that our English lexicon (CMU, 1998) is likely to contain many international words.

Once a set of "appropriate" sentences are selected from the large corpus, they can be synthesized. Since it is unlikely that a full synthesizer already exists for any given target language (though sometimes a simple one may be available), it is only necessary to generate a phoneme string and not fully synthesize the waveforms. In fact, it can be advantageous to synthesize only strings of units, which can be simply phonemes but may also be tagged with stress, tone, syllable onset/coda, and so on. This may generate a larger prompt list but will probably have better coverage.

Black and Lenzo (2001), went even further and used acoustic information to find the tag set for the expansion of the database sentences and selected those that best covered the acoustically derived unit set. Using an initial database, a cluster unit selection synthesizer (Black, 1997) was built using the techniques described below. This produced a decision tree with acoustically similar units in clusters at its leaves. Instead of simply using symbolic phonological labels to select optimal coverage, this method labeled each cluster as a type and optimized the selection of sentences that covered those types. Thus, instead of just relying on abstract phonetic and prosodic knowledge for selection, this technique used data-driven acoustic modeling to select the types to be optimized.

After converting the sentences to unit names, a greedy selection can identify the sentences that have the best unit bigram (or unit trigram) coverage. This is done by scoring each utterance on the number of new bigrams or trigrams that the sentence would add to the selected prompt set. Multiple optimal subsets can be selected from the corpus until at least 500 sentences are in the prompt set, but one to two thousand sentences are likely to give better results. Having multiple sets may be preferable, since not all recordings may be usable, and multiple examples are always useful in the final database.

7.4 Prosodic Modeling

Without proper use of prosody (e.g., phrasing, intonation, and duration), a synthesizer will sound uninteresting and monotonous. A substantial effort in early speech synthesis research went into designing and building prosodic models. With the advent of unit selection synthesis, in which sub-word units are selected from databases of natural speech, explicit modeling of prosody has declined somewhat. Given a large enough unit selection database, which itself contains sufficient prosody variation, unit selection algorithms can exploit features that influence prosody variation (such as stress and position in the phrase) rather than having explicit duration and F_0 models. The resulting prosody is often very natural but limited to the variation covered in the database. As we continue to pursue the goal of more control in speech output, explicit prosody models are returning, but they are now more data-driven than before.

There are a number of mostly language independent factors in prosodic generation, but there are also many subtle differences related to style, dialect, and so on that make it very hard to generate natural prosodic variation.

Phrasing is controlled by the amount of air one can hold in one's lungs. While speaking, air is being exhaled, and eventually the speaker must take a breath. In addition, phrasing chunks the utterance into meaningful groups that make it easier for the listener to parse the utterance. Although some languages may use punctuation (such as commas or colons) to help identify when a phrase break might occur, in text, punctuation alone is typically insufficient to determine all phrase breaks. It is interesting to note that users of text-to-speech systems typically have a very strong expectation that punctuation should correlate with breaks, despite the fact that punctuation in written text does not often lead to a phrase break when read by humans.

In building voices in new languages, it is advisable to respect punctuation but further analysis may be required. In English, for instance, it is more acceptable to insert a break between a content word and a following function word, since the function word would typically precede the content word if they formed a coherent syntactic constituent. However, other languages may have different characteristic word orders. Japanese, for example, has syntactic postmodification instead of premodification (see Section 2.2.2.3); therefore, it is more likely to insert a break between a function word and a following content word, as particles typically follow their related content words. Another general constraint is that phrases (without additional information) typically tend to have approximately the same length. Thus, a good phrase break model will typically take into account part-of-speech information as well as distance from other breaks; such a model has been proposed in Taylor and Black (1998).

The possibility of using part-of-speech information for new languages may be limited, since a trained tagger—or tagged data for building a tagger—may not be available. However, it is often sufficient to simply identify common function word classes, and to leave the remaining (unknown) classes as content words. In English, for example, we can define a small number of function word classes and (almost) exhaustively list their members. Such a tagger is still sufficient for many aspects of speech synthesis. The following table lists English function word classes (first field) and their members (PP = preposition, DET = determiner, MD = modal verb, CC = conjunction, WP = relative pronoun, PPS = personal pronoun, AUX = auxiliary verb, punc = punctuation).

```
(PP of for in on that with by at from as if
   against about before because under after
   over into while without through new between
   among until per up down)
(TO to)
(DET the a an no some this that each another
   those every all any these both neither
   no many)
(MD will may would can could should must
   ought might)
(CC and but or plus yet nor)
(WP who what where how when)
(PPS her his their its our their its mine)
(AUX is am are was were has have had be)
(PUNC "." "," ":" ";" "\"" "'" "(" "?" ")" "!")
```

Although not comprehensive, such a model can be sufficient to build phrasal models, and is easy to specify for a new language.

For duration, very simple models can be used, such as a fixed duration for all segments (e.g., 100 milliseconds [ms]) or slightly modified segment-dependent durations, assigning vowels, say, 140ms and consonants, 60ms. It is fairly easy to build simple statistical models—applying techniques such as linear regression and decision trees—if corresponding data is available. For instance, FestVox includes scripts for building CART-style (Breiman et al., 1984) decision-tree duration models from a general speech database. The results may not be the best possible models but are usually sufficient for synthesis of a reasonable quality.

Truly natural-sounding duration is hard to achieve; moreover, it is also difficult to evaluate. In spite of the apparent ease of measuring duration model accuracy by calculating the root mean squared error between generated segments and corresponding spoken segments, it is not clear that these measures actually correlate with human perception of duration.

Another duration (and intonation) model technique is to use a model from an existing supported language. This method is promising if the language is prosodically similar to the desired target language. A common duration model has been used across Germanic languages, specifically the English model for German (Macon et al., 1998) and the Dutch model for Frisian (Dijkstra et al., 2004).

Intonation can be split into two aspects: accent placement and F_0 generation. Since accents (boundaries, tones, etc.) must be predicted before durations may be predicted, and F_0 contours require durations, it is normal to view these as two separate processes. Developing new models of intonation for new languages is still very hard—especially when developmental resources are constrained. It is unlikely that sufficient time can be spent to properly model all the intonational nuances of the language. In such cases, a more general intonation model, which is understandable and not misleading, will suffice. The easiest intonation model to build is none at all, relying instead on the implicit modeling provided by unit selection.

In Black and Hunt (1996), three-point syllable-based models have been used to generate F_0 contours, in which a linear regression model over general features, such as position in phrase and word type, is trained from a natural database. Such explicit F_0 models require reliable labeling and reliable F_0 extraction programs. More recently, many corpus-based intonation modeling techniques have emerged in which the F_0 models are trained from natural databases. These are either based on parametric models as provided by HMM-based generation synthesis (Yoshimura et al., 1999) or based on unit selection techniques in which just the contours, rather than spectral components, are selected from a database of natural speech (Raux and Black, 2003).

7.5 Lexicon Building

Traditionally, building a new lexicon for a speech synthesis system has been a multiyear process. However, we often need a lexicon in a new language within a much shorter time. Therefore, a number of techniques have been developed to simplify the process of lexicon building.

There are two major components of lexicon development: (1) providing a pronunciation for each word in the lexicon, and—since we typically cannot exhaustively list all words in the language—(2) providing a model for pronouncing unknown words. The following sections describe how to generate unknown word models using letter-to-sound rules, and how to use these in a rapid bootstrapping procedure for new lexicons.

7.5.1 Letter-to-Sound Rules

Letter-to-sound rules (sometimes called grapheme-to-phoneme rules) are needed in all languages, as there will always be words in the text to be synthesized that are new to the system. For some languages, letter-to-sound rules can be written by hand, but it is also possible to learn them automatically if a lexicon of words and corresponding pronunciations are provided.

Speakers of most languages can guess the pronunciation of a word they have not yet encountered based on the words they already know. In some circumstances, their guesses will be wrong and they need to receive explicit corrective feedback—for instance, in the case of foreign words. Statistical models can be built in a similar way. To this end, we first need to determine which graphemes give rise to which phones. Looking at an existing lexicon, we will typically find that there is a different number of letters compared to the number of phones, and although we might be able to specify by hand which letters align to which phones, it is not practical to do so for tens of thousands of words.

Consider the English word *checked*, which is pronounced as /tʃɛkt/. Out of the many different ways of aligning the letters to the phonemes, one possible way is the following:

c	h	e	c	k	e	d
tʃ	_	ɛ	_	k	_	t

Such an alignment can be found automatically: first, all possible alignments are generated by adding epsilons (null positions) in all possible places. Next, the probability of each phoneme given each letter is computed, and the most probable alignment is selected. This step can be repeated until convergence, which results in a consistent alignment between letters and phonemes. Since this task can be computationally quite expensive, due to the potentially large number of possible alignments, alignments can be constrained by specifying valid letter-sound pairs. Using this set of "allowables" typically reduces the effort to a single stage of alignment selection.

It should be added that, depending on the language, there may be more than one phoneme per letter. For example, in English we require a certain number of two-phone combinations in cases in which single letters may be pronounced as more than one phone.

x → k-s, cf. “box”
u → j-u, cf. “union”

Given that we now have a letter-to-phone alignment, we can use a machine learning technique to try to learn the mapping of letters to sounds within a specific context. We can convert our aligned lexicon to a series of vectors with the letter and its surrounding context, three letters to the left and three to the right:

```
# # # [c] h e c → ch
c h e [c] k e d → _
```

(“#” is used to denote the word boundary.) Using decision-tree techniques, we can automatically build classifiers to predict phones, dual phones, or epsilons and construct an entire word pronunciation by concatenating the results of each individual letter prediction step.

This technique has been applied to a number of languages, including English (U.K. and U.S.), French, German, and Thai (Black et al., 1998a; Chotimongkol and Black, 2000). In order to give an impression of the accuracy of this method, the following table shows automatic pronunciation prediction results on an independent test set of 10% of the words in the lexicon. The numbers shown are the percentages of correctly predicted individual phones and complete word pronunciations.

		Correct	
Lexicon	Language	Phonemes	Words
OALD	U.K. English	95.80%	74.56%
CMUDICT	U.S. English	91.99%	57.80%
BRULEX	French	99.00%	93.03%
DE-CELEX	German	98.79%	89.38%
Thai	Thai	95.60%	68.76%

Although these results do reflect the relative complexity of the pronunciation of the languages, they also reflect the consistency of the lexicons themselves. For example, CMUDICT (CMU, 1998) was constructed by a number of CMU graduate students principally for speech recognition, while OALD (Mitten, 1992) was created by a small number of trained lexicographers. Therefore, OALD is much more consistent in its pronunciations. However, OALD has significantly fewer proper names than CMUDICT, which has at least 50,000 (from many different countries); thus, CMUDICT covers a larger number of harder-to-pronounce words.

7.5.2 Bootstrapping Lexicons

Traditional approaches for building pronunciation lexicons involve collecting individual words and providing pronunciations for them (see Chapter 5 for a discussion of pronunciation lexicons for speech recognition). This is a task that requires phonetic expertise but that is nevertheless tedious. Since pronunciation lexicons often need to be generated much more quickly, efficient methods have been developed to determine which words should be manually annotated with pronunciations. By using a combination of the automatic letter-to-sound rule building process and human expertise, adequate lexicons can be built more rapidly (Maskey et al., 2004). This technique assumes that a phone set has already been defined and that a large corpus of text in the target language is available (if necessary, segmented into words).

1. First select the 200–500 most frequent words and define their pronunciations by hand. Ensure basic letter coverage in this set.
2. Build a set of letter-to-sound rules from this seed set using the automatic techniques described above.
3. Select around 100 high-frequency words not already in the lexicon, and use the letter-to-sound rules to predict their pronunciations.
4. Check manually if the pronunciation is correct.

    4a. If yes, add the corresponding word (and pronunciation) to the lexicon.

    4b. If no, correct it and add the word (and the corrected pronunciation) to the lexicon.

5. Go to step 3.

This algorithm is terminated when the desired level of accuracy has been achieved.

How accurate does a lexicon have to be? In general, a lexicon will never be 100% accurate unless the vocabulary is closed (i.e., no new unknown words occur in any new, yet-to-be-synthesized text). Various different ways have been presented in the literature to assess lexical pronunciation accuracy, many of which simply provide the highest number of correct pronunciations obtained, even though this may be deceptive. Care should be taken to determine the actual definition of accuracy when comparing different letter-to-sound rule accuracies. The problem lies in the fact that high-frequency words are likely to be correct; thus, very few tokens in the text are likely to be wrong, and claims of 99.9% accuracy are therefore easy to make. However, more stringent measures are more useful to developers of speech synthesis systems, since they better indicate if and where there is room for improvement. For this reason, we use the percentage of errors on word types rather than tokens. This yields a larger error percentage but makes it easier to quantify improvements. For lexicon building, accuracy rates of around 97% are usually sufficient for general synthesis.

The above technique was tested on existing lexicons to evaluate its performance. On CMUDICT (CMU, 1998), the initial lexicon (i.e., after the letter-to-sound rules were created from 250 seed words) was 70% word type accuracy. After 24 iterations, each adding around 1,000 words, the lexicon achieved 97% accuracy. Using DE-CELEX (Linguistic Data Consortium, 1996), a German lexicon, the initial 350 seed words and letter-to-sound rules gave only 30% accuracy. After 10 iterations, an accuracy of 90% was achieved. When this technique was used to develop a new U.S. English lexicon, an accuracy of 98% was achieved with only 12,000 English entries.

In Davel and Barnard (2004a, b), a very similar technique is used to build lexicons for Afrikaans, Zulu, and Xhosa. Importantly, it was discovered that playing very simple synthesized examples of predicted pronunciations aids the creator of the lexicon and reduces errors. It was also found that nonexperts with little phonological training were able to successfully use this technique to generate useful, correct lexicons within a few weeks.

7.5.3 Grapheme-Based Synthesis

One possible simplification in building synthesizers for low-resource languages is to use the orthography directly and learn the mapping from the orthography to the acoustic representation without an explicit phone set (similar to the approach for acoustic modeling as described in Chapter 4). The success of this will depend on how closely the pronunciation of the language is to the orthographic form. In Spanish, for example, it is readily accepted that the pronunciation closely follows the written form, though it is not just a simple one-to-one mapping. For instance, in Castilan Spanish, the letter *c* can be pronounced in a number of different ways: /k/, /θ/ and /tʃ/.

Word	Castillian	gloss
casa	/k a s a/	house
cesa	/θ e s a/	stop
cine	/θ i n e/	cinema
cosa	/k o s a/	thing
cuna	/k u n a/	cradle
hechizo	/e tʃ i θ o/	charm, spell

Thus, any mapping of letters to acoustics must provide for the multiple acoustic realizations of the same letters. Conversely, some phonemes may have multiple letter realizations (e.g., /k/ as *c* or *k*).

In Black and Font Llitjós (2002), a number of grapheme-based synthesis experiments were carried out. A phonetically balanced set of 419 utterances (5,044 words of around 27,500 segments) was recorded by a Castilian and a Colombian Spanish speaker. The databases consisted of sentences selected from newspapers, using the database design technique described above. A lexicon was generated not by using letter-to-sound rules but by simply expanding each word into a list of letters, thus treating each letter as a "phone." Some initial preprocessing was added: all words were converted to lowercase, and accent characters—á, é, í, ó, ú, and ñ—were preserved as single "phones," thus yielding a phone set of 32 phones.

Each database was then labeled by building database-specific HMM acoustic models for each phone (i.e., letter). The resulting models were then used to label the acoustics using forced alignment.

Once labeled, the databases were used for cluster unit selection (Black and Taylor, 1997), as described above. In this method of unit selection, similarly named units are clustered acoustically and are indexed by high-level features such as context, position in the word, and position in the phrase.

This process allows for further automatic distinctions between acoustically different units based on context. For evaluation, two newswire stories were synthesized with both the Castilian and Colombian voices. The basic paragraphs were as follows:

> *Sevilla, Agencias. Los sindicatos UGT y CC.OO. han exigido al presidente del Gobierno, José María Aznar, que convoque la mesa de negociación de la reforma del sistema de protección por desempleo, tras reunirse con el presidente de la Junta de Andalucía, Manuel Chaves, y el de Extremadura, Juan Carlos Rodríguez Ibarra.*
>
> *El secretario general de UGT, Cándido Méndez, junto al responsable de CC.OO., José María Fidalgo, reiteró la necesidad de que sea Aznar quien convoque y esté presente en esta mesa, si bien precisó que esta reunión no servirá para nada si la cita no comienza con el anuncio del Gobierno de que retirará su actual propuesta de reforma.*

The results were then hand-checked by a Spanish speaker and judged good, poor, or bad:

Dialect	total	good	poor	bad	% good
Castilian	109	102	6	1	93.57%
Colombian	109	99	5	5	90.82%

These accuracy rates would typically be considered good, especially for a small database. However, the results are still worse than they would be for a comparable system using actual phonemes.

For other languages, such as English, in which the relationship between orthography and pronunciation is not as transparent as that for Spanish, this method might not be expected to work at all. However, it does work reasonably well. Using the CMU ARCTIC bdl voice (Kominek and Black, 2003)—a phonetically balanced database of 1,128 utterances—a synthesizer was built using the same letter-based lexicon and database-specific acoustic models. The resulting synthesis was reasonable when the relationship between the written form and the pronunciation was straightforward. Thus, sentences like "This is a pen" synthesize correctly, and even when letters or groups of letters have different pronunciations, the technique may be successful; in the sentence "We attended church at Christmas," the <ch> is properly pronounced /tʃ/ in the first two cases, and /k/ in the third. Other words, however, turn out to be errorful—for example, "meat" is pronounced /m i ae t/—and many others with more opaque pronunciations are wrong. Although grapheme-based synthesis has its advantages and is surprisingly robust, an additional mechanism to correct such errors is required. One possible technique that has not yet been tested rigorously is to allow the user to provide alternative spellings for words that are considered wrong. Thus, "meat" could be rewritten as "meet." Such corrected spelling is relatively easy for native users to specify. These rewrites can be treated as an exception lexicon, or can even be used to train further letter-to-sound rules (actually, "letter-to-letter" rules).

It is likely that such grapheme-based methods will be more advantageous for languages with fewer resources. However, their success also depends on how standardized the writing system is in the target language. To those of us who are only familiar with well-defined, standardized spelling conventions, it is often surprising to what extent such conventions are lacking in other languages. As was mentioned in Chapter 2, many languages may not even have a writing system at all.

7.5.4 Cross-Language Modeling

Designing, recording, and labeling large new databases to support new languages is expensive and requires substantial skill. Other possible routes for developing speech output in new languages are described below. Among them are (1) phone mapping in combination with acoustic data collection; (2) phone mapping without recording acoustic data; (3) adaptation, and (4) non-native spoken output.

7.5.5 Cross-Language Modeling by Phone Mapping

One of the easiest methods for building a voice in a new, unsupported language—especially if the target domain is limited—is to use an existing supported language as a basic framework and to extend it to the new language. For example, when building a talking clock for Mandarin Chinese using a simple set of 12 utterances with complete word coverage of the intended domain, it is possible to simply provide lexicon definitions for each word using approximate phonemes in English. For example, if the intended phrase is:

Chinese: xian4zai4 ling2chen2 ling2dian3 zhong1 bu2dao4
English: Now midnight 2 o'clock not yet

We can add lexical entries using the default English phone set in the form:

```
("xian4zai4" nn ( ((sh iy) 1) ((ae n) 0) ((zh ay) 1) ))
("ling2chen2" nn ( ((l iy ng) 1) ((ch eh n) 1)))
...
```

Using the voice-building processing described above, we can build the prompts with these definitions, which will generate a **very** anglicized pronunciation of the Chinese. A native Chinese may then record these utterances (preferably not listening to the anglicized versions, as they will likely distract the speaker or even make him or her laugh at the bad attempt at Chinese). However, we can still utilize the English-phoneme-based synthesized Chinese prompts to find the phoneme boundaries in the natural Chinese recordings. The results for such cross-language alignments are often surprisingly successful due to the fact that the relative differences between basic sound classes are preserved across languages—for example, a vowel in English is more like a vowel in Chinese than a consonant in Chinese. This technique has been used for a number of simple limited domain tasks in Chinese, Nepali, Catalan, Spanish, Arabic, and Japanese for both talking-clock-type applications and large weather information systems. When the final unit selection synthesizer is built from the database of the native recorded utterances, the resulting synthesized speech has no artifacts of the underlying English synthesizer and sounds perfectly native.

7.5.6 Cross-Language Modeling without Recording New Data

A more general technique to generate waveforms in the target language without having to record any target-language data is to simply use a waveform synthesizer from a closely related language.

Many minority languages are spoken within an environment of more common and more widely spoken languages. For example, many Catalan speakers also speak Spanish, and it is not unusual to hear Spanish speakers speak Catalan with a Spanish accent. Phonetically, the difference between European Spanish and Catalan is mostly the addition of two vowels (lexically, the languages are significantly different). Thus, one solution to building a Catalan voice without having to record new data is to simply use a Spanish synthesizer and map all Catalan phones to Spanish phones. Given readily available diphone databases (e.g., from projects such as MBROLA [Dutoit et al., 1996]), this technique can be an easy way to quickly develop a working synthesizer in a target language. Often this is considered simply a first step, and the synthesizer is then used to help build an improved version—for example, by generating prompts for the alignment of actual native recordings or by aiding in building a lexicon.

In Tomokiyo et al. (2005), several experiments were carried out using phoneme mapping to develop a deliberately accented English voice. Using a Spanish unit selection database, three levels for crosslingual voices were built. The first simply used the Spanish pronunciation model, the second used the U.S. English model and mapped the U.S. phonemes directly to the Spanish model, and the third built new letter-to-sound rule mappings to more tightly couple the mapping of English phones to the smaller number of Spanish phones. In listening tests, the third voice produced the most understandable output, in spite of having a significant Hispanic accent.

7.5.7 Cross-Language Modeling through Adaptation

The third and perhaps most exciting (though also the least well developed) method for cross-language modeling is to directly follow the philosophy pursued in speech recognition (Schultz and Waibel, 2001), that is, to use data from multiple languages and a small amount of natively recorded data to perform adaptation (see Chapter 4). The principal technique currently used in this context is **voice transformation**, in which a mapping is learned between aligned spectral parameters in the source and target languages. Voice transformation was initially developed to allow the building of new voices from existing voices in the same language. It has also been used for style conversion for the same voices, but without requiring large databases to be recorded in different styles (Langner and Black, 2005; Tsuzuki et al., 2004).

Several ways of learning the spectral mapping have been investigated; the most popular of these is to train a Gaussian Mixture Model (GMM) to map between source and target languages (Stylianou et al., 1995; Kain, 2001). As a first step, parallel waveforms are created—that is, real or synthesized source waveforms paired with waveforms recorded by native speakers in the target language. The amount of data required is around 20–30 utterances with a reasonable phonetic coverage. The next stage is to align the spectral frames, which typically consist of either mel frequency cepstral coefficients (MFCC) or line spectral pairs (LSP). This can be achieved by dynamic time warping (DTW) alignment of each parallel sentence. Finally, a GMM mapping function between source and target spectral frames can be built as follows:

$$\begin{aligned}\hat{y} &= \boldsymbol{E}[\boldsymbol{y}|\boldsymbol{x}] \\ &= \sum_{i=1}^{m} p(C_i|\boldsymbol{x})[\boldsymbol{\mu}_i^{(y)} + \Sigma_i^{(yx)} \left(\Sigma_i^{(xx)}\right)^{-1} (\boldsymbol{x} - \boldsymbol{\mu}_i^{(x)})] \end{aligned} \qquad (7.1)$$

$$p(C_i|\boldsymbol{x}) = \frac{w_i N(\boldsymbol{x}; \boldsymbol{\mu}_i^{(x)}, \Sigma_i^{(xx)})}{\sum_{j=1}^{m} w_j N(\boldsymbol{x}; \boldsymbol{\mu}_j^{(x)}, \Sigma_j^{(xx)})}, \qquad (7.2)$$

where $\boldsymbol{x}$ and $\boldsymbol{y}$ denote a target and source frame, respectively. $\boldsymbol{\mu}_i^{(x)}$ and $\boldsymbol{\mu}_i^{(y)}$ denote the mean vector of the i-th mixture for $\boldsymbol{x}$ and that for $\boldsymbol{y}$, respectively. $\Sigma_i^{(xx)}$ and $\Sigma_i^{(yx)}$ denote the covariance matrix of the i-th mixture for x and that for x and y. $N(\boldsymbol{x}; \boldsymbol{\mu}, \Sigma)$ denotes the Gaussian distribution with the mean vector $\boldsymbol{\mu}$ and the covariance matrix Σ. w_i denotes a weight of the i-th mixture, m denotes the number of mixtures.

Thus, for language transformation, we can select a closely related source language and synthesize a set of utterances with appropriately close phonemes from the source language. We then record these utterances from a native speaker, align them with the synthesized examples, and build the GMM mapping function as described above. The results can produce acceptable waveform quality in the target language with a minimal amount of recorded data.

Latorre et al. (2005), use such a technique to build Spanish voices from Japanese (and the reverse). As these languages are phonetically relatively close, the results are certainly understandable and actually quite good.

Their technique used HMM-based generation synthesis and around 10–20 minutes of speech in the target language.

7.6 Non-native Spoken Output

Non-native or accented speech also plays an important role in speech synthesis. It has become obvious that accented speech (either non-native or from an unfamiliar dialect) can actually be more acceptable to some listeners.There is an advantage to using non-native synthesized speech in that speech errors are often attributed to the accent of the speaker rather than to the synthesizer errors themselves.

One specific example of this is the AWB CMU ARCTIC voice (Kominek and Black, 2003). The speaker is a Scottish English speaker, yet the basic build process uses the U.S. English lexicon and phone set. Scottish English has more vowels than U.S. English, thus some of the speaker's variation is confounded by the U.S. English front end. Therefore, the synthesizer makes a number of gross phonetic errors in pronunciations. However, U.S. English speakers do not immediately notice these errors and attribute almost all of the unusualness of the voice to it being a somewhat unfamiliar English accent. For example, Scottish English (like Southern British English) distinguishes between the vowels in *Mary*, *marry*, and *merry*, but the speaker's native variation of these are all treated as the same vowel /ɛ/. Although the clustering technique partially models this variation, the "wrong" vowel is sometimes selected during synthesis. Therefore, to a Scottish listener, confusions might exist between *Harry* and *hairy*, however no such confusion can be heard by the U.S. English listener. It should also be noted that some standard Scottish English phonetic variations are properly modeled by the system, particularly palatalization in words like *new* and *student*, and trilled and flapped /r/. Thus, although this voice does not actually provide a native Scottish accented synthesizer (native Scots immediately notice problems with it), U.S. English listeners find the voice acceptable and even find its novelty entertaining.

Apart from its entertaining qualities, however, the fact that accented synthesizers are often more acceptable than native synthesizers may actually make it easier to build communicating machines. As the human listener is more willing to adapt to a perceived non-native or nonlocal accent, taking advantage of this disposition facilitates the process of synthesizer building. In addition, in some applications—particularly speech-to-speech translation (see Chapter 10)—an accented voice is probably a very good idea. Since the translation will never be of perfectly native quality, it is not just acceptable but even expected that the output voice of a translation engine will have an accent. This may also encourage the replies from the target language speaker to be appropriately simplified, thus allowing easier communication through a limited translation device.

7.7 Summary

This chapter has presented the basic process of building speech synthesizers for new languages. Although there is still a substantial amount of art in building a new voice, it is a much better-defined task than it was even a few years ago. Widely available tools, such as those provided in the FestVox suite (Black and Lenzo, 2000a), have helped to increase the number of experts trained in speech synthesis and have thus paved the way for successful research ad commercial systems. In the near future we are likely to see greater use of voice conversion techniques, which, in the long run, will further facilitate the process of generating speech output in any desired language.

Chapter 8

Automatic Language Identification

Jiří Navrátil

Automatic Language Identification (LID) is the task of automatically recognizing language from a spoken utterance. In view of current globalization trends in communication technology, LID plays an essential part in providing speech applications to a large, multilingual user community. These may include multilingual spoken dialog systems (e.g., information kiosks), spoken-document retrieval, and multimedia mining systems, as well as human-to-human communication systems (call routing, speech-to-speech translation). Due to the challenge posed by multiple (and possibly unknown) input languages, interest in automatic LID has increased steadily, and intensive research efforts by the speech technology community have resulted in significant progress over the last two decades. This chapter surveys the major approaches to LID, analyzes different solutions in terms of their practical applicability, and concludes with an overview of current trends and future research directions.

8.1 Introduction

The core problem in solving the Language Identification (LID) task is to find a way of reducing the complexity of human language such that an automatic algorithm can determine the language identity from a relatively brief audio sample. Differences between languages exist at all linguistic levels and vary from marked, easily identifiable distinctions (e.g., the use of entirely different words) to more subtle variations (e.g., the use of aspirated versus unaspirated syllable-initial plosives in English versus French). The latter end of the range is a challenge not only for automatic LID systems but also for linguistic sciences themselves. It is, for instance, often unclear where the line between a language and a dialect should be drawn (see the discussion in Chapter 2). The fact that differences between varieties of a single language may be larger than those between two distinct languages must be borne in mind when developing LID systems. In this chapter, the term "language" is broadened to include dialects as distinct classes.

The ultimate role model for automatic LID systems is the human listener. Human listeners are capable of recognizing a language even from extremely short audio samples, provided they have a certain degree of familiarity with the language. Familiarity can vary from full knowledge of the lexicon, grammar, and pronunciation with native or near-native proficiency to a simple acoustic experience—that is,"familiarity from listening." Even with little previous exposure, listeners can often identify the language in question, though their accuracy may deteriorate or longer test samples may be required (see the detailed discussion in Section 8.2).

Figure 8.1 depicts various levels of abstraction from the raw speech signal and the type of information typically conveyed at those levels. Whereas the lowest level includes only purely acoustic signal variation, determined by channel and background noise, all other levels provide information relevant to distinguishing languages from each other. At the elementary-unit level, languages may be identified by their inventory of phonemes (e.g., French and Italian do not have the phoneme /x/, whereas Dutch and German do), by the acoustic realization of phonemes (word-initial stops such as /p,t,k/ are aspirated in English and German but not in French or Spanish, or the phone /S/ [as in '*shine*'], will sound differently in English and in Russian, with the latter being produced at a slightly more postalveolar position), and by phonotactic constraints (e.g., with few exceptions, Japanese does not allow two adjacent consonants, but Danish

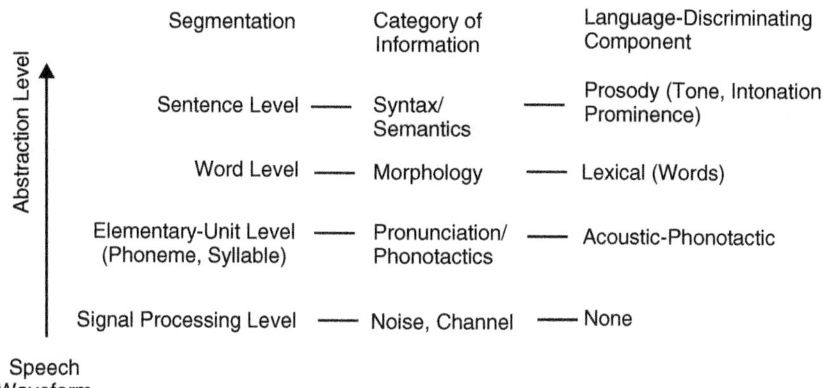

Figure 8.1: Levels of signal abstraction by acoustic analysis along with components of information.

and Swedish do). In addition to the mere existence of sounds or sound combinations in a language, their *frequency* is another key indicator of language identity (see Section 8.6). At the word level, languages are naturally distinguished by their different vocabularies. Finally, at the prosodic level, we find characteristic differences in intonation, tone, or prominence; variations in tone, for instance, are often used to identify Asian languages such as Chinese and Thai, and a word-final stress pattern distinguishes French from Hungarian, which has a word-initial stress pattern. In the ideal case, information from all these levels is used to perform language identification.

8.2 Human Language Identification

Since the performance of human listeners on LID tasks is the performance target for automatic LID algorithms, it is interesting to look at the question of what perceptual cues are used by human listeners to identify a language. The literature on human perceptual LID experiments provides several answers to this question.

8.2.1 Perceptual Benchmarks

A study on human LID was carried out in 1994 at the Oregon Graduate Institute (Muthusamy et al., 1994b) with the goal of creating a comparative perceptual baseline for automatic LID algorithms, as developed in Muthusamy (1993). In this experiment, 28 participants were presented with 1-, 2-, 4-, and 6-second excerpts of telephone speech in 10 languages: Chinese (Mandarin), English, Farsi, French, German, Japanese, Korean, Spanish, Tamil, and Vietnamese. The set of participants included 10 native speakers of English and at least 2 native speakers of each of the remaining languages. The average LID accuracies (the percentage of correctly identified test samples) for individual languages are shown in Figure 8.2. The accuracy rates shown are those achieved in the first and last quarters of the experiment. Several interesting results were obtained from this study. First, since participants were allowed to obtain the correct answer after every individual test sample during the experiment, a learning effect occured, as can be seen in the diagram: the accuracy improved from the first to the fourth quarter in all cases. Second, in spite of an initial training

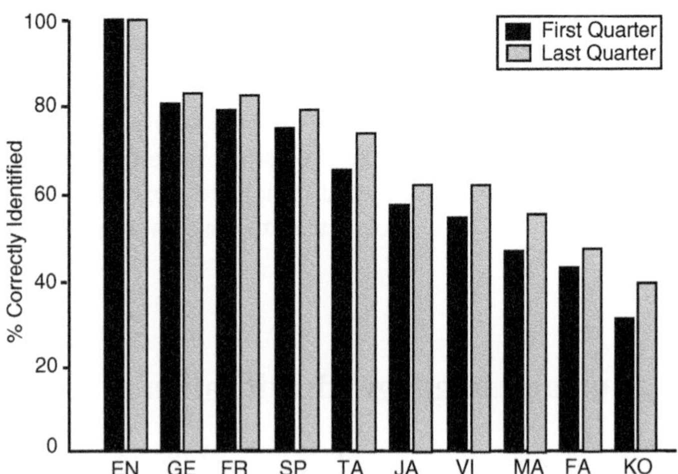

Figure 8.2: Human language identification rates for individual languages in the first (dark) and last (gray) quarters of the experiment, averaged over all participants (Muthusamy et al., 1994b).

phase during which participants were acquainted with acoustic samples of all languages, performance varied significantly for different languages. The more common languages, English, French, German, and Spanish, were recognized with an above-average accuracy, while Korean and Farsi were the least accurately identified languages.

Additional experiments studied to what extent human accuracy depended on the test signal duration as well as on the participants' general knowledge of foreign languages. In both cases, a positive correlation was observed. Finally, participants were surveyed in postexperiment interviews in order to determine the types of cues used during the LID process. The most frequently mentioned features included "typical sounds" (syllables and short words) occurring in each language (e.g., nasalization in French, the word *ich* in German) as well as prosodic features (e.g., the "sing-song" rhythm of Vietnamese).

With respect to the type of listening experiment, a similar study using excerpts of audio recordings was also done, involving five languages: Amharic, Romanian, Korean, Moroccan Arabic, and Hindi (Maddieson and Vasilescu, 2002). Identification accuracy as well as a similarity/discrimination performance were recorded. The study focused on cognitive factors that might help subjects in their ability to perform LID tasks. While the outcome was inconclusive as to how much linguistic knowledge was a factor in LID tasks, the authors concluded that such knowledge plays a positive role in language discrimination and similarity judgments.

8.2.2 Listening Experiments with Modified Waveforms

In order to obtain an assessment of the importance of different types of features (phonotactic, acoustic, prosodic) for human language identification, a more recent study (Navrátil, 2001, 1998) conducted a listening experiment based on telephone speech signals that were modified to selectively emphasize different information components. Whereas the experiments described in the previous section established perceptual benchmarks in terms of absolute performance and were based on unmodified speech signals, this experiment aimed at measuring the *relative* importance of lexical, phonotactic-acoustic, and prosodic features for the human listener.

The experiment involved five languages (Chinese, English, French, German, and Japanese) and consisted of three test sections. Listeners were asked to identify the language in each test section. In the first section (A), the test person heard speech excerpts (3 and 6 seconds long) from each language without any signal modifications. The listener thus had access to complete signal information; this experiment therefore served as a reference baseline. The test signals in the second section (B) consisted of short segments (syllables) of 6 seconds, which were concatenated in a random order. By shuffling syllables, the meaning of the original sentence as well as its prosodic contour were destroyed, leaving only the syllable-level phonotactic-acoustic information intact. The signals in the third section (C) were filtered by an adaptive inverse LPC filter, thus flattening the spectral shape and removing the vocal tract information. Only the F_0 and amplitude components—that is, the prosodic information—were audible. During the data collection period of three months, a total number of 78 participants of 12 different nationalities contributed to this study. The overall identification rates for the individual test sections are shown in Table 8.1.

The overall LID rates seem to indicate the different relative importance of phonotactic-acoustic and prosodic information. The weakest feature appears to be prosody—presumably due to its linguistic complexity and the influence of various factors (speaker, emotion, sentence modus, semantics). These results are confirmed by the different performance rates obtained by automatic systems using prosodic versus phonotactic information (Hazen and Zue, 1997; Yan et al., 1996; Schultz et al., 1996a; Hazen and Zue, 1997; Navrátil and Zühlke, 1998), which are described in more detail in the following sections. Prosody appears to be useful mainly for distinguishing between certain broad language groups, such as tonal versus nontonal languages. In spite of the relatively strong importance of

Table 8.1 Language identification accuracy rates over all test persons for test sections A (full speech), B (syllables), and C (prosody).

Section	German	English	French	Japanese	Chinese	Overall
A	98.7%	100.0%	98.7%	81.7%	88.7%	93.6% (3s) 96.0% (6s)
B	79.7%	98.7%	79.1%	54.6%	57.7%	73.9% (6s)
C	32.1%	34.3%	69.4%	45.3%	65.9%	49.4% (6s)

the phonotactic-acoustic features (section B), the word-level information available in section A seems to be a decisive factor for high performance. This is an indicator for the necessity of (at least a partial) lexical decoding to be included in an LID system, as discussed in Section 8.8.

8.2.3 Psycholinguistic Study Using Speech Resynthesis

A systematic study focusing on the importance of various prosodic and rhythmic features in distinguishing between English and Japanese was carried out by Ramus and Mehler (1999). Motivated by psycholinguistic questions of language learning of children in bilingual environments, Ramus and Mehler developed an experimental paradigm involving speech resynthesis with various modifications of the speech signal in order to selectively remove certain information components. Specifically, the speech signals in the two languages were transformed to contain (a) broad phonotactics, rhythm, and intonation; (b) rhythm and intonation; (c) rhythm only; and (d) intonation only. This was achieved by a thorough manual speech analysis followed by resynthesizing certain segments to sound as other phonemes. For example, condition (a) entailed replacing all fricatives with the phoneme /s/, all stop consonants with /t/, all liquids with /l/, all nasals with /n/, all glides with /j/, and all vowels with /a/. This would remove nonprosodic syntactic information from the speech while preserving the broad phonotactic-prosodic structure. In the other conditions, similar but even more drastic replacements were performed, such as replacing all consonants with /s/ and all vowels with /a/, or replacing all phonemes with /a/ (preserving the fundamental frequency only).

A total of 64 native French speakers participated in the listening experiments. The listeners did not have prior knowledge about which languages were being used in the tests. After a controlled training phase, allowing for adaptation to the nature of the modified signals, each listener was asked to discriminate between two languages in a series of tests. The average accuracy in the discrimination experiments ranged between 50% (chance) and 68% (rhythm). The main result showed that test subjects could not distinguish the languages well in the prosodic-only mode, while their performance hovered around 67% in all other modes, which involved the broad phonotactics and rhythmic information. While Ramus and Mehler present many more derived results and discuss psycholinguistic aspects, the main conclusion—relevant for perceptual LID—is a confirmation of the relatively weak discrimination power of prosodic information alone (in the sense of fundamental frequency), as compared to rhythmic, phonotactic, or even underlying higher-level syntactic features in the speech signal.

8.3 Databases and Evaluation Methods

The availability of sufficiently large speech corpora is one of the major driving factors in speech research in general. The other is the existence of generally agreed on standards for evaluating LID systems. We will first survey existing corpora for LID research and then discuss the two main evaluation paradigms.

8.3.1 Corpora

On the basis of work by Yeshwant Muthusamy et al. (1992), the first multilingual database for LID research was created by the Center for Spoken Language Understanding at the Oregon Graduate Institute (OGI)[1] and released to the research community in 1993. This database, a collection of telephone speech in 11 different languages, provided a unified way of comparing and evaluating LID algorithms and replaced the then common practice of reporting isolated results on different proprietary data sets. As a result, the volume of technical publications on LID increased noticeably after 1993, which proved the significance of the "data factor" (Muthusamy et al., 1994a). The release of the first OGI database was followed by a series of subsequent data collection efforts, the most important of which are listed below.[2] The OGI-11L corpus consists of speech data from English, Farsi, French, German, Hindi, Korean, Japanese, Mandarin Chinese, Spanish, Tamil, and Vietnamese. Thus, the language repertory includes languages from the same families (such as French and Spanish) but also distinctly different languages (e.g., tone languages like Vietnamese and Mandarin versus stress languages like German and English). An average of 100 speakers were recorded in each language over a real-world telephone

[1]This is now known as the OHSU-OGI School of Science and Engineering.

[2]All of the cited corpora are distributed through the Linguistic Data Consortium (LDC) (LDC, 2005).

channel. Utterances fell into three categories: (1) those with a fixed vocabulary (e.g., the days of the week, or numbers from one through ten); (2) those with topic-dependent vocabulary (answers to requests such as "Describe your last meal!"), and (3) spontaneous, topic-independent monologues from each speaker. Details on the collection protocol can be found in Muthusamy et al. (1992). An important feature of the OGI-11L database is the availability of manual segmentation and labeling information at both the phone and the word level for a subset of six languages. These annotations can be used for training various multi- and monolingual phonetic recognizers, which are an essential component in many LID systems—especially those based on the phonotactic approach described in Section 8.6. The OGI-11L database was later released through the LDC as OGI-TS, with minor modifications to the original OGI distribution.

In a subsequent OGI project, multilingual data was collected from at least 200 speakers in 22 different languages (OGI-22L) (Lander et al., 1995). The languages included are Arabic, Cantonese, Czech, English, Farsi, French, German, Hindi, Hungarian, Italian, Japanese, Korean, Malaysian, Mandarin (Chinese), Polish, Portuguese, Russian, Spanish, Swahili, Swedish, Tamil, and Vietnamese. In comparison to the OGI-11L, the repertory was expanded by languages from additional language families, such as Semitic and Slavic languages. Similar to OGI-11L, phonetic time-aligned transcriptions for a subset of the languages as well as romanized transcripts for Arabic, Cantonese, Farsi, Hindi, Japanese, Korean, Mandarin, and Tamil were provided. The database was released in 1995.

In 1996–1997, the Linguistic Data Consortium (LDC) organized a series of LID-relevant data collections of unscripted telephone conversations between friends in 6 (CallHome) or 12 (CallFriend) languages. There are about 60 conversations per language, each lasting between 5 and 30 minutes (see Chapter 3 for a more detailed description of these corpora). The CallFriend corpus was used by the National Institute of Standards and Technology (NIST) as a data source to conduct an evaluation of LID algorithms, organized in 1996 and 2003.[3]

8.3.2 Evaluation Methods

Most of the earlier published studies in the LID area view the quality of an algorithm in terms of its accuracy at identifying languages—that is by

[3] http://www.nist.gov/speech/tests/lang/index.htm

measuring an identification rate or an error rate (typically averaged over all languages). In recent years, an increased interest in the related task of **language *detection*** (or verification) can be observed. Unlike the LID task, the goal of language detection is to either confirm or reject a language hypothesis about the incoming speech. As such, this task only involves providing a yes/no answer to the question "Is X the language of this test signal?," in contrast to identifying one out of N languages as the correct one. Language detection is desirable mainly because of the fact that the possible language population in LID can quickly become too large in real-world applications and cannot be modeled exhaustively. Therefore, by performing a series of language detection tests to scan the input for (a few) relevant languages, open-set LID is achieved.

Typical tools to assess the performance of a language detection system include the **Receiver Operating Characteristics (ROC)** curve—or, equivalently, the **Detection-Error Trade-Off (DET)** curve (Martin et al., 1997). The ROC/DET representation allows for an immediate assessment of the system's behavior in terms of its two types of errors, the false alarm rate and the miss rate. The detection performance is often sampled at a specific operating point to obtain a single-valued measure of accuracy. NIST, which conducted evaluations of LID algorithms in 1996 and 2003, defined the key operating point for their evaluation to be the **equal-error rate**—that is, an operating point at which the two error types are in balance.

8.4 The Probabilistic LID Framework

The most successful LID systems use a statistical framework (see, for example, Hazen (1993) and Hazen and Zue (1997)), which is similar to state-of-the-art approaches to speech and speaker recognition. The core of this approach is the Bayes classifier.

Identifying a language L^* out of a repertory of possible languages $L_1, \ldots, L_N$ involves an assignment of the most likely language L^* (out of all N languages) to the acoustic evidence A from the speech signal, such that

$$L^* = L_j \text{ when } P(L_j|A) > P(L_i|A) \ \forall i = 1 \ldots N, i \neq j \tag{8.1}$$

or, in a concise form:

$$L^* = \arg\max_i P(L_i|A). \tag{8.2}$$

The acoustic evidence, originally available as a speech waveform $s(t)$, is typically transformed into a sequence of acoustic feature vectors for further processing $v = v_1, \ldots, v_T$. This process generally aims at removing redundancy and irrelevance and usually involves a short-term spectral analysis followed by a series of linear and nonlinear transforms. In addition to the spectral information contained in v, a "prosody contour" is also taken into account, represented by the fundamental frequency and amplitude components jointly denoted by $f0$. Given these representations, (8.2) becomes

$$L^* = \arg\max_i P(L_i|v, f0). \tag{8.3}$$

In Hazen and Zue (1997), Hazen decomposes the probability (8.3) into several components, as follows: it is assumed that the speech input can be segmented into a sequence of phones $a = a_1, \ldots, a_M$ (via phonemic analysis), with their corresponding segment boundaries $s = s_0, \ldots, s_M$. Considering such segmentation, (8.3) can be represented as

$$L^* = \arg\max_i \sum_a \sum_s P(L_i, a, s|v, f0) \tag{8.4}$$

and rewritten as

$$L^* = \arg\max_i \sum_a \sum_s P(L_i|a, s, v, f0)P(a|s, v, f0)P(s|v, f0). \tag{8.5}$$

Equation (8.4) is an exact equivalent of (8.3), as no intermediate assumptions were made up to this point. In practice, however, a probability estimation for all possible segmentations and phone sequences is not tractable. Replacing the summation by a calculation of a single optimum segmentation and phone hypothesis can reduce the complexity significantly. This step is commonly and successfully applied in speech recognition algorithms as well. Hence, the summation reduces to a search for optimum phone segmentation and phone assignment (Hazen, 1993):

$$L^* = \arg\max_{i,s,a} P(L_i|a, s, v, f0)P(a|s, v, f0)P(s|v, f0) \tag{8.6}$$

In order to simplify this form further, it is assumed that the search for an optimum hypothesis $\{s^*, a^*\}$ can be performed independently of the language identification task itself, meaning the utterance decoding is language universal. Hence, the search becomes a self-contained decoding (phone recognition) step:

$$\{s^*, a^*\} = \arg\max_{s,a} P(a|s, v, f0)P(s|v, f0) \tag{8.7}$$

and can be implemented, for example as a Viterbi decoder. Once $\{s^*, a^*\}$ is known, LID is performed based on

$$L^* = \arg\max_{i} P(L_i|a^*, s^*, v, f0), \tag{8.8}$$

whereby this form may be conveniently rewritten as

$$L^* = \arg\max_{i} \frac{P(L_i, a^*, s^*, v, f0)}{P(a^*, s^*, v, f0)} = \arg\max_{i} P(L_i|a^*, s^*, v, f0). \tag{8.9}$$

A further decomposition of the latter term in (8.9) leads us toward indentifying models for individual components of information in speech signals, as discussed in Hazen (1993):

$$L^* = \arg\max_{i} P(L_i)P(a^*|L_i)P(s^*, f0|a^*, L_i)P(v|a^*, s^*, f0, L_i) \tag{8.10}$$

Besides the fact that it may be easier to estimate the individual probabilities in (8.10) as compared to (8.9), they also correspond to essential speech components, namely:

- $P(a^*|L_i)$ describes the frequency of phone co-occurrences. Statistical dependencies between phonemes in languages are termed *phonotactics*; thus, this probability represents the *phonotactic* model. In practice, the probability is decomposed into a product of conditional terms, whereby each term is approximated by a Markov process of a certain order. These are commonly referred to as N-grams. Phonotactic approaches are highly effective due to their relatively uncomplicated training procedures and will be discussed extensively in Section 8.6.

- $P(v|a^*, s^*, f0, L_i)$ is the *acoustic* model. The acoustic model captures patterns of pronunciation contained in the feature vector sequence v. In the literature, this probability has been considered both dependent as well as independent of the phonemic (a) and prosodic ($f0, s$) information. More details are discussed in Section 8.5.
- $P(s^*, f0|a^*, L_i)$ represents the *prosodic* model. The segmentation information $f0$ (amplitude and F_0 contours) are modeled language specifically and ideally depend on the underlying phone sequence. The latter condition, however, is often dropped in favor of extracting simpler and more robust features. A question of how to best represent the $f0$ contours represents a great challenge in prosodic LID and remains an active problem. Section 8.7 discusses the prosodic model in more detail.

The following sections survey approaches to automatic language identification from the aspect of the components and information types delineated above, recognizing the fact, however, that no strict boundaries between the various methods can be drawn. Most successful systems implement more than one such component model by joint modeling of several information types or by combining isolated components at various levels.

8.5 Acoustic Approaches

Purely acoustic LID aims at capturing the essential differences between languages by modeling distributions of spectral features directly. This is typically done by extracting a language-independent set of spectral features from segments of speech and using a statistical classifier to identify the language-dependent patterns in such features. Thus, the evidence for the language-dependent classifier in Equation 8.10 consists of one or more acoustic feature vectors, $v = v_1, \ldots, v_T$, which may represent an unordered set of vectors or—in more advanced systems—a temporal sequence. Considered independently from other components contained in Equation 8.10, the acoustic classification rule, given the acoustic evidence A, becomes

$$L^* = \arg\max_L P(L|A) = P(L|v_1, \ldots, v_T). \tag{8.11}$$

Since there is no intermediate mapping to explicit linguistic units (such as phones or syllables), differences in acoustic features may represent a variety of phenomena, such as pronunciations of individual sounds across languages or coarticulation of sounds in sequence.

8.5.1 Acoustic Pattern Matching

Acoustic approaches are among the earliest studies of automatic LID (Cimarusti and Ives, 1982; Sugiyama, 1991). An example is the 1982 study by Cimarusti and Ives, in which an LID system for eight languages (Chinese, Czech, English, German, Farsi, Korean, Russian, and Vietnamese) was developed on the basis of LPC analysis. The entire feature set consisted of approximately 100 measured values (LPC-, ACF-, Log-Area-Coefficients, formant positions, etc.) followed by an iterative training procedure of a **polynomial classifier** in order to achieve a full separation of all training samples. The authors reported an accuracy of 84% based on 3-minute speech recordings. Although this result was probably not speaker independent due to the relatively small number of speakers in the data collection, it appeared promising.

Improvements to this simple spectral pattern matching technique can be obtained by using an acoustic codebook with more than one reference pattern per language. An example of this technique is a study presented by Foil (1986), which uses **vector quantization** (VQ) of acoustic and prosodic features. In the first part of the study, prosodic features (energy, zero-crossing rate, and derived values) were extracted from a short-time analysis of speech. In the second part of the study, formant positions and other derived features were calculated from voiced speech segments. Models in the form of VQ codebooks (sets of reference vectors placed throughout the acoustic space) were trained for each language with the objective of minimizing the VQ error (the loss obtained in recovering each individual vector from the nearest reference vector). The minimum accumulated VQ error served as a basis for the LID decision. While the formant-based feature set achieved an accuracy of 64% (with an 11% rejection rate), the prosodic feature set led to an accuracy only slightly higher than random choice, indicating that there are significant challenges associated with the extraction and modeling of prosodic features for LID. This has been confirmed by subsequent work (as discussed in Section 8.7).

8.5.2 Acoustic Modeling with Gaussian Mixture Models

Gaussian Mixture Models (GMMs), used extensively in speech recognition (see Chapter 4), have also successfully been applied to modeling the acoustic probability $P(v_1, \ldots, v_T|L)$ for LID. Although the specific form varies by implementation (depending on the type of phonetic information modeled or the inclusion of temporal structure through Hidden Markov Models [HMMs], for example), the core modeling principle is the same: an acoustic feature vector v is assumed to be generated by a multivariate stochastic source with a probability density function:

$$p(v|L) = \sum_{i=1}^{M} \pi_i N(v, \mu_i, \Sigma_i), \tag{8.12}$$

where N denotes a multivariate normal density function (Gaussian) with mean parameter μ and covariance Σ. M densities form a mixture distribution, in which each component is weighted by its prior probability π. Due to the fact that GMMs can approximate any arbitrary density function and that powerful training algorithms for GMM structures exist, these models are the preferred choice in many state-of-the-art LID systems. Compared to the VQ-based pattern-matching approach described above, the advantage is that the distribution of acoustic vectors given the language is no longer approximated by a discrete set of reference values, but is modeled in a continuous fashion, which typically results in higher accuracy. An example of the use of GMMs for LID is found in Zissman (1993), which used GMMs trained on relative spectra (RASTA), followed by a maximum-likelihood based decision rule.

In their segment-based approach, Hazen and Zue used a phonetic-class GMM with $M = 15$ mixture components, trained on vectors of MFCC features (Hazen, 1993; Hazen and Zue, 1997). There were up to 59 such phonetic classes, derived from an automated clustering technique. As in many other practical applications, diagonalized covariances were used to reduce the model complexity and to allow for robust parameter estimation. Hazen and Zue reported LID accuracy rates of about 50% measured on the NIST-1994 evaluation dataset, compared to about 60–70% achieved by a phonotactic component (see Section 8.6) on the same data. Similar observations were reported by Zissman (1996), and Navrátil and Zühlke (1998), who used GMMs modeling individual phonemes out of a broad language-independent repertory. The acoustic performance in isolation, measured on the NIST-1995 evaluation dataset, averaged about 50% while the same data could be identified with an accuracy of up to 85% using a phonotactic component. Despite a significant gap in performance, however, a combination of both components yielded improvements.

All acoustic modeling approaches presented so far only model the static distribution of acoustic features given the language, although much language-discriminating information resides in the dynamic patterns of acoustic features changing over time. Temporal acoustic modeling is most often performed using HMMs, which are not only the preferred modeling technique in speech recognition systems, but also frequently used for acoustic decoding in language and speaker identification (Corredor-Ardoy et al., 1997; Dalsgaard et al., 1996; Lamel and Gauvain, 1994; Pellegrino et al., 1999; Teixeira et al., 1996; Ueda and Nakagava, 1990; Zissman, 1996). An example of an HMM structure is shown in Figure 8.3 consisting of a set of states with transition probabilities (a_{ij}) and a set of observation probability functions ($b_i(\cdot)$). The state structure is a Markov chain of

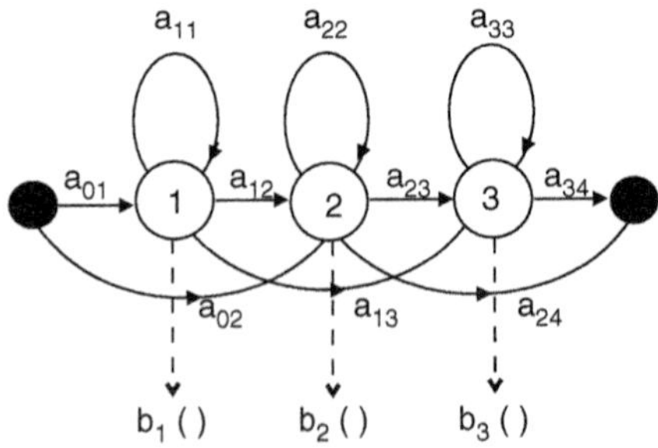

Figure 8.3: An illustrative example of an HMM structure.

1-st order, with nonemitting and emitting states (shown in dark and white, respectively, in Figure 8.3) accounting for the sequential time behavior of the modeled speech, while the observation probability functions (typically GMMs) capture temporal acoustic patterns. Due to their capability of capturing temporal information in speech, HMMs represent a natural bridge between the purely acoustic approaches described in this section and the phonotactic approaches described in the next section.

More recently, the GMM/HMM classification framework has been applied on a frame-by-frame basis whereby acoustic vectors are calculated as differentials in a sequence of cepstral vectors (Singer et al., 2003; Torres-Carrasquillo et al., 2002b). A high-order GMM consisting of 1,024 components was used by Torres-Carrasquillo et al. (2002b) to model the differential features, termed "Shifted-Delta Cepstra," generating an acoustic classification score but also tokenizing the speech into a discrete sequence of indices that were then modeled by a phonotactic component. Further refinements include channel compensation and gender-specific modeling (Singer et al., 2003); vocal-tract length normalization (Wong and Sridharan, 2002); and adaptation from large background models (Singer et al., 2003; Wong and Sridharan, 2002), resulting in GMM systems with a performance comparable to that of phonotactic systems.

8.5.3 Acoustic Modeling with ANNs and SVMs

There are alternative ways of modeling the distribution of acoustic feature vectors given the language; other popular choices besides GMMs are **artificial neural networks** (ANNs) and **support vector machines** (SVMs).

In an ANN, elementary functional units (neurons) are ordered in a layered manner such that each unit connects to a subset of others via weighted arcs (synapses). A neuron is characterized by a weighted summation of its inputs followed by a nonlinear warping (sigmoid) function to obtain its output. The networks predominantly used in LID are the feed-forward structured ANNs, that is, those with neuron outputs of one layer connecting to the next layer only. ANNs can approximate any arbitrary mappings and therefore present a powerful modeling tool (Funahashi, 1989).

Muthusamy (1993) utilized an ANN in the form of a multilayer perceptron, which was trained to map perceptual linear prediction (PLP) feature vectors to one of seven broad phonetic classes, followed by alignment and segmentation. These broad classes were then used to perform LID; results ranged between 65 and 70% for 10 languages of the OGI-11L corpus. In a 1998 study, Braun (Braun and Levkowitz, 1998) described the use of recurrent neural networks (RNNs) applied to perceptually "important" segments of test utterances that were detected in an interactive manner by human listeners. Recent literature on LID with acoustic ANNs is relatively sparse, possibly due to the relative success of other methods and the fairly large number of training samples required by ANNs.

Support vector machines (SVMs) were introduced by Vapnik (1999) as an analytic result of the empirical risk minimization problem—a basic principle in machine learning. The essence of SVMs lies in representing the class-separating boundaries in a high-dimensional space in terms of few crucial points obtained from the training sample, termed support vectors. The selection is carried out to attain a maximum-margin solution best separating the classes. The transition from an original input space to the high-dimensional space is achieved by applying a certain kernel function. Campbell et al. (2004) describe an application of support vector machines to the task of LID. Similar to a successful implementation in speaker detection (Campbell, 2002), the SVM system uses a monomial kernel expansion from the input to a high-dimensional feature space, in which a linear classification on averaged vectors is carried out. Measured on the NIST-1996 and the 2003 LID evaluation datasets, the SVM system provides results better or comparable to the GMMs and the phonotactic components (Campbell et al., 2004; Singer et al., 2003), and hence represents one of the promising novel additions to state-of-the-art LID techniques. As a summary, Table 8.2

Table 8.2 Some acoustic LID systems and their performance rates.

System	Task	Test Duration	ID Rate	Ref.
GMM (40 comp.)	OGI-10L	10s/45s	50%/53%	Zissman, 1996
GMM (16 comp.)	OGI-11L	10s/45s	49%/53.5%	Hazen and Zue, 1997
Phone-HMM	OGI-10L	10s	59.7%	Lamel and Gauvain, 1994
Syllabic-NN	OGI-10L	10s	55%	Li, 1994
Vowel-GMM	5L (read speech)	21s	70%	Farinas et al., 2002

lists selected representatives of acoustic LID implementations in terms of their performances, ordered roughly chronologically.

8.6 Phonotactic Modeling

Phonotactics—the constraints on relative frequencies of sound units and their sequences in speech—is one of the most widely used information sources in LID. Its popularity is due to its relatively simple and scalable implementation on the one hand, and its relatively high language-discriminating power on the other. The phonotactic LID method uses a probabilistic framework and builds on the inherent property of every language to exhibit strong language-specific frequencies and dependencies between individual phones[4] in an utterance. This phenomenon can be easily illustrated by an example: Whereas the combination of the phone /i/ followed by /ç/ occurs frequently in German, it is nonexistent in English. It is interesting to point out that similar relationships between the fundamental units of written texts (graphemes) were already observed by Markov in 1913 and the resulting differences among languages have been analyzed in numerous linguistic and information-theoretic reports (Küpfmüller, 1954). As Küpfmüller demonstrates, by modeling sequences of letters in English and German texts using a Markov model of an increasing order, random strings can be generated such that their resemblance to the corresponding language becomes striking despite the lack of meaning in such texts. Retaining this basic idea, it is only a small step from graphemes in written texts to phone units in spoken language.

Similar to other identification and recognition techniques, phonotactic modeling in LID involves both a training and a test phase. During training, a set of conditional probabilities is estimated from training data in each language; in the test phase, observation probabilities of a test sample are calculated given all models, typically followed by a maximum-likelihood decision rule. The probabilistic phonotactic models can have a variety of structures. The most popular of these is that of an ergodic model of $N-1$-st order, where $N = 1, 2, \ldots$, also known as N-grams (Hazen and Zue,

[4]We use the broader term *phone* instead of *phoneme*, which has a linguistically stricter definition (see Chapter 2).

1997; Muthusamy et al., 1994a; Zissman and Singer, 1994). Alternative techniques to achieve more flexible N-gram modeling include binary decision trees (Navrátil, 2001) and various clustering and smoothing schemes (Kirchhoff et al., 2002b; Schukat-Talamazzini et al., 1995).

8.6.1 Basic N-gram Modeling

Let $\mathbf{A}$ denote the set of defined sound units (e.g., all phones), $\mathbf{A} = \{\alpha_1, \ldots, \alpha_K\}$. Based on an observed sequence of length T, $A = a_1, a_2, \ldots, a_T$, the discriminant used in a maximum-likelihood classifier is the probability of A given a hypothesized language $L \in \{L_1, \ldots, L_M\}$, where M is the number of languages in the system. This can be written in terms of conditional probabilities as follows:

$$P(A|L) = P(a_1, \ldots, a_T|L) = P(a_1|L) \prod_{t=2}^{T} P(a_t|a_{t-1}, \ldots, a_1, L) \quad (8.13)$$

The use of N-grams involves the following approximation to (8.13) (see also Chapter 6):

$$P(a_t|a_{t-1}, \ldots, a_1, L) \approx P(a_t|a_{t-1}, \ldots, a_{t-N+1}, L), \quad (8.14)$$

that is, a unit at time t is modeled as dependent on the $N-1$ units immediately preceding it and independent of the units beyond $N-1$. The set of the approximating probabilities over $\mathbf{A}$ can be also seen as a discrete stationary Markov chain of $(N-1)$-st order. As in language modeling, the N-gram models for $N=2$ and $N=3$ are referred to as *bigrams* and *trigrams*, respectively. Given a set of such language models, the Bayes classifier makes a decision based on the maximum-likelihood rule:

$$L^* = \arg\max_i P(A|L_i). \quad (8.15)$$

Using our previous example of the phone sequence /iç/ modeled in two languages—English (L_1) and German (L_2)—the observation probabilities would be calculated as

$$P(/\text{iç}/|L_i) = P(/\text{i}/|/\text{ç}/, L_i) \cdot P(/\text{ç}/|L_i), i \in 1, 2 \quad (8.16)$$

that is, as a product of the bigram probability of phone /ç/ following /i/ and the unigram probability of /i/ occurring in the respective language. Given well-estimated language models, a relationship $p_1 \ll p_2$ should be the result, as /iç/ occurs frequently in German but not in English.

Phonotactic LID is still an active research area with many problems and challenges. The first challenge is to achieve modeling of a sufficiently high statistical order. As previously discussed, the modeling accuracy grows with the model order. Language-characteristic dependencies may be found that span several phonemes; thus, a restriction to bigrams, for example, will generally lead to suboptimal results. On the other hand, the model complexity in terms of the number of free parameters grows exponentially with its order. For a phone repertory of 60 units covering several languages, a bigram model will have $60^2 = 3{,}600$, and a trigram model will have $60^3 = 216,000$ free parameters that need to be estimated from data. If we assume that on average 10 observations are needed to estimate one parameter with sufficient reliability, and that on average 10 sound units are spoken per second, then a bigram would need about an hour and a trigram about 2.5 days worth of continuous speech for it to be estimated with sufficient robustness. Obviously, a transition to a 4-gram model would cause a further exponential increase in the number of parameters, thus rendering full model estimation impractical. Since language-specific information exists across contexts longer than trigrams, many research efforts have addressed parametric complexity, including the statistical smoothing and context-clustering techniques described below.

The second challenge is posed by the recognition (decoding) process of the sound units from continuous speech, which often needs to be performed in adverse acoustic conditions. Speech decoding is an unsolved research challenge in itself, and its accuracy is far from ideal. On telephone-quality speech signals, for instance, the error rate of typical HMM-based phone decoders is around 50%—that is, on average, every second phone in the decoded sequence is incorrect. Decoding errors are a significant source of noise for phonotactic analysis and need to be addressed (Navrátil, 1996; Zissman, 1996, 1997).

8.6.2 *N*-Gram Estimation and Smoothing

The phonotactic models are constructed by estimating the individual conditional phone probabilities in each language from a large training corpus using a maximum-likelihood approach. Using the example of a bigram, the probability is calculated as

$$\begin{aligned} P(\alpha_i|\alpha_j) &= \frac{P(\alpha_i, \alpha_j)}{\alpha_j} \\ &= \frac{c(\alpha_i, \alpha_j)/T}{c(\alpha_j)/T} \\ &= \frac{c(\alpha_i, \alpha_j)}{c(\alpha_j)}, \end{aligned} \tag{8.17}$$

where $c(.)$ denotes an occurrence count, and T, the total number of training samples. A disadvantage of the maximum-likelihood (ML) estimate is its requirement for relatively large amounts of data, whereas in practice, data sparseness is a common problem—for example, some possible phone combinations may never be observed in the training set. In such cases, the ML estimate assigns a zero probability to the corresponding parameter, which in turn causes zero probabilities for any such observations occurring in the test set and results in classification errors. This is analogous to the situation in statistical language modeling for word sequences (see Chapter 6), though the problem is somewhat less severe for smaller inventories of modeling units, such as phone sets. As in word-based language modeling, however, smoothing can be applied to adjust the ML estimate and thus prevent errors due to data sparseness.

In LID, the most common smoothing technique is linear interpolation between the N-gram and its lower-order $(N-1)$-gram marginal (Jelinek et al., 1997). In the case of bigrams, the interpolated probability is composed as follows:

$$P_{interp}(\alpha_i|\alpha_j) = \lambda_2 P(\alpha_i|\alpha_j) + (1 - \lambda_2)P(\alpha_i), \tag{8.18}$$

with λ_2 being the bigram interpolation parameter, which can be a function of α and is trained jointly in an ML fashion. In LID practice, λ is often a constant common to all α_i (Zissman, 1996; Hazen and Zue, 1997; Yan et al., 1996; Navrátil, 2001), with its value being determined heuristically. Similar findings apply to trigrams with an additional constant λ_3.

8.6.3 Extended Phonotactic Context

The exponential relationship between the length of the phonotactic context and the number of N-gram parameters presents a practical difficulty preventing a context length of more than two or three units (that is a length beyond the context of bigram or trigram). While the entropy of the co-occurrence of two units naturally decreases with their growing mutual distance in a sequence, significant information will still exist in contexts beyond phone triples.

The basic structure of N-grams can be easily modified to access some of the unused information. A simple example is a bigram probability defined between two units with a time gap between them, such as $P(a_t|a_{t-2})$. This idea was studied in a more general light as "permugrams" (Schukat-Talamazzini et al., 1995) and was also used under the term "skip-ngrams" in Navrátil and Zühlke (1997). Recent work of Kirchhoff et al. (2000b) generalized such extensions in the framework of Mixed-Memory Markov Models representing an N-gram model as a mixture of k lower-order bigrams:

$$P(a_t|a_{t-1},\ldots,a_{t-k}) \approx \sum_{i=1}^{k} w_i P(a_t|a_{t-i}), \quad \text{with } w_i \geq 0, \sum_i w_i = 1 \tag{8.19}$$

whereby the weights w_i can be estimated via the Expectation Maximization (EM) algorithm. The exponential parametric complexity of the N-gram is thus reduced to quadratic complexity in the mixture model. The above-mentioned techniques were shown to improve the performance of a baseline system that used a comparable number of free parameters.

Another type of structural modification studied extensively in the literature is context clustering. Here, the probability of the current unit a_t is conditioned on a set of phone histories (a "cluster of histories") rather than on a single context. Since the clusters may include histories of arbitrary lengths, information from longer contexts can be modeled while the model complexity is only determined by the number of such clusters. The latter may in turn be chosen flexibly to maintain estimation robustness. Context clustering represents a general view of N-gram modeling, which includes most of the existing modeling alternatives for extended phonotactic context as special cases. A probability with clustered context can be written as an approximation

$$P(a_t|a_{t-1},\ldots,a_1) \approx P(a_t|C(a_{t-1},\ldots,a_1)), \tag{8.20}$$

with C being a discrete function of the context. Since $C(.)$ may have an arbitrary form, the standard bigram, trigram, as well as the previously mentioned models can be included as special cases by choosing $C(.)$ appropriately. Obviously, the choice of the clustering function is a decisive factor in the model design and may involve a variety of criteria to select the clusters, ranging from heuristic methods to fully data-driven approaches. The latter make use of held-out data sets to create an optimum mapping of histories to clusters in an unsupervised manner. One particular method based on binary decision tree structures in combination with a maximum likelihood objective (Navrátil, 2001) is described below.

Context Clustering with Maximum-Likelihood Binary Decision Trees

Binary decision trees are a common machine learning technique that has been successfully applied to various modeling problems in speech (see Chapter 4) and language processing, such as for predicting words from a word history (Hartmann et al., 1982). Since they use a minimum prediction entropy rule during training (which can be shown to be equivalent to maximizing the likelihood of the data) and are not limited to a particular history length, tree-based models represent a very flexible way of modeling context information.

A tree model consists of nonterminal and terminal nodes (see Figure 8.4). Each nonterminal node is associated with a binary question that leads to one of two child nodes. To answer the question, a certain phone from the history, the so-called predictor, is compared to a node-dependent subset of phones. If the predictor is in the subset, the result is positive; otherwise, it is negative. When a terminal node (leaf) is reached, the probability of the phone a_t can be obtained from the associated distribution. It is clear that different histories result in reaching different terminal nodes—that is, different distributions. In this way, the context, represented in terms of predictors leading to different leaf nodes, is exploited in a very flexible manner.

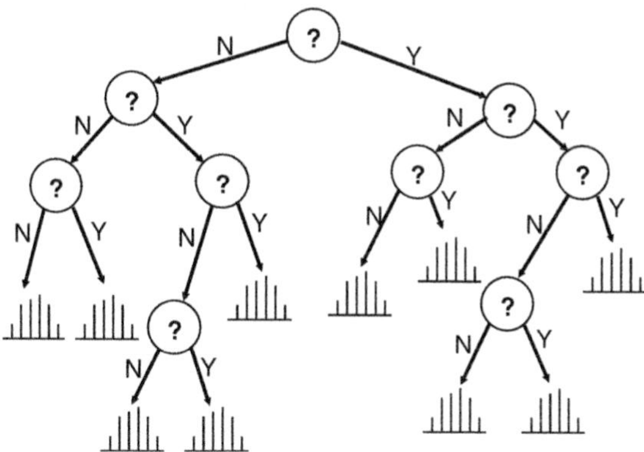

Figure 8.4: Example of a binary decision tree.

The parameters of the tree model are the binary questions as well as the leaf-dependent distributions, which need to be estimated for each language. In Bahl et al. (1989), a tree-growing algorithm was described that aims at minimizing the overall entropy of the phone distribution Y at each node. The advantage of this algorithm, described below, is the unsupervised construction of the tree on a training set, which eliminates the need for linguistic experts.

1. Let c be the current node of the tree. Initially, c is the root.
2. For each predictor variable $X_i (i = 1, \ldots, m)$, find the subset $\mathbf{S}_i^c$, which minimizes the average conditional entropy at node c

$$\overline{H}_c(Y \mid \text{“}X_i \in \mathbf{S}_i^c\text{?”})$$

$$= -\Pr(X_i \in \mathbf{S}_i^c \mid c) \sum_{j=1}^{K} \Pr(\alpha_j \mid c, X_i \in \mathbf{S}_i^c) \cdot \log_2 \Pr(\alpha_j \mid c, X_i \in \mathbf{S}_i^c)$$

$$- \Pr(X_i \notin \mathbf{S}_i^c \mid c) \sum_{j=1}^{K} \Pr(\alpha_j \mid c, X_i \notin \mathbf{S}_i^c) \cdot \log_2 \Pr(\alpha_j \mid c, X_i \notin \mathbf{S}_i^c), \tag{8.21}$$

where $\mathbf{S}_i^c$ denotes a subset of phones at node c; K, the size of the phone repertory; and α_j, the j-th phone from this repertory.

3. Determine which of the m questions derived in step 2 leads to the lowest entropy. Let this be question k:

$$k = \arg\min_i \overline{H}_c(Y \mid \text{“}X_i \in \mathbf{S}_i^c?\text{”})$$

4. The reduction in entropy at node c due to question k is

$$R_c(k) = H_c(Y) - \overline{H}_c(Y \mid \text{“}X_k \in \mathbf{S}_k^c?\text{”}),$$

where

$$H_c(Y) = -\sum_{j=1}^{K} \Pr(\alpha_j \mid c) \cdot \log_2 \Pr(\alpha_j \mid c).$$

If this reduction is “significant,” store question k, create two descendant nodes c_1 and c_2, pass the data corresponding to the conditions $X_k \in \mathbf{S}_k^c$ and $X_k \notin \mathbf{S}_k^c$, and repeat steps 2–4 for each of the new nodes separately.

The principle of tree growing is obvious: if the data at a given node may be divided by a question in two sets that together have a smaller entropy than that of the undivided data, two new nodes are created. The entropy reduction is considered significant relative to some threshold.

In order to determine the phone subset $\mathbf{S}_i^c$ in step 2, a “greedy” algorithm can be applied, as suggested in Bahl et al. (1989). The search for **S** can be done through the following steps:

1. Let **S** be empty.
2. Insert into **S** the phone $\alpha \in \mathbf{A}$, which leads to the greatest reduction in the average conditional entropy (8.21). If no $\alpha \in \mathbf{A}$ (**A** being the phonetic set of size K) leads to a reduction, make no insertion.
3. Delete from **S** any member α if doing so leads to a reduction in the average conditional entropy.
4. If any insertions or deletions were made to **S**, return to step 2.

An example of a node-question for a particular content $a\_\{t = 1\},\ldots,$ $a\_\{t = m\}$ may look like “$a_{t-1} \in \{[s], [sh], [f]\}$?”—that is, all phones that were preceded by the phones /s/, /sh/, or /f/ would be passed to the "yes"-node and vice versa (taking a_{t-1} as the optimal predictor determined in step 3 of the tree growing algorithm in this particular example). An essential structural parameter in the training is the significance threshold. Smaller thresholds will result in large trees with a great number of terminal nodes, whereas higher values will cause the tree to stop growing after a few nodes. Further on, the number of predictors considered—that is, the history length—will be chosen. Experimental results comparing the binary-tree models to standard N-grams and their combination is presented in Navrátil (2001).

In a similar vein, binary decision trees find their application in language modeling for speech recognition. A noteworthy recent work in this area utilizes combined sets of random trees, termed "random forests" (Xu and Jelinek, 2004).

8.6.4 System Architectures in Phonotactic LID

Monolingual versus Multilingual Phone Decoders

From the above discussion, a natural question arises: what phonetic repertory should be chosen to represent the spoken multilingual utterances as sequences of discrete "tokens" $\alpha \in \mathbf{A}$? Zissman and Singer (1994) used a single English phone recognizer and proved that modeling phonotactic constraints in terms of the phones of one language is feasible for identification of other languages. As illustrated in Figure 8.5, the incoming speech is decoded (tokenized) into a sequence of discrete units by means of a phone

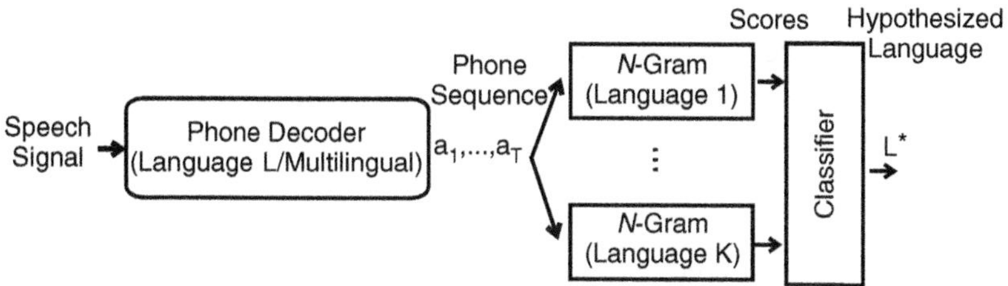

Figure 8.5: Phonotactic architecture with a single mono- or multilingual phone decoder.

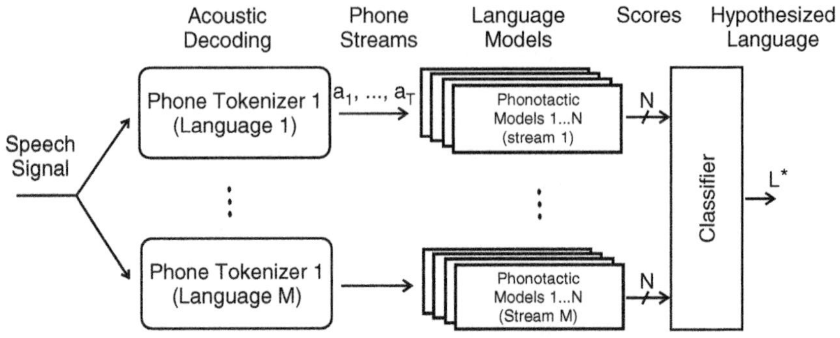

Figure 8.6: Phonotactic architecture with multiple phone decoders.

recognizer, which then undergoes a phonotactic analysis (for example, a bigram modeling). For better phone coverage across languages, Hazen describes a single decoder with a *multilingual* phone repertory and a variable number of phone units (Hazen and Zue, 1997). Yan and Barnard (1995) employed six language-dependent phone recognizers to better represent the wide phone inventory and to increase robustness of the classifier by combining the outputs of the individual decoder systems. This parallel-decoder architecture also appears in Zissman (1996), and is termed PPRLM (Parallel Phone Recognition followed by Language Modeling). As shown in Figure 8.6, in the PPRLM architecture, the incoming speech is processed simultaneously by a number of language-specific tokenizers, resulting in multiple phone streams, which, in turn, are modeled by a separate set of phonotactic models. The generated scores are combined in a linear fashion (Zissman, 1996) or via a nonlinear classifier (Yan and Barnard, 1995). In order to reduce the computational overhead due to the multiple acoustic decoding processes, Navrátil and Zühlke studied a single multilingual decoder with a multipath decoding strategy, using multiple language-specific bigram models within the Viterbi decoder (1998), which compares favorably to the parallel-decoder approach without increasing computation much beyond that of a single tokenizer. This system was further simplified in Caseiro and Trancoso (1998), to remove the need for manual labeling. An alternative way of generating a multitude of streams is via phone lattices (Gauvain et al., 2004).

Generally, the multistream PPRLM architectures seem to outperform single tokenizer systems, which is likely due to increased robustness in the multiple sets of phonotactic models created for each of the decoders.

The typical number of sound units ranges roughly between 20 and 40 in language-dependent decoders, and between 30 and 100 in multilingual decoders, and the most frequently used sound unit is the phone.

Modeling Cross-Stream Dependencies

An interesting additional dimension of modeling within the PPRLM architecture was proposed by Kirchhoff and Parandekar (2001; 2003). Here, a multistream system with explicit modeling of phonotactic constraints within as well as *across* multiple streams was proposed. Instead of phone units, the cross-stream architecture involved articulatory features—discrete sets of events describing an acoustic observation in terms of voicing, place, and manner of articulation. The articulatory features were grouped into mutually exclusive subsets, and an acoustic decoding of the speech signal was performed in each such subset, thus producing multiple streams of tokens. As in other PPRLM architectures, phonotactic models are applied to each stream separately; however, additional pairwise dependencies across the streams are also considered. A general idea of cross-stream modeling involving dependencies from various units is illustrated in Figure 8.7. To reduce the relatively large number of possible dependencies, pairwise and time-context restrictions were placed on their choice via a greedy search (2001) and via a genetic algorithm (2003). Using cross-stream dependencies in addition to modeling the phonotactics within the same stream

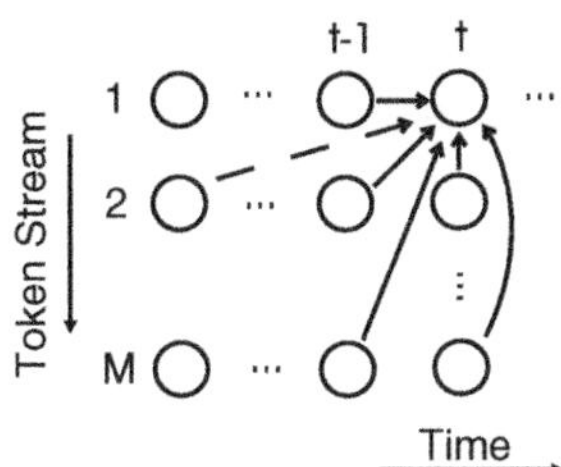

Figure 8.7: Illustration of modeling cross-stream dependencies for a specific token.

was shown to improve the performance, as did the combination of the articulatory-feature system and the baseline phone-based system.

In summary, Table 8.3 lists selected LID implementations of phonotactic-only components and their performances spanning the last decade, in chronological order. Note that despite the common use of the OGI database, the exact subset used to obtain the error rates may vary in each study. Furthermore, baseline systems with a variable level of complexity were used, therefore the listed error rates may not be seen as an outline of progress in phonotactic modeling in an absolute sense.

8.7 Prosodic LID

Prosodic information—such as tone, intonation, and prominence—is primarily encoded in two signal components: fundamental frequency (F_0) and amplitude. Thus, properties of F_0 and amplitude contours can intuitively be expected to be useful in automatic LID. In Eady (1982), for instance, two languages with different prosodic properties were studied: English, which belongs to the category of languages marking different levels of prominence by means of F_0 and amplitude, and Chinese, in which tonal patterns are lexically distinctive (see Chapter 2). The author compared the time contours of the fundamental frequency and the energy extracted from sentences of the same type (e.g., all declarative sentences) and found unique language-specific differences—such as a higher rate of change within each contour—and a higher fluctuation within individual syllables for Chinese.

However, the use of prosody for the purpose of identifying languages is not without problems. The difficulty of obtaining a clear assessment of the usefulness of a prosodic component for LID derives from the large number of additional factors influencing F_0 and amplitude. These include:

- speaker-specific characteristics (voice type, characteristic speaking rate, emotional state, health, etc.)
- lexical choice (word characteristics, such as word stress and lexical tone)
- syntactic content of the utterance (statement, question)
- pragmatic content/function in discourse (contrastive emphasis, given versus new distinction).

Table 8.3 Examples of phonotactic LID systems and their recognition rates.

System	Task	Test Signal Duration	ID Rate	Ref.
Interpolated Trigram Multilingual Tokenizer	OGI-11L	10 s/45 s	62.7%/77.5%	Hazen and Zue, 1997
Bigrams (1 tokenizer) Gender-Dependent	OGI-10L	10 s/45 s	54%/72%	Zissman, 1996
PPRLM (3 tokenizers)	OGI-10L	10 s/45 s	63%/79%	Zissman, 1996
PPRLM (6 tokenizers)	OGI-6L	10 s/45 s	74.0%/84.8%	Yan and Barnard, 1995
Extended *N*-grams (PPRLM with 6 streams)	OGI-6L	10 s/45 s	86.4%/97.5%	Navrátil, 2001
Cross-Stream Modeling	OGI-10L	mixed	65%	Parandekar and Kirchhoff, 2003
GMM-Tokenizer	CallFriend (12L)	30 s	63.7%	Torres-Carrasquillo et al., 2002a
GMM-Tok.+PPRLM	CallFriend (12L)	30 s	83.0%	Torres-Carrasquillo et al., 2002a

In many cases, particularly in languages from the same family (or the same prosodic type; see Section 8.2), the language-specific prosodic variation is overridden by other, more dominant factors, rendering its extraction and recognition extremely difficult. In order to successfully exploit prosodic information, the question of how to best separate language-dependent characteristics from speaker-dependent or other irrelevant characteristics needs to be addressed and indeed remains one of the open challenges in current LID technology. A potentially useful application of prosodic features might be the preclassification of language samples into broad, prosodically distinct categories, followed by more fine-grained subclassification using other methods.

LID systems based solely on prosody are relatively rare. One such approach was described in Itahashi et al. (1994), and Itahashi and Liang (1995). In this system, fundamental frequency and energy contours were extracted from the speech signal and approximated via a piecewise-linear function. In order to preserve prosodic language characteristics, a heuristically selected set of variables derived from the line-approximated representation was calculated (mean slope of the line segments, mean relative start frequency of a line segment with positive and negative slope, the correlation coefficient between the fundamental frequency and the energy contour). Vectors of these features were then processed using principal components analysis (PCA) to perform a dimensionality reduction. The PCA-reduced samples were stored and compared to test samples via the Mahalanobis distance. The system was evaluated on speech material from six languages (Chinese, English, French, German, Japanese, and Korean); LID rates ranged between 70 and 100%. The number of speakers used in the experiments, however, was rather small (5 per language). An identical approach was tested with data in six languages taken from the OGI-11L corpus, producing accuracy rates of about 30% (Dung, 1997). Similar prosodic results were reported by Hazen and Zue (1994). A recent study used similar modeling of F_0-based and rhythmic features from pseudo-syllabic segments of read speech in ten languages, utilizing a Gaussian Mixture Model (GMM)–based classifier (Farinas et al., 2002).

Probably the most thorough study of prosodic features for LID was presented in Thyme-Gobbel and Hutchins (1996). The authors compared 220 prosodically motivated features that were calculated from syllable segments represented as histograms, and used for classification with a likelihood ratio detector. The experiments involved a series of pair-wise language classification tasks on four languages: English,

Mandarin Chinese, Japanese, and Spanish, resulting in accuracies of up to 70–80% for certain language pairs. Several conclusions were drawn from this study. First, prosodic features can be useful in LID to an extent comparable to other methods (e.g., acoustics). Second, F_0 information seems generally more useful than amplitude information. Third, the usefulness of individual features and their combinations varies strongly with the specific language pair thus, one open question in prosodic LID is how the extraction of complex prosodic information can be tailored to specific language pairs or combinations.

Motivated in part by the above study, Cummins et al. (1999) pursued a more data-driven prosodic feature extraction approach to determine the usefulness of individual features. Using five languages from the OGI-11L corpus (the same four languages as in Thyme-Gobbel and Hutchins plus German), a recurrent ANN was trained on a temporal derivative of the F_0 and energy contours. No manual feature selection was done, leaving the task of finding relevant time patterns in these contours in a data-driven fashion by the ANN. Cummins et al. reported results comparable to those of Thyme-Gobbel and Hutchins in terms of performance as well as the relative usefulness of F_0 versus energy. They concluded that the automated selection of suitable prosodic features is feasible; however, a purely prosodic set of features may lead to merely moderate results if used as the only information source.

In addition to the small number of purely prosodic LID systems, many systems used certain prosodic features, such as segment length variation, as a component in combination with other evidence, such as acoustic or phonotactic information (Farinas et al., 2002; Hazen and Zue, 1997; Navrátil and Zühlke, 1998; Zissman, 1996).

Table 8.4 summarizes example results obtained by purely prosodic systems.

Table 8.4 Some prosodic components and their recognition rates.

System	Task	Test Duration	ID Rate	Ref.
Duration	OGI-11L	10 s/45 s	31.7%/44.4%	Hazen and Zue, 1997
Rhythm	5L (read speech)	21 s	22%	Farinas et al., 2002
F_0 features	OGI-2L (averaged)	45 s	70%	Rouas et al., 2003

8.8 LVCSR-Based LID

As previously discussed, the most effective approach to LID ideally utilizes complete knowledge of the lexical and grammatical structure of a language. The automatic approach closest to the ideal situation of a human listener familiar with a given language is represented by LID systems using fully fledged large vocabulary continuous speech recognizers (LVCSR) to decode an incoming utterance into strings of words with a subsequent analysis for LID-specific patterns. The advantages of this approach are obvious: since more higher-level constraints (in particular, a lexicon and language model) can be utilized, results can be expected to be more accurate and less noisy than those of a phonotactic or purely acoustic system, for example. On the other hand, the requirement that the LVCSR be trained with properly transcribed and labeled data in each language implies a considerable effort during training—not to mention that such training resources may not be available for many languages. An LVCSR-based LID system also requires more computational resources (memory and CPU time) during testing, which makes it a less attractive solution for applications on portable and handheld devices.

Simply put, the basic principle of LVCSR-based LID system is a "brute-force" approach: a test sample in an unknown language is processed using all available LVCSR systems in all hypothesized languages. Since each component is assumed to produce a normalized score expressing the degree to which the hypothesis matches the signal, a maximum-score (typically, maximum-likelihood) decision is taken over all outputs to select the final language hypothesis.

Schultz and Waibel report on LID experiments carried out in the framework of a machine translation project involving four languages: English, German, Japanese, and Spanish (Schultz et al., 1996a). The results, measured on a scheduling task with spontaneous speech, show an improvement of the LID performance through the lexical processing. Among the four languages, an LID error rate of 16% was measured on tests with variable length signals of less than five seconds, whereas the baseline performance of the same systems without the use of the lexicons (i.e., using the acoustic and phonotactic information only) increases to 17.4%.

Further results are presented by Hieronymus and Kadambe (1996, 1997), who applied five LVCSR systems to English, German, Japanese, Mandarin Chinese, and Spanish, achieving an error rate of 15.6%

for 10-second utterances. This was subsequently improved by introducing a normalization of the LVCSR output scores based on phone-only decoding in each language, resulting in a final LID error rate of 7% (3% after further optimization). Based on the reported results on the five languages, Hieronymus and Kadambe's system seems to belong to the most accurate LID system (compare, for example, the result of 9.8% with six languages and 10-second utterances achieved by a phonotactic system on similar data (Navrátil and Zühlke, 1998)). However, when making this comparison, one needs to consider computational costs during training and testing. LVCSR processing typically involves dictionary sizes of 20,000–100,000 (and often more) words, resulting in considerable memory and CPU requirements. This cost multiplies with the number of supported languages. Therefore LVCSR-based LID seems appropriate and justified primarily in scenarios in which the effort invested in developing and running the LVCSR components will be reused, such as in a translation system.

A compromise between full lexical decoding via the LVCSR and no utilization of word-level information whatsoever is sought in systems implementing algorithms for keyword spotting. Particularly in application scenarios that allow for a reliable prediction of word usage in each language (e.g., airport information tasks narrowing down the dictionary usage to flight scheduling), a system browsing the speech utterances for language-specific versions of certain objects or actions can achieve highly accurate LID while reducing the computational load by avoiding a full decoding. Another possibility is to integrate lexical information into a phonotactic system by incorporating models for the most frequently occurring words (Matrouf et al., 1998).

Table 8.5 summarizes results obtained by some LVCSR-based LID implementations.

Table 8.5 Some LVCSR LID components and their error rates.

System	Task	Test Duration	ID Rate	Ref.
LVCSR	4L	<5 s	84%	Schultz et al., 1996a
LVCSR	6L	10 s	97%	Hieronymus and Kadambe, 1997
LVCSR	2L (air control)	mixed	98.4%	Fernandez et al., 2004

8.9 Trends and Open Problems in LID

Over the past 10–15 years, language identification has become a well-established branch of speech research, with regular conference sessions dedicated to its problems, developments, and applications. Recent LID evaluations organized by the National Institute of Standards and Technology (NIST) in 2003 and 2005, as a continuation of previous such evaluations from 1995–96, contributed to continued attention to this problem and spurred new activity in this area.

We have discussed how the main approaches can be described in terms of the signal features they seek to exploit as well as in terms of their modeling techniques. However, the various degrees of difficulty in the LID experiments, such as test duration, amount of training data and data annotation, channel quality, and test training mismatch, render a consistent comparison of the individual algorithms difficult. Based on a somewhat rough comparison, a performance-oriented ranking will place the LVCSR-based systems among the most powerful methods, followed by the phonotactic, the acoustic, and finally the prosodic approaches.

A major trend in current LID methodology is system and component fusion. While a system fusion aims at combining entire LID systems, the component fusion exploits the fact that different sources of discriminative information within one LID system provide partially decorrelated output and therefore may lead to improved performance when integrated into one decision process. Most common fusion implementations combine outputs of the individual components, such as phonotactic, acoustic, and prosodic modeling, into a final hypothesis by means of a higher-level classifier. This classifier can be a simple linear function (Hazen and Zue, 1997) or a nonlinear classifier, such as a multilayer perceptron (Navrátil, 2001; Yan et al., 1996).

An interesting and equally important aspect of comparing LID approaches is the intended application. Different applications pose different demands due to their choice of languages, typical test duration, availability of training material, and so on. Table 8.6 summarizes some of the advantages distinguishing individual methods from several application-oriented aspects.

While LVCSR-based LID systems can be considered the most accurate solution to LID today (Hieronymus and Kadambe, 1997), their high computational cost as well as the requirement for sufficient

Table 8.6 Comparison of basic LID approaches from an application development aspect.

Basic Approach	Strength	Constraint	Example Application
Prosodic	Robust in channel mismatch	Mostly suitable for distinguishing language groups	Preclassifier of tonal versus stress languages
Acoustic	Low cost in training and testing (data and computation)	Usable in combination with other components	Language and accent ID in a multiapproach system
Phonotactic	Good performance--to-cost ratio, No linguistic knowledge required to train	Useful for tests with duration >5 seconds	LID system with large language population, including rare languages without linguistically labeled training data
LVCSR/ Keyword Spotting	High accuracy, Short tests	Significant training effort, Linguistic input required	Multilingual dialog systems with LVCSR components, Audio mining systems

word-transcribed training data, limits their area of use. Potential application scenarios are multilingual dialog or translation systems, in which a multilingual LVCSR infrastructure is already in place, independently of the LID task. In order to reduce computational demands, speech signals can first be decoded by all LVCSR systems in parallel, followed by a process of hypothesis pruning and eventual convergence on a single language hypothesis. In the extreme case of systems with limited vocabularies and finite-state grammars, the LID task can be implemented as an implicit by-product of word recognition by unifying all words in a common (multilingual) dictionary.

In terms of performance-to-cost ratio, the phonotactic approaches seem to fare best. Their main advantage is that they are capable of modeling arbitrary languages in terms of sound units of other languages (Zissman, 1996) and hence do not require manual labeling and segmentation. They therfore represent an acceptable solution for many practical applications—especially those involving large numbers of languages or rare and resource-poor languages. The many improvements in the area of N-gram model smoothing (Section 8.6.2), context clustering

(Section 8.6.3), and combining multiple streams (Section 8.6.4) have boosted the performance of phonotactic systems to a level close to that of LVCSR-based systems. However, the typical minimum test duration required for phonotactic systems lies between 5 and 10 seconds. Phonotactic systems should therefore be used with test signals of sufficient length in tasks such as audio mining, indexing, archiving, broadcast transcription, and speaker diarization. Phonotactic and LVSCR components can be efficiently combined into a high-accuracy system, with the phonotactic component performing a preclassification, followed by a more fine-grained analysis using a small number of LVCSR systems.

The greatest potential for acoustic approaches (i.e., those modeling pronunciation patterns of sound units) and prosodic approaches seems to be in multicomponent LID systems, providing supporting information to other core components (typically phonotactic components). Recent developments in acoustic modeling using GMMs and SVMs (Section 8.5) improve performance significantly and make such component combinations highly desirable.

Future development of LID algorithms will certainly be driven by application demands and, as such, will most likely need to be focused both on improving the performance of its core techniques and on developing solutions to new problems related to LID. A secondary factor will be further development of multilingual databases.

A new task of **speaker diarization**, introduced in the recent rich transcription evaluations conducted by NIST, focuses on enhancing speech transcriptions by various sets of "metadata" annotations. This opens up an interesting new direction for research in LID. For a given speaker, not only the language and the words spoken are of interest but also the dialect or accent. The information sources used in an LID system could be leveraged to detect divergences from a standard with respect to a given speaker. For instance, by simultaneously processing speech from a given speaker via a prosodic, an acoustic, and a phonotactic component, it would be possible to detect a divergence in scores from those of a model trained on a standard speaker group. Such divergences may indicate that the speaker is a multilingual non-native individual; whereas the phonotactic hypothesis will tend to indicate the actual language spoken (say, English), the acoustic-prosodic blocks might hypothesize an alternative, conflicting language (e.g., French). Given an appropriate confidence measurement indicating the degree of the divergence, an enhanced output reflecting the hypothesis that English is being spoken by a French native speaker can be of high value in the diarization task. Dialog systems may benefit from such additional information by adapting the acoustic models appropriately, thereby increasing their accuracy. Some work on accent and dialect recognition was done in Kumpf and King (1997), and Angkititrakul and Hansen (2003); however, due in part to a lack of sufficiently large data corpora, this research direction remains largely open.

The task of speaker diarization further leads to other LID-related tasks, such as language segmentation, or language change detection. In this case, a continuous audio stream is segmented not only according to different speakers but also according to language, dialect, or accent. Although various techniques are available from speaker and topic segmentation, and some preliminary studies have been carried out (Chan et al., 2004), a broader investigation of such a problem remains yet to be done.

Another currently unsolved problem is language rejection. LID systems are always developed for a finite number of languages. Detecting and rejecting input signals that do not belong to any of the supported languages rather than assigning a wrong hypothesis is still an open research issue (e.g., at the 2003 NIST-conducted benchmark evaluation of LID systems, none of the participating systems solved this task satisfactorily).

Finally, in spite of several decades of research activity, utilizing prosodic features in LID systems still remains a considerable challenge. As discussed in Section 8.7, the relatively high complexity of the linguistic information encoded in prosodic features is certainly the primary reason for the relatively low success rate so far. Furthermore, due to their suprasegmental nature, a robust modeling of prosodic features typically requires a minimum test signal duration, thus limiting its scope of potential applications. Important open issues include the problem of how to best separate language-relevant prosodic features from those related to semantics and speaker idiosyncrasies.

Another area of strategic interest for application developers is voice-based user interface and dialog system design. Obviously, any dialog system expected to handle more than one language needs either to have an appropriate LID component or an alternative system setup as described in Chapter 11. A simple but challenging problem is the question of how a dialog should be initiated by the machine without assuming a known language. Furthermore, how should the dialog be designed to achieve the LID step in the shortest time and with the highest possible accuracy?

Simple applications today often implement a system-initiative strategy and use an initial (explicit) prompt for determining the language. However, when a large number of possible input languages needs to be considered, explicit prompts will likely become tedious for users.

As proven in the past, one of the major driving forces in LID is the development of large speech corpora with features relevant to the task or the application of interest. Since a typical LID solution is of a statistical nature and is parametrically complex, certain tasks will only be solved by experimentation involving large amounts of data. For instance, for studying accent recognition for speaker diarization, a corpus comprising a number of different languages, each with a collection of foreign accents recorded from a large amount of speakers, is necessary in order to train such systems and to enable researchers and developers to draw statistically significant conclusions. Language resources organizations, such as the LDC and ELRA, have in the past served as collectors and distributors of data to the speech community; future LID research will certainly be closely tied to their data collection efforts.

Chapter 9

Other Challenges: Non-native Speech, Dialects, Accents, and Local Interfaces

Silke Goronzy, Laura Mayfield Tomokiyo, Etienne Barnard, Marelie Davel

9.1 Introduction

There has been much progress in the past few years in the areas of large vocabulary speech recognition, dialog systems, and robustness of recognizers to noisy environments, making speech processing systems ready for real-world applications. The problem of dialect variation, however, remains a significant challenge to state-of-the-art speech recognition systems. Whether these variations originate from regional variations or from other languages, as is the case with non-native speech, they tend to have a large impact on the accuracy of current speech recognition systems. This is the case for small vocabulary, isolated word recognition tasks as well as for large vocabulary, spontaneous speech recognition tasks. In an example of the first case, the word error rate (WER) rose from 11.5% for a native reference test set to 18% for non-native speakers (which corresponds to an increase in WER of 56.5% relative) (Goronzy, 2002). In an example of the latter case, a relative increase in WER of more than 300% from 16.2% for native speakers to 49.3% for non-native speakers was observed (Wang and Schultz, 2003). These two examples indicate the severity of the problem for non-native speech; similar results have been reported for cross-dialect recognition (Fischer et al., 1998; Beaugendre et al., 2000). However, practical speech recognition systems need to be robust to non-native speech as well as a certain amount of regional variability. Although there exist databases that allow well-tuned recognizers to be designed for many languages and their major dialects, it is often not practical to integrate recognizers for more than a few languages within one system. This is especially true in embedded environments such as mobile or automotive applications, in which resource restrictions prohibit the use of multiple recognizers—that is, one recognition engine using different sets of acoustic models (AMs), grammars or language models (LMs), and dictionaries. As a result, systems are restricted to the most common languages. Consequently, many speakers will need to use these systems in a language that is not their native language. In other environments (e.g., telephone-based services), it may be possible to employ several recognizers, but finite resources and training data nevertheless require the recognition of non-native and cross-dialect speech.

We also encounter the problem of non-native speech in various content-selection tasks. When designing speech interfaces for applications in which the user wants to access multimedia content or some other kind of information, this often involves the pronunciation of foreign music and movie titles, city and street names, and so on. This means that even if the system language is the speaker's native language, some names and phrases may be pronounced according to the phonological system of a different language or dialect.

One of the main causes for the performance degradation of recognizers is that they are usually trained for one particular language and for limited dialect variation within that language. This means that the AMs are trained on native speech, that the pronunciation dictionaries contain native pronunciations as well as mostly native variants, and that the LMs are trained on native texts—in each case "native" refers to a limited set of dialects of the language.

When discussing dialect variation and its impact on speech recognition, we will not attempt a rigorous definition of the range of variability between different idiolects, on the one extreme, and unrelated languages, on the other. In practice, certain forms of variability (e.g., between midwestern and Atlantic dialects of U.S. English) are relatively benign, whereas other forms (e.g., between U.S. and British English, or for strong cross-language variability) can have a severe impact on recognition; the focus here will be on the latter type of variation.

In this chapter, we focus on the challenges of non-native speech and briefly review how these same concepts apply to other types of dialect variation. We will use the terms *source language/dialect* or *L1*, to refer to the native language or dialect of the speakers, and *target language/dialect*, or *L2*, to refer to the language or dialect that the system is designed to recognize. The term *accent* will be used to refer specifically to the features of pronunciation that distinguish non-native from native speech (or different dialects), most notably, phonetic realization, but also prosody, allophonic distribution, and fluency.

We will begin this chapter with a description of the manifold characteristics of non-native speech in Section 9.2. This includes theoretical models as well as descriptions obtained from corpus analysis along with a description of non-native databases. Almost all investigations in non-native speech require non-native speech data. Obtaining sufficient non-native speech data is one of the biggest problems. While there are plenty of databases for many different languages, they usually contain only native speech data. Very few databases containing non-native speech are publicly available, as will be described in Section 9.3.4.

If speakers speak the target language, they will do so with more or less significant deviation from the acoustic realization modeled by the target-language AMs. Of course, variation in acoustic realization and pronunciation is a rather general problem in speech recognition. We will examine in the following sections whether the standard approaches to this problem can also be helpful for non-native speech. We will look at approaches that use special acoustic modeling for non-native speakers in Section 9.4. Speaker adaptation techniques have also proven valuable for adapting AMs to the current speaker; this is true for both native and non-native speakers. Speaker adaptation techniques in the special context of non-native speakers are described in Section 9.5.1.

Non-native speech differs from native speech not only in acoustics (realization of individual phonemes) but also in the phoneme sequences themselves. As we have seen in Chapter 5, deviations at the lexical level may need to be explicitly represented in the dictionary for improved recognition rates. This is true for both native and non-native variation. Since the adaptation of the pronunciation dictionary, henceforth called **pronunciation adaptation**, plays an important role in the case of non-native speakers, we will discuss this problem in more detail in Section 9.5.2. Speaker adaptation as well as pronunciation adaptation are likely to improve recognition rates. Furthermore, it can be expected that these improvements are additive if both methods are combined; the performance of this combination is reviewed in Section 9.6. In Section 9.7 we summarize some results on the performance of these techniques for cross-dialect recognition for native speech.

Finally, Section 9.9 discusses other factors relevant to the development of multilingual speech-recognition systems. We show that cultural factors can have a significant impact on user interface design, and then discuss the specific challenges of the developing world.

9.2 Characteristics of Non-native Speech

The differences between native and non-native speech can be quantified in a variety of ways, all relevant to the problem of improving recognition for non-native speakers. Differences in articulation, speaking rate, and pause distribution can affect acoustic modeling, which looks for patterns in phone pronunciation and duration and cross-word behavior. Differences in disfluency distribution, word choice, syntax, and discourse style can affect language modeling. And, of course, as these components are not independent of one another, all affect overall recognizer performance.

In this chapter, we discuss both theoretical and corpus-based descriptions of non-native speech.

9.2.1 Theoretical Models

It is a common assumption that pronunciation of an L2, or target language, is directly related to the speaker's L1, and that the path to native-sounding speech is a straight trajectory through phonetic and phonological space.

Research in **second language acquisition** (SLA), however, has shown that the reality is far more complex. Learners of a language are on a journey to proficiency, affected by developments in motor control, perception of the L2, familiarity with languages outside the L1-L2 pair in question, stress level, and a myriad of other influences that make the speech of any one speaker at any one moment a dynamic idiolect. The broadly-shared *perception* of accent—that is, the ability of native speakers across demographic lines to identify a non-native accent as French or Spanish or English—is difficult to reconcile with quantitative evidence from SLA research showing that speakers of a particular L1 differ from one another in more dimensions than they share when speaking an L2.

The theoretical model of SLA probably closest to the common lay view of foreign accent is known as *Contrastive Analysis* (CA) (James, 1980). CA theory claims that "speakers tend to hear another language and attempt to produce utterances in it in terms of the structure of their own language, thus accounting for their 'accent' in L2" (Ferguson, 1989). While CA is intuitively very attractive, it has not been an effective method for predicting all mistakes language learners make. Non-native speakers may approximate with a sound from outside the phonetic inventory of either L1 or L2 (see, for example, Brière [1996]). Other attempts to diagnose, explain, and predict pronunciation errors in non-native speech include *Error Analysis* and *Transfer Analysis*.

Contrastive Analysis Standard realization of L2 is compared with standard realization of L1 to predict deviations.

Error Analysis Standard realization of L2 is compared with an intermediate language (IL)—which represents the speaker's model of L2 at any given time—to predict deviations.

Transfer Analysis A variant of CA that only claims to predict those deviations that have their root in influences from L1, recognizing that other deviations will occur.

Production of a foreign language, of course, is about much more than phonetic similarity between L1 and L2. As speakers gain experience in a language, certain pronunciations become fossilized while others continue to change. Learners also change in their approach to the pragmatics of speaking the language. Tarone et al. (1983) describe strategies that language learners use to overcome difficulties in four major areas: phonological, morphological, syntactic, and lexical. They identify the strategy classes of *transfer*, *overgeneralization*, *prefabrication*, *overelaboration*, *epenthesis*, and *avoidance*, with most having an application in all four domains. This model points out that phonological transfer is only one of the techniques learners use in speaking an L2 that can affect speech recognition. Avoidance, for example, is highly relevant for data collection. Is it desirable to elicit data in such a way that speakers cannot avoid words and constructions that are difficult for them, potentially introducing disfluencies and capturing expressions that will not occur in real-world use?

When listeners hear a foreign accent, they are basing their characterization partly on perceptual recovery strategies specific to their native languages. For example, a frequently cited feature of Spanish-accented English is the substitution of long and tense /i/ for short and lax /ɪ/, making "win" sound like "wean." But what sounds to an English listener like the same /i/ from five different Spanish speakers may actually be five completely different phonetic realizations. Beebe (1987) shows that the majority of L2 pronunciation errors are phonetic rather than phonemic as may appear to native listeners. For example, in Japanese-accented English, Beebe found that although the pronunciation error rate of /l/ attempts was 46%, the actual rate of /r/ substitution was only 3%. Most of the perceived errors were actually phonetic deviations from [l], not phonemic substitutions of /r/.

Research in SLA seems to be consistently telling us that a foreign accent cannot be neatly described. Does this matter to an automatic speech recognition (ASR) system, however? It could be argued that the recognizer perceives nonstandard pronunciation much as a monolingual native listener does—by forcing unexpected acoustic events into the sound system it knows. It could also be argued that as the recognizer models sounds at an even finer level than was used in studies such as Beebe's, even small phonetic deviations will cause a recognition error. The experiments described in Sections 9.5.1 and 9.5.2 will attempt to answer these questions.

9.3 Corpus Analysis

Although collecting a non-native speech corpus for every L1-L2 pair is impractical, we can broaden our understanding of the general problem of non-native speech production as it relates to ASR by examining the limited set of corpora that have been collected and transcribed.

9.3.1 Elicitation of Non-native Speech

Capturing non-native speech presents many challenges aside from those faced in transcribing and recognizing it (Mayfield Tomokiyo and Burger, 1999). In planning the elicitation, it is necessary to consider how speakers will handle unfamiliar words (for read tasks) and rely on the scenario that is presented to them (for guided spontaneous tasks), and even whether they will be willing to complete the task at all (for many non-native speakers, exposing a lack of proficiency in the language is very uncomfortable). This adds to the processing cost of speaking, increasing the potential for disfluencies and errors that might not occur in real use. One of the most important goals of data collection for ASR is to capture the acoustics and expressions that would be found in naturally-occuring speech; researchers must strike a delicate balance, requiring speakers to generate speech in their own unique way without putting them in a stressful situation.

9.3.2 Transcription of Non-native Speech

In order to perform supervised training or adaptation with a non-native speech corpus, transcriptions of the speech data are needed. Transcriptions can be very broad, indicating only the words that are supposed to have been spoken in a read speech task, or very narrow, showing time stamps, classifying speech errors, and even providing phonetic annotation. Table 9.1 shows some of the annotations that are useful to include in transcriptions of non-native speech, with examples.

Phonetic transcription of pronunciation errors and phonetic deviations can be very difficult to generate because the articulatory productions often do not fit neatly into the International Phonetic Alphabet (IPA) (cf. Chapter 2), much less into approximations like SAMPA. For example, in the transcriptions discussed in Mayfield Tomokiyo (2001), one speaker was experimenting with sounds of /r/ in English and intermittently applied rhoticity to vowels that are not ordinarily observed to take it. The challenge for the transcribers was to consistently identify when the realization had crossed the threshold into rhoticity, and then use an ASCII approximation to encode it.

Table 9.1 Targets for annotation with examples.

Insertions	Will <;ins the> Fox's film sell many action figures?
Deletions	Only a hundred <;del years> ago salmon were plentiful.
Substitutions	Will all this effort <;1 &effect> be worth it?
Repairs	The books are -/upsteam/- upstairs in the study.
Repeats	The +/government/+ government must approve the plans.
Neologisms	The ruins lay buried in layers of seaweed <;1 ˜seawood>.
Mispronunciations	Near the *Willamette [w i l j ah m eh t] river.
Unintelligible regions	I have so many ((wines?)) in my Excel spreadsheet.
Word fragments	Baby orangutans were wedged into cra= crates.
Meta information	Some travel -/hundred/- <;meta I'm sorry> hundreds of miles.

9.3.3 Characteristics of Non-native Speech

Understanding how native and non-native speech differ is an important first step in attacking the problem of non-native recognition. There are many possible ways to quantify non-native speech characteristics.

Phonetic Confusion

Phonetic confusion, or the inability of a speech recognizer to identify the correct phones in the input speech signal, is one indicator of the degree of phonetic deviation from the standard against which the recognizer was trained. By comparing phonetic confusion for native and non-native speakers, we can determine more closely where the two types of speech differ.

Figure 9.1 shows phonetic confusions for native speakers of English and Japanese speaking English.[1] In this experiment, the recognizer was used to identify the vowels in speech. This was essentially a forced alignment task, with consonants and vowel presence known but vowel identity unknown. This is one of many ways in which phonetic confusion can be calculated.

Figure 9.1 (top) provides a global view of the distribution of phonetic confusion for native and non-native speech, with the size of the bubble

[1] Detailed descriptions of the experiments described in this section can be found in Mayfield Tomokiyo (2001).

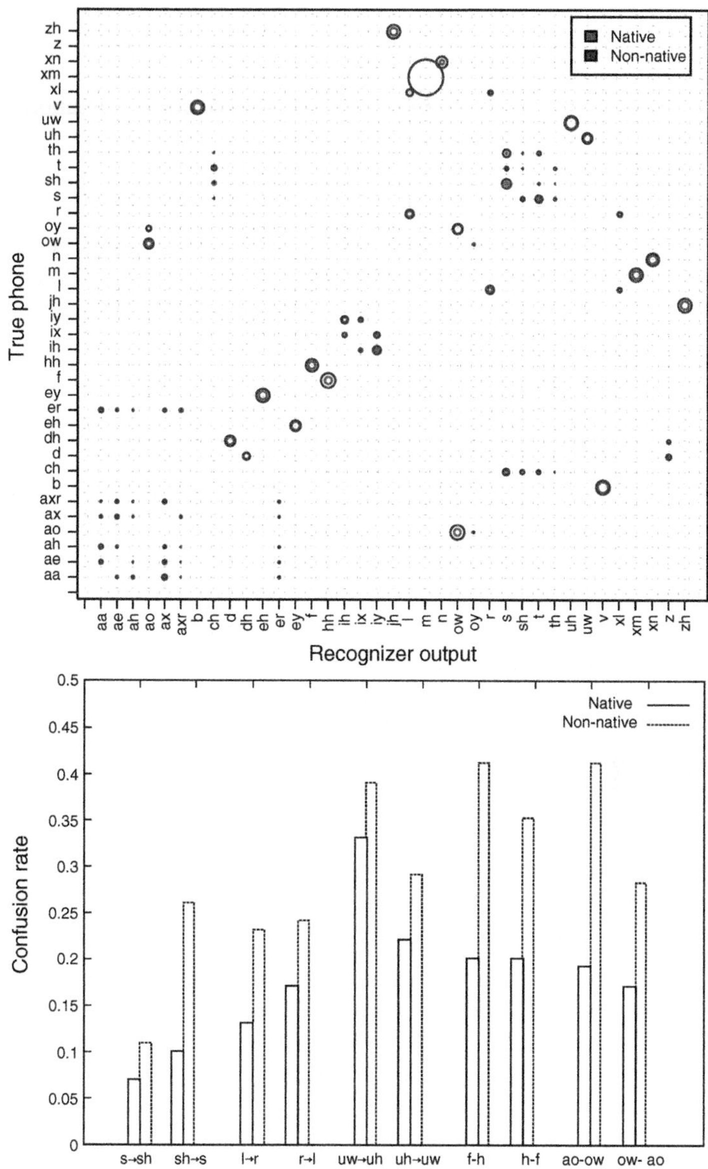

Figure 9.1: Phonetic confusions for native English and native Japanese speakers, full phone set (top), specific phones (bottom).

representing the magnitude of the confusion. The non-native speech here is primarily characterized by greater degrees of acoustic confusion between the same pairs of phones that are confusable in native speech. A more detailed view, presented in Figure 9.1 (bottom), compares phonetic confusion in native and non-native speech for a few selected phones. Confusion between [ʊ] and [u], for example, is high for non-native speakers, but it is also high for native speakers. [l]-[r] confusion, the subject of much study in Japanese-accented speech, is also fairly high in native speech. We do see notable differences in confusion between [h] and [f], and between [ʃ] and [s], which would be predicted by a transfer-based analysis of Japanese-accented English.

Fluency

When speaking a foreign language, one must concentrate not only on the meaning of the message but also on getting the syntax right, articulating the sounds, capturing the cadence of the sequence of words, speaking with the right level of formality, and mastering other elements of spoken language that are more or less automatic for a native speaker. The additional cognitive load can result in slower speech, with more pauses as the speaker stops to think. The fluidity of speech is called *fluency*, and offers a number of opportunities for quantification.

Table 9.2 compares fluency for native speakers of English, Japanese, and Chinese speaking English in read and spontaneous speech tasks. The overall word rate (number of words per second) is much lower for the non-native speakers for both types of speaking tasks. The main factor in the decrease, though, seems to be the number of pauses inserted

Table 9.2 Speaking rate and pause distribution statistics for native and non-native speakers of English in spontaneous versus read speech. Average phone duration and pause duration are measured in seconds. The pause-word ratio is the number of pauses inserted per word in speech.

	words per second		**phone duration**		**pause duration**		**pause-word ratio**	
speaker	spont	read	spont	read	spont	read	spont	read
Japanese	2.42	2.33	0.11	0.11	0.10	0.09	0.17	0.49
Chinese	2.70	2.28	0.11	0.11	0.10	0.12	0.18	0.47
Native	4.01	3.84	0.08	0.07	0.10	0.11	0.10	0.22

between words, rather than the actual length of phonemes or silences. This has important implications for both acoustic modeling and language modeling. The pauses between words suggest that there is far less cross-word coarticulation than would be found in native speech. For example, "did you" might be pronounced as /dɪʤu/ in native speech, but with a pause inserted would be pronounced /dɪd<pau>ju/. Word sequences, too, may be generated with less reliance on common and felicitous combinations, making them poorly modeled in an N-gram LM trained on native speech.

Disfluencies, including pauses as well as filler words, stumbles, and other speech errors, have been shown to have a significant effect on speech recognition accuracy (Shriberg and Stolcke, 1996). Understanding and modeling the disfluency characteristics of the particular speaker population and task of concern may enhance the performance of the system.

Lexical Distribution

Non-native speakers, in general, do not have available to them the range of words and expressions that native speakers do. It would not be accurate, though, to say that their speech is more *restricted.* On the contrary, depending on the task, native speakers may rely on fixed expressions, while non-native speakers may each come up with a unique way of presenting the same message. Accurate modeling of lexical distribution is an essential component of an effective language model.

Some differences in lexical distribution can be attributed to L1 transfer. German speakers of English, for example, make much heavier use of the past perfect than native U.S. English speakers do. Depending on context, this could either mirror simple differences in English dialect (*I've been to the parade today* versus *I went to the parade today*) or cause misunderstanding (*Have you been to the party?* versus *Did you go to the party?*). This pattern can be seen as reflective of the use of past and past perfect in German.

Other differences can be linked to the educational system under which speakers learned the L2. Japanese speakers of English offer an excellent opportunity for study of this phenomenon, as English education in Japan is highly standardized, and there is relatively little opportunity to deviate in one's practice of English outside the educational system during the years of compulsory formal study. The corpus of English tourist-domain queries by native speakers of Japanese (henceforth called the "Japanese tourist corpus") studied in Mayfield Tomokiyo (2001), contains many examples of usage and lexical choice that can be traced to expressions commonly taught in Japanese secondary schools. The word *which*, for example, ranks 25th in a corpus of Japanese-accented tourist information dialogues, and only 306th in the native U.S. English component of the corpus. The Japanese speakers showed a strong tendency to use non-restrictive relative clauses, whereas native speakers would omit the relative pronoun or use a modifier, as in *Please give me the name of the restaurant which is near my hotel* (L2) versus *Are there any good restaurants near the hotel?* (L1). This tendency cannot be tied to direct transfer of Japanese syntactic patterns. It can be linked, however, to strategies that are taught to simplify generation of English sentences. For example, most English nouns take prepositional adjectives (*new mail*), but some take postpositional adjectives (*something new*). With the relative clause construction, it is not necessary to remember where the adjective goes: *mail which is new; something which is new.*

Distribution of N-grams in native and non-native speech can also be contrasted to get quantitative measures of the difference between native and non-native speech. Table 9.3 shows the top N-grams in the Japanese tourist corpus. We see evidence of preference in the non-native speech for simple questions (*Where is the Empire State Building?*), whereas the native speakers prefer embedded forms (*Could you tell me where the Empire State Building is?*). We also see evidence of grammatical errors. Looking further at the discrepancy in frequency of *is there any*, it can be observed that the Japanese speakers used the term as a general query of existence (*Is there*

Table 9.3 Frequent trigrams in native and non-native speech.

Frequent trigrams in non-native speech			Frequent trigrams in native speech		
N-gram	Non-native rank	Native rank	N-gram	Native rank	Non-native rank
where is the	1	did not occur	you tell me	1	6
can I get	2	52	could you tell	2	18
do you know	3	52	get to the	3	50
does it take	4	32	I'd like to	4	12
how long does	5	77	tell me how	5	41
you tell me	6	1	to get to	6	164
is there any	7	130	I need to	7	94

any good restaurant around here?), whereas the native speakers reserved *is there any* for mass nouns and paired plural nouns with *are* (*Are there any good restaurants around here?*).

Task-Dependent Factors

It should be noted that task is an important factor in distinguishing native from non-native speech, and that task does not affect all L2 speakers and speaker groups equally. Referring back to Table 9.2, we see that the silence insertion rate, while always greater for non-native speaker groups, is much lower in spontaneous speech. Read speech forces readers to speak word sequences that they might not naturally produce, and allows them to speak without concentrating on meaning. The result is that while the prescribed N-grams may fit nicely into a native language model, the pronunciations do not fit well into the acoustic model; moreover, reading errors may diminish the effectiveness of the native language model.

In spontaneous speech, we see that anxiety level impacts language performance greatly. In a stressful situation, speakers can forget words and constructions that they ordinarily control well. Pronunciation may start off poor and improve throughout the interaction as the speaker gains confidence (a problem, incidentally, for speaker adaptation). If the spontaneous speech task is one that the speakers are familiar with (often the case with directions and navigation), their speech may be closer to native speech than in a task that is entirely new for them, such as medical triage.

Other Measures

In this section, we have examined only a few of the possibilities for quantifying differences between native and non-native speech. Others include perplexity, distribution of contractions,vocabulary growth, reading errors, and phonetic confusion with manual transcriptions.

9.3.4 Publicly Available Databases

Recorded and transcribed corpora of non-native speech, particularly those that have a parallel native component, are necessary for understanding and modeling non-native speech for ASR. The database on which the observations in Section 9.3.3 were based is proprietary, as are many others. There are a number of excellent publicly available speech databases, however.

- Translanguage English Database (TED): Contains non-native speakers of English having a variety of different L1s. They were recorded in a lecture scenario in which academic papers were presented. Recorded at Eurospeech 1993 (Kipp et al., 1993).
- Strange Corpus 1 and 10: Contains speakers of various source languages/L1s reading the German "Nordwind und Sonne" and other read and nonprompted sentences (Strange Corpus, 1995).
- ISLE Corpus (Interactive Spoken Language Education): The speakers are German and Italian learners of English. The project for which this database was recorded aimed at developing a language learning tool that helps learners of a second language to improve their pronunciation (ISLE, 1998).
- Hispanic-English database containing read as well as spontaneous speech (Byrne et al., 1998a).
- NATO Native and Non-Native (N4) Speech Corpus: The recordings include NATO naval procedures in English. The native language of the speakers are Canadian English, German, Dutch, and British English (NATO Non-Native Speech Corpus, 2004).
- Broadcast News has a non-native component, and is useful for comparing native to non-native speech in an unrestricted domain (LDC, 1997).
- *Wall Street Journal* — Spoke 3: Contains read speech (English) of 10 non-native speakers with different L1s.
- Non-native city names: A cross-lingual speech database for the analysis of non-native pronunciation variants of European place names was recorded for five languages (English, French, German, Italian, and Dutch) in which each speaker spoke all five languages, only one of which was their mother tongue (Schaden, 2002).

These databases are very specialized (e.g., to cover city names or naval procedures only) and thus might not be suitable for other tasks. All databases have in common the fact that they can only capture a kind of "snapshot" of certain speaker groups. It will always be particularly difficult to formally measure how much the selected speaker group is really representative of speakers of a certain source language. This might be the reason some investigations of different databases put more emphasis on certain factors while others do not. In general, we can state that the number of potential tasks and source-target language pairs is very large, and it is not expected that non-native databases will ever be available for more than a few of the possible combinations.

9.4 Acoustic Modeling Approaches for Non-native Speech

The variation in acoustics across speakers is a common problem in speech recognition. With appropriate acoustic modeling, much of this variation can be captured. However, when considering dialects and non-native speech, the variations can become too large to be treated in the traditional way.

In native dialects of the same language, we expect shifts within one phoneme set. In non-native speech, we can also expect the use of phonemes from other languages, that is other phoneme sets, or even sounds that cannot be categorized at all. The majority of current speech recognizers use Hidden Markov Models (HMMs) as briefly introduced in Chapter 4. High-complexity AMs, such as triphones with many Gaussian mixture components, are already capable of modeling pronunciation variation and coarticulation effects to a certain extent (Holter and Svendsen, 1999; Adda-Decker and Lamel, 1999; Riley et al., 1999). Jurafsky et al. (2001) investigated this issue more closely and found that on the one hand, some of the variation, such as vowel reduction and phoneme substitution, can indeed be handled by triphones, provided that more training data for the cases under consideration are available for triphone training (which is difficult to obtain for non-native speech, as was outlined in the previous section). On the other hand, there are variations like syllable deletions that cannot be captured by increased training data; therefore, other approaches are needed to cover these kinds of variation.

Witt found that for non-native English speakers, triphones perform worse than monophones if trained on native speech, meaning that less detailed (native) models perform better for non-native speakers (Witt and Young, 1999). Witt's observation was shared by He and Zhao (2001) and Ronen et al. (1997). This might indicate that high-complexity triphones can be a good choice for covering native pronunciation variation; for the non-native case, this is questionable.

Since the lack of data for non-native speakers is omnipresent, in some system designs all native and non-native data, regardless of the speaker's source language or L1, is pooled to build a general AM.

Uebler et al. investigated the task of recognizing accented German of various accents (Uebler and Boros, 1999). They found that pooling data from native and non-native speakers to build one global AM yielded quite uniform performance across non-native speakers. However, further results suggest that accent-specific models yield better performance.

Livescu (1999) trained AMs by pooling data of different source languages spoken by native and non-native speakers. She applied these AMs to non-native speech recognition and found that they outperformed AMs, which had been trained solely on either native or non-native speech.

Mayfield Tomokyio compared many different methods of combining native and non-native data in acoustic model training (Mayfield Tomokiyo and Waibel, 2001). With 80 hours of native data and 3 hours of non-native data, it was found that the most effective use of the non-native data was to first train native models, then run additional Viterbi training iterations on the fully trained models using the non-native data. A further improvement was achieved by interpolating the resulting models with the original native models to counteract overfitting, as in Huang et al. (1996). In the best case, WER was reduced from 63.1 to 45.1%—a relative reduction of 29%.

It is to be expected that there are more factors to be considered when considering LVCSR applications. Here, apart from differing pronunciations and acoustic realizations, limited vocabularies and ungrammatical sentences as described in Section 9.3.3 might need to be considered.

9.5 Adapting to Non-native Accents in ASR

Before adapting speech recognizers to non-native speech, the difference between non-native speech and the *models* that were created for native speech need to be quantified. In the case of native speech recognition, we build models that describe native speech in a more or less accurate way. While we cannot deny that there is an influence of source language on how the target language is spoken, we have seen in Section 9.2.1 that this relationship is very complex. This means that speakers can, but do not necessarily have to, pronounce target phones as "something between" the true target phone and some L1 phone that they see it as corresponding to. Other factors, such as articulatory mishaps, experimentation, or individual preconditions, have a large impact on the target language's phone realization. Such preconditions may be a result from when (at what age), how (by reading, by listening), and where (social background) the speaker acquired the target language.

Livescu (1999) found proof for both cases on a small subset of the corpus of the JUPITER system (Zue et al., 2000). Some effects on the target language could be explained by constraints of the source language of the speaker. For example, speakers with a source language that has a simpler syllable structure than the target language tend to use simple consonant clusters when speaking the target language. It was also found that speakers of source languages that do not reduce vowels in unstressed positions substitute full-quality vowels for schwas in the target language. At the same time, she observed many unpredictable pronunciations occurring inconsistently among speakers. Even the same speakers did not pronounce the same words consistently. However, it is difficult to find out what exactly was the reason for these unpredictable variations.

Another corpus-based analysis of non-native speech was performed by Kim and Flynn (2004). They studied the phonological phenomena that occur when first-language Korean (L1) speakers produce English (L2). In particular, 48 native Korean speakers were asked to imitate a number of utterances produced by a native speaker of midwestern American English. The results show similarities between L2 and production of L1—for example, limited stress reduction, which they attribute to the influence of a syllable-timed L1. Additionally, some evidence for over-generalization was seen: speakers learn certain phenomena that occur in L2 but not in L1 (e.g., rising pitch in noun phrases), but also extend these productions to contexts in which a native speaker would not do so. Cognitive load was seen to play a substantial role, meaning that deviations from native speech were more significant when longer or more complex utterances were required.

Articulatory mishaps or social background, of course, are not generally considered when modeling native speech. So when we want to adapt to non-native speech, we should focus on those parameters that are explicitly modeled in our ASR system. Generally speaking, we usually have AMs, pronunciation dictionaries, and language models as main building blocks in our ASR system. Most of the approaches that try to deal with non-native speech adapt one or several of these models.

Ignoring all effects in non-native speech that are not modeled in the native case means that at a minimum we have to deal with the following two phenomena:

- Acoustics that will be different from native ones. This phenomenon can be tackled by speaker adaptation techniques, that will be described in the next section.
- Pronunciations that deviate from native ones. Methods to model this will be explained in Section 9.5.2.

Since speaker and pronunciation adaptation modify different models in our system, we might expect that improvements, if any, yielding from adaptation of the AMs are at least partly additive to those from the adaptation of the pronunciation dictionary. Whether or not this can be confirmed is shown in Section 9.6.

9.5.1 Speaker Adaptation for Non-native Accents

In the past years, speaker adaptation has become a standard method to improve the recognition rates of speaker-independent (SI) ASR systems. One problem one is almost always faced with is the mismatch between the available training data and the current test situation. Often there is a mismatch between the environments and microphone used for training and those found in real-world use. Furthermore, the training speaker set might not be representative of the current test speaker. As a consequence, the recognition rates are not satisfying. To close the gap between the SI AMs and the acoustics of the current speaker, state-of-the-art systems apply speaker adaptation techniques.

We can distinguish between supervised and unsupervised, as well as between batch and incremental adaptation. In supervised adaptation, speakers have to provide speech data prior to using the system; usually they are required to read a predefined text. In batch adaptation, this speech data as a whole is processed and the acoustic models are adapted before the user starts to use the system. By using predefined enrollment text, the AMs can be reliably adapted because we can ensure that the AMs are adapted with speech data that really "belong" to the AMs—assuming that the user spoke the prompted text without mistakes.

Incremental and unsupervised speaker adaptation does not, in contrast to batch and supervised adaptation, need any prerecorded enrollment speech data but directly exploits the user's speech input during system deployment. Starting with a set of SI models, the AMs are adapted after each (set of) utterance(s). While unsupervised, incremental adaptation is far more comfortable for users, it is a much more challenging task from a technological point of view since it means that each adaptation step has to be performed on very limited amounts of data. Furthermore, we have to rely on the recognized words being what was actually spoken. Depending on how well the SI models match the acoustics of the current speaker, we will have more or less misrecognized utterances that will be used for adaptation. As a consequence, some AMs will be adapted using speech data that actually does not "belong" to these AMs. To avoid such misadaptation, confidence measures are often used to make sure that only those utterances are used for adaptation that have been recognized with high confidence, (Goronzy, 2002). Byrne et al. (1998b) found a positive effect of supervised speaker adaptation for non-native speakers. However, adapted models from one non-native speaker did not improve the performance for other speakers. This means that enrollment data would have to be collected for each speaker, which is, as they also state, problematic.

The two most popular methods for speaker adaptation are maximum-likelihood linear regression (MLLR) and maximum a posteriori (MAP) adaptation. Both methods adapt the means and variances of the AMs. In MLLR, adaptation parameters are pooled and updated with transformation matrices, as will be described in the following section; the MAP approach, however, conducts adaptation on the phoneme level. Although yielding a phoneme specific and thus finer adaptation than MLLR, the disadvantage of MAP is that it only updates the parameters of those phonemes that were observed in the adaptation data. As a consequence, it needs a lot of adaptation data to reliably reestimate the parameters for all phonemes used in the system, which takes a rather long time. Therefore, MLLR might be advantageous in systems of higher complexity, since it can potentially provide fast adaptation with a limited number of utterances. Woodland describes the basics of these two methods and their recent extensions in more detail (2001).

Even though we assume that a phoneme-specific adaptation might be particularly relevant for non-native speakers, in the following example, MLLR was chosen because a fast approach for unsupervised, incremental adaptation was required. MAP was considered impossible because of the lack of non-native data.

Standard MLLR, as described by Leggetter and Woodland (1995), performs a linear transformation of the model parameters by estimating a transformation matrix for each regression class—that is, for each set of parameters. While faster than MAP, it still requires several sentences of adaptation data before the transformation matrices can be reliably estimated. For many applications, the available amount of adaptation data is still insufficient for this method. For such applications, the standard MLLR approach was extended by Goronzy (2002) to include a MAP-like weighting scheme. Using this extension, improvements in WER of more than 17% on a command-and-control task were possible for a set of eight non-native speakers (English native speakers speaking German), with very limited amounts of adaptation data. For each update of the HMM parameters, only 1,000 frames of speech, which roughly corresponds to four words, were used. Such kinds of adaptation is well-suited for non-native speakers. When they start using the ASR system, the deviation in AMs can be expected to be high, which is why a fast adaptation is necessary. However, since less data can be expected (as compared to native speakers in the same time frame), an adaptation method that can handle such limited amounts of data is required.

Ikeno et al. (2003) experimented with Hispanic-accented English. They collected a corpus of about 25 hours of speech from 18 participants originally from South and Central America, who have Spanish as their source language but who had been living in the United States for at least one year. Data from 14 speakers was used to adapt the AMs and language models of a state-of-the-art HMM-based large-vocabulary recognizer, which had originally been trained on the *Wall Street Journal* corpus. On this large-vocabulary task, adaptation was found to reduce word error rates from 68 to 39%, with most of the benefit derived from improvements in the AM. A particularly significant difference between the native and non-native speech was found in vowel reduction: non-native speakers were less likely to reduce vowels than their native counterparts.

Liu and Fung (2000) used MLLR adaptation to adapt AMs that were trained on native English speech to a Cantonese accent. They used native Cantonese speech that was aligned with a phoneme recognizer using HMMs trained on native English after an appropriate mapping of the phoneme sets was applied. The English models were then adapted using this aligned data. A comparison with adaptation using Cantonese-accented English speech data showed very similar improvements over the baseline model.

Although yielding improvements, the application of speaker adaptation by itself is insufficient. While adapting the AMs to non-native realizations, non-native phoneme sequences are not taken into account. We will show in Section 9.5.2 why this could be problematic.

9.5.2 Pronunciation Adaptation

Since we expect that non-native speakers pronounce words of the target language differently from native speakers of the language, the straightforward approach would be to simply add these non-native pronunciations to the dictionary.

The problem, however, is how to determine the non-native pronunciations. As seen in Section 9.2.1, pronunciation deviations are not easy to predict. One way would be to manually transcribe at least a part of the non-native data. To be exact, a phonetic transcription would be necessary. This, however, apart from the omnipresent lack of non-native data, raises additional questions concerning the phone set. If only the native phoneme set was used for the transcriptions, valuable information about the "real" pronunciation might be lost. On the other hand, if we additionally used the non-native phonemes for transcribing the data, how do we transcribe phones that are somewhere "in between" a native and a non-native phoneme (a case that we consider likely to happen)? How do we transcribe productions that fall outside even the IPA? Do we, as a result of adding to the phone set used for transcription, need to change the model set of the recognizer? This might lead to a potentially huge phone set if we consider speakers with different source languages and thus seems infeasible for real applications.

Pronunciations are often derived automatically from corpora (see Chapters 5 and 7). In this case, a phoneme mapping of source-language phonemes to the closest target-language phonemes often becomes necessary. Schaden (2003a) raises the question of how to do this mapping. Simply considering the acoustic or articulatory proximity does not seem to be a reliable indicator of the phoneme substitutions that really occur in non-native speech. He observes that often the choice of the substituted phone is influenced by the phonetic or graphemic context. In the rewrite rules he generated, this contextual information was taken into account.

As described in Chapter 5, traditional approaches to pronunciation modeling mainly consider pronunciation variants that occur in native speech. Two kinds of approaches can be distinguished, namely knowledge-based and data-driven approaches. Knowledge-based approaches apply expert knowledge to develop rules that generate pronunciations variants from canonical pronunciations. In contrast, data-driven approaches try to derive the pronunciation variants directly from the speech signal. While the rule sets resulting from the knowledge-based approaches are rather general and can be used in any application, they also tend to overgenerate and provide no indication of how likely the different variants are. Adding all generated variants is often counterproductive since too many pronunciations increase the confusability and thus lower recognition rates. If the variants are learned from speech databases, overgeneration can be avoided and application likelihoods can be computed; however, the observed variation is very database specific and might not be relevant for other tasks. Of course a combination of both approaches is also possible. Both techniques have been applied for non-native pronunciation modeling. Sahakyan (2001) investigated non-native databases and manually generated pronunciation rules that generated German- and Italian-accented English pronunciation variants occurring in the ISLE database. Humphries et al. (1996) chose a data-driven approach and grew decision trees that were supposed to learn a phoneme-to-phoneme mapping from canonical pronunciations to pronunciation variants, with the variants for decision tree training derived from a corpus. Schaden (2003b) manually generated rewrite rules for three different language pairs in both directions (English, German, French) to transform native canonical pronunciations into accented variants. He distinguished four different levels of accent, ranging from a "near native" pronunciation over "minor allophonic deviation," "allophone/phoneme substitution," and "partial transfer of source language spelling pronunciation to the target language" to "almost full transfer of source language spelling pronunciation to the target language," where the respective higher levels always include the phenomena from previous levels. Amdall et al. (2000) also uses rewrite rules, which were obtained from co-occurrence statistics between canonical pronunciations and non-native variants obtained by a phoneme recognizer.

Non-native speakers are known to have difficulty acquiring context-conditioned phonetic contrasts when the L2 phoneme is perceived as corresponding to an L1 phoneme that is not subject to, or does not trigger, the same variation (Fox and Flege, 1995). Allophonic variation is an important part of the acoustic model, however, and if the allophonic distributions are estimated from native speech, the polyphone modelpolyphone model may not be appropriate for non-native speech. This situation has some similarities to multilingual ASR bootstrapping, in which an existing recognizer is augmented with a set of phonemes and allophones from a new language, and for which Polyphone Decision Tree Specialization (PDTS) (Schultz and Waibel, 2000) has been successful. PDTS does not appear to be as effective for non-native speech, however, which exhibits far less consistency in allophonic variation than does native speech (Mayfield Tomokiyo, 2001; Wang and Schultz, 2003). One limitation of PDTS is that it only considers adding branches to the polyphone decision tree that represent contexts not found in the original data. If two phones have already been clustered together based on what is seen in native speech, PDTS does not provide a mechanism for separating them to represent the different allophonic distribution over the same phoneme set found in non-native speech.

Cremelie and ten Bosch used grapheme-to-phoneme (g2p) converters for multiple languages to generate non-native pronunciations for the special task of recognizing foreign names (Cremelie and ten Bosch, 2001). Depending on the weighting scheme used to weight the non-native pronunciations, they could improve the WERs by almost 60% relative. However, Schaden assumes that such g2p rules are only partially applied because most of the speakers are aware of target-language pronunciation rules (which was confirmed by Fitt (1995)). He furthermore assumes that even speakers with no formal background in the target language seem to be aware of at least some of the pronunciation regularities of these languages. However, the task of pronouncing foreign city names is certainly of particular difficulty, so that a fallback to the source-language-guided pronunciations seems more probable than for "normal" words.

9.5.3 Simulating the Reproduction of Foreign Words

It is often difficult for a non-native listener to perceive and produce phonetic differences that are not relevant in their native language (Compernolle, 2001). For example, French does not have the glottal fricative /h/. Native speakers of French may experience difficulty producing it when speaking English or German, and may not be able to hear it when spoken in words like "here." Japanese does not have a rhotic /r/ phoneme, and native Japanese speakers' pronunciation of words like English "bright" can vary greatly in accuracy of the /r/.

Japanese speakers of English show a tendency to break up consonant clusters with epenthetic vowels. "Table" (canonically /te͡ɪbl̩/) might be pronounced as something like [te͡ɪbɯɾɯ] or [te͡ɪbɯlɯ]. Italian speakers of English also exhibit a similar tendency, pronouncing "team" (canonically /ti:m/) as [ti:mə] (Sahakyan, 2001). This might be traced to the consonant-vowel (CV) syllable structure of these two L1s. It can also be argued, however, that there is a universal tendency of the articulators toward a CV syllable, and that as cognitive cycles are taken up by articulating unfamiliar phonemes and constructing L2 sentences, some degree of syllable simplification can be observed from speakers of any L1 (Tarone et al., 1983). Regardless of the reason for the epenthesis, however, it is clear that phone-based speaker adaptation algorithms can be insufficient for capturing the variation found in non-native speech, as they do not consider alternative phone sequences. In the "table" example, acoustic adaptation would try to adapt the HMM for /l/ to [ɯɾɯ], since it uses the native canonical pronunciation as a reference for adapting the AMs. In reality, however, three phones instead of one have been produced.

Goronzy attempted to predict pronunciations that would be generated by speakers of English speaking German (Goronzy et al., 2004). The experiment was meant to be a simulation of a person learning a target language by listening to a native speaker of that language. To simulate the English-accented reproduction of German words, a British English HMM-based phoneme recognizer was used to decode German speech. The German speech consisted of several repetitions of German command words uttered by several native German speakers. Because the HMMs used for the recognition were trained on British English, the resulting phoneme transcriptions simulate the way a native speaker of British English would perceive German. An English phoneme N-gram model was applied, representing English phonotactic constraints. It should be noted that in the case of native pronunciation modeling using a phoneme recognizer, the recognizer should be as unconstrained as possible to find unknown variants. In this situation, however, we are particularly interested in those phoneme sequences that share characteristics of the source language, so we apply the phoneme language model. Ideally, by using a data-driven approach, unpredictable variants (which result from factors different from the source language) can be covered as well. The resulting English-accented phoneme transcriptions were then used to train a decision tree that learns a mapping from the German canonical pronunciation to an English-accented variant. The resulting English-accented variants were added to the recognizer's dictionary to then recognize the accented speech. The experiment was repeated for German native speakers who spoke French with the corresponding settings (German phoneme recognizer that decoded French words) (Goronzy and Eisele, 2003). In addition to the dictionary modification, acoustic adaptation using weighted MLLR was applied (Goronzy, 2002). The results in WERs of the experiments are shown in Table 9.4. "Base" stands for the baseline dictionary, which only contained native German and French pronunciations, respectively. Reference tests on corresponding German and French native test sets yielded WERs of 11% (German) and 12.3% (French). Using these dictionaries for the non-native (English-accented German and German-accented French) test sets of 10 English and 16 German speakers resulted in WERs of 28.91% and 16.17%, respectively.[2] Applying weighted MLLR ("Base+MLLR") improves the results. In the first case, using the extended dictionary ("extdDict") alone slightly improves the results but in the second case results are even slightly worse. However, in combination with weighted MLLR ("extd+MLLR"), the best results can be achieved in both cases.

Preliminary experiments on Spanish speakers speaking German show the same tendency. The only model that required modification in these experiments was the dictionary, which is a feasible solution if several dialects and accents are to be covered in real systems. Of course, the AMs

Table 9.4 WERs for accented speech using adapted dictionaries and weighted MLLR; baseline results on native speech are 11% for German and 12.3% for French.

Language	Base	Base +MLLR	extdDict	extd +MLLR
Eng-acc German	28.91	24.06	27.42	23.66
Ger-acc French	19.63	16.17	20.56	15.69

[2] The difference in WER increase can probably be explained by the fact that some of the English speakers could speak almost no German (they arrived in Germany just two weeks before the recordings were made), while in the German speaker set, all speakers rated their knowledge of French as beginner or intermediate level.

are also changed by the speaker adaptation, but this is considered a standard procedure in state-of-the-art speech recognizers. Another advantage is that Goronzy's method uses only native German and English corpora; this is a desirable approach for the many source-target-language combinations for which native language speech corpora already exist and are relatively easy to obtain. Because of the immense variability in non-native speech, however, the above experiments show a simplified simulation of the perception and production of a language, and one pronunciation is not enough if a wider range of non-native speakers should be covered. This means that ideally, multiple-accented pronunciations covering different non-native effects would be generated to then select and add only the optimal pronunciation for each speaker (when adding several pronunciations to the dictionary, one always risks an increase in confusability). This is clearly an issue for future research.

9.6 Combining Speaker and Pronunciation Adaptation

Because of the expected different acoustic realizations of phonemes on the one hand and the expected differing phoneme sequences on the other, both adaptation of the AMs and adaptation of the dictionary should be applied, with the speaker adaptation accounting for different realizations of certain phonemes, and the pronunciation adaptation accounting for the "wrong" phoneme sequences.

We have seen in the previous example that the results were indeed (partly) additive and that up to 20% relative improvement could be achieved when both acoustic and pronunciation adaptation are applied. Woodland (1999) also underlines the necessity of combining both techniques for improved recognition results. Huang et al. (2000) were able to improve the baseline results for a Mandarin speech recognizer faced with accented speech by 13.9% relative when the dictionary was adapted, by 24.1% when MLLR was used, and by 28.4% when both were used. Wang and Schultz (2003) used polyphone decision trees specialized to non-native speech in combination with MAP speaker adaptation. They considered German-accented English speech and improved the recognition rates by 22% relative when applying MAP and an additional 6% relative when using PDTS in combination with MAP. Livescu used one set of AMs that were interpolated from native and non-native training data and combined them with special lexical modeling that took into account phoneme confusions specific to non-native speakers (Livescu and Glass, 2000). She also found additive improvements of lexical and acoustic modeling.

Tian et al. (2001) used bilingual pronunciation modeling in combination with online MAP adaptation of the acoustic models. This method yielded relative improvements between 55% and 71% in clean and noisy environments, respectively, in an isolated word recognition task.

9.7 Cross-Dialect Recognition of Native Dialects

As discussed in the introduction to this chapter, the recognition of speech in different native dialects introduces many of the same challenges as non-native speech. (In an early study of non-native speech recognition [Kubala et al., 1991], one of the most difficult "non-native speakers" for a system trained to recognize American English was in fact a first-language speaker of British English.) In this case, it is common practice to simply train different recognizers for different dialects, but some of the same factors that apply to non-native speech (limitations of computational resources and speech data, the occurrence of proper names) also cause cross-dialect recognition to be of practical importance. Since many of the same techniques and principles that apply to non-native speech recognition are relevant to cross-dialect recognition, we will now summarize how well those techniques have performed in selected cross-dialect experiments.

Mandarin Chinese has been the focus of numerous experiments in dialect adaptation; it is the most widely spoken language on earth, and has at least eight major dialects and as many as 40 minor dialect variants (Ramsey, 1987). In fact, adaptation to different dialects of Mandarin Chinese represent two classes of challenges: (1) many speakers have different dialects of Mandarin Chinese as their first language, and (2) others who speak a language such as Shanghainese as their first language produce Mandarin Chinese with a dialect influenced by that first language.

Beaugendre et al. (2000) experimented with the Beijing and Taiwan dialects of Mandarin Chinese, using the Beijing dialect as source and the Taiwan dialect as target. They found cross-dialect recognition accuracy to be significantly inferior to that of a dialect-specific recognizer (word error rates of 28.3% compared to 7.1%, respectively, on a large-vocabulary corpus of read speech). MLLR adaptation with a global transformation matrix had only limited benefits, reducing the word error rate to 24.3%. However, combining MLLR with MAP improved matters significantly, and by further interpolating the HMM weights of the adapted models with the source models, they were able to reduce the error rate to 15.3% with as few as 250 adaptation utterances, and 11.0% with 2,500 utterances. (These results were obtained without explicit tone recognition; if tone modeling is included, the best adapted accuracy is 8.7%, compared to 5.9% for models built specifically for the target dialect.)

At the 2004 Johns Hopkins University summer workshop, a research group focused on the recognition of Mandarin Chinese as spoken by first-language Wu speakers. Three forms of adaptation—acoustic, pronunciation, and language model—were considered in the context of large-vocabulary character recognition. Although systematic variability in pronunciation and language use were found to exist, acoustic adaptation was found to be the only significant source of improved accuracy. A somewhat more sophisticated form of MLLR was found to reduce character error rates from approximately 60% to about 50%, using around 160 minutes of adaptation data (from 20 different speakers). This compares to a character error rate of 44% that was obtained with a more extensive adaptation data set.

Different dialects of native German speakers were studied by Fischer et al. (1998). In particular, a large-vocabulary continuous-speech-recognition system was constructed from about 90 hours of speech by 700 German speakers from Germany, and tested on both German and Austrian speech. (Even within Germany, there are significant dialect differences, but the large number of training speakers were assumed to provide adequate coverage of these variations.) Whereas a test-set accuracy of 13.2% was obtained for speakers from Germany, the error rate for speakers from Austria was 20.1%. Two approaches to compensate for these differences were studied.

- The training data was clustered into a number of subsets based on acoustic similarity, and different recognizers were constructed for the various clusters. During recognition, the most appropriate cluster for a test speaker's utterances was determined and used for recognition. Six clusters were found to be optimal for this set, and resulted in an error rate of 17.8% on the same test set as above. (With this set of clusters, the error rate for the speakers from Germany was also marginally reduced.)

- Adaptation data (about 15 hours of data from 100 speakers) was used to refine the acoustic models—either by inclusion of the Austrian data in the German training data or by MAP-like Bayesian adaptation of HMM parameters. Combination of the training data reduced the Austrian speakers' word error rate to 15.6%, while increasing the German speakers' rate to 13.7%. Bayesian adaptation, in comparison, produced error rates of 14.2% and 17.1%, respectively, for the speakers from Austria and Germany at the most aggressive setting of the smoothing parameter.

  If the Austrian data is used by itself to train the recognizer, the lowest error rate for Austrian test speakers (12.2%) is obtained, at the cost of a significant increase in error rate for the speakers from Germany (22.3%). To optimize the combined recognition accuracy across the two test populations, the best approach was to simply combine the training data.

Overall, then, acoustic adaptation for dialect variations is seen to be both important and successful, even with a relatively small amount of training data. To date, these successes have not been mirrored by pronunciation adaptation for cross-dialect recognition. The results are generally similar to those found by Mayfield Tomokiyo for non-native speech: compared to acoustic adaptation, adaptation of pronunciations offers relatively little benefit, and the combination of the two methods is dominated by the contribution from adapting the acoustic models (Mayfield Tomokiyo and Waibel, 2001).

9.8 Applications

In the preceding sections we have seen many perspectives on how non-native speech—and recognition of non-native speech—differ from native speech. In designing a speech recognition application that is expected to see any volume of non-native users, differences in language ability, task understanding, and language use are important considerations if non-native users, particularly less proficient non-native users, are to have a satisfying experience.

Many approaches to improving recognition accuracy for non-native speakers assume a given L1-L2 pair. This information is not necessarily available in a real-world application. Accent identification can be accomplished using acoustic features (Liu and Fung, 2000), competing AMs (Teixeira et al., 1996), and recognizer output (Mayfield Tomokiyo and Jones, 2001), but these methods all rely on some amount of training data.

Sometimes it is even necessary to identify the language that is being spoken. Code-switching, or switching back and forth between languages or dialects, is an important pragmatic strategy for multilingual speakers. Code-switching is seen in many different forms. Francophone Canadians may speak of French may speak English street names with perfect English pronunciation when giving directions. Latino speakers may pepper their English with Spanish expressions that are widely understood in their community. Speakers in Hong Kong may switch between Cantonese and English almost without noticing it (Fung and Liu, 1999). In most cases of code-switching, speakers are highly fluent in both languages, but whether or not they fully switch to the pronunciation of the new language or adapt it to the primary language of the interaction depends on the nature of the interaction. Speakers of very low proficiency may also switch languages mid-interaction as a last resort when they feel unable to continue in the system language.

The following sections describe some of the types of applications that are likely to encounter non-native users.

9.8.1 Speech-Based User Interfaces

Imagine that you are a non-native speaker of English attempting to find out whether a particular procedure is covered under your health insurance. Before, calling the insurance company, you study the medical terminology that is likely to come up during the conversation and are prepared to talk to an agent at the company. When you call the number on the back of your insurance card, however, you are confronted with a speech-driven interface.

System: Thank you for calling American Health Insurance! You can speak your answer at any time. If you are a provider, press "one." If you are a patient, press or say "two."

User: Uh... **one** more time please?

System:	You have reached the automated provider information access line. To authorize a procedure, say "authorize." To confirm benefits, say "benefits."
User:	Hello? I would like to ask about my contract.
System:	I'm sorry, I must have misunderstood. To authorize a procedure, say "authorize." To confirm benefits, say "benefits."
User:	Authorize.
System:	Please speak the patient's ID number.
User:	123-456-789.
System:	Please speak the code for the procedure.
User:	... <click>

Although at the time of this writing, speech-driven interfaces such as this are generally disliked by the public, speech application developers have made great strides in providing a user-friendly experience for the native speaker, including expeditious and effective error recovery. The improvements do not necessarily extend to the non-native user, however. Use of industry lingo, like "provider" instead of "doctor," may be trendy and in some sense more accurate, but it throws up an immediate barrier to those users who do not know the terms. A human operator would immediately be able to sense that rephrasing or simpler language was necessary to accommodate the user. In a system like this, however, there is no recourse for speakers having difficulty other than to wait for any error handling procedure to be activated (if they are even aware that this is an option). Non-native speakers may also not realize as quickly as native speakers that they are not speaking with a human.

There are two primary considerations in making a speech application non-native-speaker friendly: making the task itself easier for non-native speakers, and tailoring error recovery strategies to non-native speakers.

Task Design

Simplification of the terminology and sentence structure would seem to be an obvious first step in improving access for non-native speakers, but native user interface (UI) designers do not necessarily have accurate intuitions about what is easy and hard for a non-native speaker to understand. Working directly with non-native users of varying proficiencies and language backgrounds is very important in streamlining the task. Even an expression as seemingly straightforward and polite as "I must have misunderstood" can cause confusion. The use of the modal "must have," the term "misunderstand" and the conjugation "misunderstood" may all be unfamiliar to the listener, causing him or her to spend extra time processing the expression and possibly missing what comes next. "I did not understand" would be much more straightforward, but UI designers may feel that this is too direct for native listeners. This is just one example of the delicate balance that must be struck to achieve optimal comprehensibility for the non-native speaker and optimal satisfaction for the native speaker. If the goals of user satisfaction and task success are viewed as linked, however, surely there is room for compromise.

Providing a top-level option for people having trouble understanding ("If you don't understand, press 9") could be a very useful route for non-native speakers. They could further be offered options for speaking to a human or being taken to a simplified interface that would allow options to be repeated without having to listen to the entire menu.

This all assumes that speakers have some familiarity with the concept of speech interfaces in general, which cannot be said for many speakers in the developing world. Section 9.9.2 explores the challenges of extending the benefits of spoken language technology to speakers with limited access, limited exposure, and a limited cultural basis for understanding.

Error Recovery

Most speech applications have some implementation of error recovery—for example, transferring to a human agent if there are three consecutive recognition failures. The system does not always realize that there has been a failure, however, and the user is not always aware that if they persist, the interface may change.

It is not always the case that a speech interface is a bad option for a non-native user. Native speakers are notoriously bad at repeating a sentence when asked to; intending to be helpful, they may change the sentence, further confusing the listener. Some of the stress of speaking a foreign language may also be avoided when speaking to a computer rather than a human.

Some possibilities for improving error recovery strategies for non-native speakers include acoustically detecting that the speaker is non-native and applying some of the adaptation techniques discussed in this chapter; tracing the path through the task hierarchy to assess whether the user is showing different patterns from native users; offering users the ability to repeat or slow down individual sentences; and asking periodically if the user is having trouble understanding. The field of speech-based user interface design is active and rapidly expanding. Although algorithms and advances will not be covered here, speech-based systems appear poised to become a primary means for dissemination and transfer of critical information, and the system designers should be encouraged to incorporate the needs of speakers of limited language abilities and exposure to technology.

9.8.2 Computer-Aided Language Learning

As technologies for processing human language have matured, it has become possible to view them as pedagogically valuable tools. Advances in speech recognition and parsing have been enthusiastically received in the field of computer-aided language learning (CALL). Common roles of speech recognition in CALL systems fall into four categories.

Interactive: record and playback functions, adding variety to otherwise tedious drills

Quantitative: provide feedback regarding acoustic features such as duration and formants F1/F2

Probabilistic: estimate the likelihood of an acoustic model having produced the acoustic event provided by the speaker

Communicative: incorporate speech with natural language understanding to act as a conversation partner

In an *interactive* context, speech is used to give learners instant and repeated access to their own pronunciations and to those of native speakers that they wish to emulate. Critical issues include monitoring (If learners have full control over the interaction, will they proceed in the way that is most beneficial to them?) and feedback (Without evaluation from a teacher, will learners know what they are doing wrong?) as well as authenticity, individual learning styles, and limitations in the hard-coded processing domain (Garrett, 1995).

At least one of these concerns can be addressed by providing *quantitative* feedback to the users, so that deficiencies and improvements in their speech are clearly visible. Speaking rate and pause frequency are known to have significantly different distributions in native and non-native speech (Mayfield Tomokiyo, 2000) and correlate well with fluency ratings given by speech therapists and phoneticians (Cucchiarini et al., 1998). Eskenazi and Hansma (1998) have found that prosodic features that can be extracted directly from the speech signal are good indicators of fluency and pronunciation quality.

While systems that offer this kind of quantitative feedback without requiring the user to utter isolated phones do need an acoustic model to generate a time-phone alignment, they are not making a statement about the relationship between the learner's speech and the information about native speech contained in the model. Many CALL systems use *probabilistic* output from the acoustic model to derive a pronunciation score. The scores themselves are then evaluated by comparing them to scores given by human listeners; a scoring algorithm is considered effective if it produces scores that correlate well with scores that experienced humans, such as language teachers, give the speakers. Pronunciation scores can also be given at different levels; a sentence-level score would give speakers an idea of how good their overall pronunciation is, whereas a phone-level score would be useful for training articulation of specific phonemes. Metrics used for probabilistic feedback include phone log-likelihood and log-posterior scores (Franco et al., 1997), competing AMs (Kawai, 1999; Ronen et al., 1997), and competing pronunciation models (Auberg et al., 1998).

Using probabilistic scores directly from the ASR engine does carry some risk. It is not clear that speech recognition technology has reached the point at which it can make judgments as to correctness of pronunciation that correspond to human judgments at a satisfactory level (Langlais et al., 1998). Different ASR engines accumulate and report different kinds of statistics, and some may correlate better than others with human judgments (Kim et al., 1997).

Communicative systems address relevance and authenticity concerns about CALL by not only evaluating but also understanding and responding to what the user says. The SUBARASHII Japanese tutoring system (Ehsani et al., 1997) allows beginning learners of Japanese to interact with a fictitious Japanese person to perform simple problem-solving tasks. As the goal of SUBARASHII is not to correct speakers' mistakes but rather to give speakers experience using the language, significant flexibility is allowed at the syntactic and lexical level. Within the context of four constrained situations (one of the situations involves asking whether the fictitious character would like to go see a movie on Friday), the model of acceptable responses from the user is augmented with probable errors and variations in phrasing. This allows users flexibility in what they are allowed to say (correct sentences are not rejected just because they are not exactly what the model predicted), and even with some errors, users are able to interact with the system, just as they would in real life with a human listener.

During recognition, monophone acoustic models are used, and the search is constrained by the response model. It would not be possible to take advantage of these restrictions in a full conversational system, but in a system in which the topic and direction of the conversation can be highly *constrained* (as is often the case in language classrooms), Ehsani et al. found that "meaningful conversational practice can be authored and implemented and that high school students do find these encounters useful." The recognition accuracy for grammatically correct and incorrect utterances that were in the response model were 80.8% and 55.6%, respectively. Recognition accuracy for utterances that were not in the response model was not reported.

9.8.3 Speech-to-Text

The CALL research described above focused not on improving recognition quality but on using speech recognition, in some form or another, to aid language learning. Accurately recognizing heavily accented and poorly formed non-native conversational speech has not been a priority in CALL, perhaps because even with high-quality recognition, analyzing and providing feedback on conversation is very difficult.

In large vocabulary continuous speech recognition (LVCSR), the objective is to improve the system's recognition of the speaker, not the speaker's language skills. The acoustic, lexical, and language models in an LVCSR system can all be adapted to more accurately represent non-native speech, as has been described in earlier sections of this chapter. The better the representation, the better the recognition (or so one would hope).

An early study of non-native speakers in LVCSR focused on Hispanic-accented English (Byrne et al., 1998a). Initial word error rates were extremely high, averaging 73% in an unrestricted-vocabulary task-based test. It is interesting to note how speaker proficiencies were evaluated in this study. An *intermediate* skill level implied only some reading knowledge of English, yet the speakers were expected to answer questions such as "What is going on here" and "What will happen next," requiring nontrivial conversational skills. *Advanced* speakers required solid reading comprehension, and were assumed to be able to participate in an argumentative dialog. It is doubtful that the same correspondence between reading and speaking skills would apply to Japanese speakers. Most Japanese learners of English study the language in Japan before coming to the United States, and can have a high level of competency in reading but an extremely limited ability to carry on a conversation. The correspondence between reading and speaking competencies will be different for different target populations, and the data collection protocol and ultimate system design may benefit from reflecting this.

Studies using more constrained tasks or higher-proficiency speakers have had more success in bringing word error rate to a reasonable level. Witt and Young (1999) have shown that for a simple task, fully-trained source and target language model parameters can be interpolated to form a new set of accent-dependent models that perform well on speakers of different native languages. For high-proficiency speakers and speakers of regional dialects, adaptation using source-language data is effective to the point of being sufficient (Beaugendre et al., 2000; Schwartz et al., 1997), and target-language data may also contribute to WER reductions in some cases (Liu and Fung, 2000).

In designing an LVCSR system for non-native speech, an important consideration is whether the objective is to model non-native speech in general or to focus on a particular L1 group. Another consideration is the size of the lexicon; allowing multiple variants for each word can have an enormous impact on a system with a 100,000-word lexicon. Amdall et al. (2000) collected possible transformations by aligning reference to automatically generated pronunciations and showing how small gains in accuracy for the WSJ non-native speakers can be achieved by pruning the list of word variants based on the probability of the rules invoked for the individual phone transformations. Livescu and Glass (2000) use a similar alignment-to-phone-hypothesis method to derive pronunciation variants for speakers in the JUPITER weather query system. Their objective, like Amdall's, is to model non-native speech in general as opposed to focusing on a particular L1 group. Fung and Liu (1999), on the other hand, concentrate on English spoken by Cantonese native speakers. They use a knowledge-based approach to predict the likely phone substitutions for improved recognition on the Hong Kong TIMIT isolated phone database.

9.8.4 Assessment

If one views LVCSR as the "acceptance" model of non-native speech recognition, where any pronunciation is valid and the system must adapt to it, assessment applications could be considered the "rejection" model. In assessment applications, user speech is compared to a gold standard and given a rating. Unlike CALL applications, the objective is not to interact with the speakers or improve their pronunciation.

Development of ASR-based assessment systems is in its infancy. Automated scoring of language proficiency is very promising in theory. In the United States, the number of non-native speakers of English is growing just as mandates for testing and provisions for special needs make providing fair assessments more critical. Human assessors can be biased, rating speakers higher in intelligibility for L2 accents they have been exposed to, or subconsciously giving lower scores to speakers with an L2 accent that is negatively marked. Human assessors may grow more or less tolerant over time, or they may tire. Automatic assessment offers a remedy for these problems.

The risks of automatic assessment, however, are serious. A poor score on an important test can make educational dreams unreachable or destroy a career. ASR is notoriously sensitive to certain features of speakers' voices that make them "sheep" (recognized well) or "goats" (recognized poorly), with a distribution that corresponds to neither human ratings nor measurable features of speech. There are known discrepancies in recognition accuracy on male and female speech mostly due to unbalanced data. To date, automatic speech recognition has not advanced to the point where it can be an alternative for human scorers on important evaluations of spoken language proficiency.

9.9 Other Factors in Localizing Speech-Based Interfaces

Accounting for variations in local dialects and accents is crucial in the development of multilingual speech-based interfaces for local markets.

Several additional factors related to the characteristics of the user population also influence the quality and effectiveness of these interfaces. In this subsection we discuss two such influences that should be considered in the design of speech interfaces, namely cultural factors, and the specific challenges that prevail in the developing world.

9.9.1 Cultural Factors

The importance of cultural factors in human computer interfaces is well understood (Marcus and Gould, 2000). A number of guidelines have, for example, been developed for the design of culturally appropriate Web sites. Less is known about the role of cultural factors in the design of spoken interfaces, but it is likely that many of the theoretical constructs that have been useful in other aspects of interface design are also applicable when the mode of communication is speech. We discuss two such theoretical constructs, namely Nass's perspective from evolutionary psychology (Nass and Gong, 2000) and Hofstede's "dimensions of culture" (Hofstede, 1997). To illustrate these principles, we summarize an experiment on spoken interfaces that was carried out in rural South Africa.

During the past decade, Nass and collaborators have established a significant body of findings to support the following statement: When humans interact with a speech-based device, their responses are strongly conditioned by human-human communication. Factors such as gender, personality, or level of enthusiasm are perceived as salient in human-human communication. These factors are therefore also surprisingly influential in speech-based systems. For example, Nass and Gong (2000) describe an experimental system that used spoken output to advertise products in an electronic auction. Even though participants professed neutrality with respect to the gender of the voice used and had insight that the speech was electronically generated, they nevertheless acted as if "gender-appropriate" voices were more persuasive. That is, products such as power tools, which are generally associated with male users, were more successfully marketed with a male voice. Similarly, products such as sewing machines were more readily accepted when advertised with a female voice.

Nass and collaborators have documented a range of such influences of generated speech on user behavior (Nass and Brave, 2005). Besides gender, they have studied such factors as emotion, informality, ethnicity, and empathy in spoken content. In each case, significant effects were demonstrated.

Each of these factors is clearly highly variable across different cultures; it is therefore reasonable to expect that the details of the observed effects will also depend on the specific nature of the culture within which the system is used. (Nass and collaborators performed the majority of their research in the United States and have found similar results in Japan [Tan et al., 2003]; however, they do not claim that identical results would be obtained elsewhere.) This argument suggests that the design of spoken interfaces will have a significant cultural dependency, and that appropriate guidelines for interface design should be developed based on norms and expectations within a particular culture.

Some understanding of the expected variability across cultures can be obtained from the work of Hofstede (1997), who used statistical analyses of interviews with employees of a multinational corporation to derive salient differences between cultures. Although the particular setting of these studies limits the generality of the derived cultural differences, the most important dimensions are likely to be essential for interface design (Nass and Gong, 2000).

In particular, Hofstede found the following five variables to be most descriptive of the cultures in his studies:

- Power distance: The extent to which individuals expect there to be an unequal distribution of power within a culture.
- Collectivism versus individualism: The amount of group cohesion that is expected and practiced, as opposed to a focus on individual rights, goals, and achievements.
- Gender distinction: In some cultures, differences in gender roles are strongly maintained, whereas others allow for more of an overlap between the gender roles.
- Uncertainty avoidance: Measures the extent to which different cultures demand informality and structure—that is, cultures with high uncertainty avoidance tend to value conformity and certainty.
- Long-term or short-term orientation: The balance between long-term and short-term goals that is prevalent in a society (which, somewhat counter-intuitively, correlates with the balance between practice and belief [Hofstede, 1997]).

Marcus has shown that these cultural differences can be related fairly directly to different paradigms for graphical user-interface design (Marcus and Gould, 2000). It is less clear how to relate these factors to the design of speech-based interfaces, but the significant overlap between these factors and those identified by Nass suggests that careful consideration is required as well as much additional research.

Some hints on the variable effects of culture on voice interfaces can be gained from recent experiments that investigated the feasibility of telephone-based services in rural South Africa (Barnard et al., 2003). These experiments involved users from a traditional African culture, which is thought to differ significantly from modern Western cultures along most of Hofstede's dimensions. These differences demonstrated themselves in a number of ways, some of which were expected (e.g., the need for clear instructions and the need to establish significant user trust before usage of the system). Other observed differences were less predictable. For example, users found that the prompts were not sufficiently loud; this can possibly be explained by the higher collectivism index of many African cultures, which leads to the expectation that individuals should speak loudly in order to share the conversation with all bystanders. Also, users did not respond to silences after prompts as indications that their responses were expected. Although this is partially explicable in terms of the limited experience that these users have with telephone-based services, the high uncertainty avoidance of their culture probably further increases the need for explicit measures to elicit responses.

Overall, little is known about the role of cultural factors in speech-based interfaces. Theoretical frameworks such as those of Hofstede and Nass suggest a number of useful guiding principles, and Nass's work contains much relevant information for the specific case of speech output in Western societies. Significant progress in this domain is expected in the coming decade.

9.9.2 Multilingual Speech Systems in the Developing World

Given its heritage as a high-technology showcase, it is natural that the field of spoken-interface design should have originated as a tool for literate, technologically sophisticated users in the developed world. Early applications of spoken-dialog systems—such as airline reservations and stock-trading systems—were clearly aimed at such users. It has, nevertheless, long been hoped that such systems would be of particular use in bringing the power of information technology to users in the developing world who have not had the benefit of formal learning and technological mastery (Barnard et al., 2003). To date, this hope has not materialized, due to the many challenges inherent in bringing speech-based interfaces to an extended user population.

There are a number of reasons to believe that speech-based interfaces can play a significant role in bringing the information age to the developing world. On the one hand, such interfaces are accessible through conventional telephone networks. Thus, they can be used in areas where computers and computer networks are not available as long as telephone service is present—and this is true for a rapidly expanding proportion of the globe (Anon., 2005). On the other hand, the fact that such interfaces (potentially) do not require literacy or computer skills is an obvious benefit in countries where many citizens do not have these capabilities.

These potential benefits have long been understood and have played a role in the attention that developing countries such as India and South Africa have given to speech and language technologies. However, surprisingly little work has been done to date to address the practical realities of speech-based interfaces for the developing world. These realities fall into four broad categories, related to the availability of (1) basic linguistic and cultural knowledge, (2) linguistic resources, (3) guidelines for the design of user interfaces, and (4) software tools.

The development of speech-based systems requires a diverse set of location-specific knowledge sources. These include linguistic knowledge about aspects such as phonology, phonetics, prosody, and grammar as well as cultural aspects—for example, an understanding equivalent to Hofstede's indices (listed above). In addition, this knowledge must be codified in a manner that is suitable for the development of technology. For the majority of the world's languages, the basic facts have never been documented, and even for relatively widespread languages (such as the Nguni languages of southern Africa, some of which are spoken by tens of millions of speakers), surprisingly fundamental linguistic issues are not fully resolved (Roux, 1998).

When these hurdles are overcome, developers of speech technologies are faced with the need for linguistic resources. These resources include basic and annotated corpora of text and speech, pronunciation dictionaries, grammars, morphological and syntactic parsers, and the like. In the major languages of the developed world, such resources are often available for free, or can be purchased for a nominal fee from institutions such as LDC and ELRA (see Chapter 3). In stark contrast, these resources do not exist at any price for the vast majority of developing-world languages. Given the diversity of socially important languages in the developing world, it seems as if novel techniques (e.g., highly efficient bootstrapping [Schultz and Waibel, 1997b; Davel and Barnard, 2003]) will be required to address this state of affairs. By embodying these techniques as tools that are easily used by L1 speakers with limited computer literacy, it is possible to tap into the vast pool of linguistic skills that are present in every human population (Davel and Barnard, 2004b). It is thus possible to gather diverse linguistic resources efficiently.

The design of speech-based interfaces for developing-world audiences has received virtually no scientific attention. Aside from the poorly understood cultural factors described above, the specific requirements of users with limited technological and general exposure need further study. These requirements have a number of obvious implications (e.g., the need to avoid terms such as "*menu*" or "*operator*," which are likely to be unfamiliar to such users), but also some that may not be so obvious—for example, the ubiquitous DTMF feedback requires careful design for users who may recognize numbers from their visual shapes rather than knowing them by name.

With respect to software tools, the state of affairs is much more encouraging. The movement toward open-source software has produced tools such as Festival, Sphinx, and HTK, which are readily adapted to a wide range of languages. There is every indication that these tools are playing a key role in enabling the creation of developing-world speech-based interfaces and will continue to do so. The deployment of multilingual systems requires much beyond the capabilities offered by these tools (for example, run-time environments and authoring tools that make it possible to develop applications without detailed knowledge of spoken dialog systems). The same open-source approach is likely to support these requirements as well but is currently less developed. The success of the tools cited is also highly dependent on the availability of linguistic resources, such as pronunciation dictionaries, recorded speech, and basic linguistic knowledge, which cannot be ensured for many of the languages of the world.

Taken together, the obstacles listed above may seem insurmountable. Fortunately, there are a number of examples in which such problems have been solved by a combination of technical and sociological means. Two recent examples are the widespread uptake of cellular telephony, and several successful AIDS-awareness campaigns. In both cases, the critical success factor was the importance of the intervention to the lives of the users by providing dramatically improved communications in the one case, and limiting the spread of a debilitating disease in the other. If spoken interfaces can be shown to have comparable value in the developing world, one can be confident that the hurdles to their development will be overcome.

9.10 Summary

In this chapter, we have examined characteristics of and approaches to automatic recognition of non-native, accented, and dialectal speech. This topic has received comparably little attention in the past. For the real-world use of speech-driven interfaces, systems need to be able to handle speech other than purely native speech; the number of languages for which ASR systems can be built will always be limited, and non-native words will occur in many domains. To date, most ASR systems show significant increase in word error rates if faced with non-native speech.

In order to appropriately model non-native speech, we first have to determine how non-native speech differs from native speech. We have tried to give some insight into these differences and their origins. We have shown that influences on non-native speech are complex, far exceeding the commonly assumed sole influence of the mother tongue.

A particular challenge in non-native and dialectal speech is that a huge number of permutations exist in terms of which language was spoken with the accent of which other language. Detailed investigations are often constrained by the lack of sufficient amounts of non-native or dialectal data.

We have also described how standard methods in native speech recognition, such as specialized acoustic modeling, speaker adaptation, and adaptation of the pronunciation dictionary, apply to the non-native case, and what the particular challenges are for these methods. We have shown that these methods can indeed help non-native speech if they are tailored to this special case.

Apart from the modeling issues described above, we have also investigated how the design of different interfaces needs to be modified in order to increase usability for non-native speakers, who often have particular needs and problems when using automated systems. This is of particular importance for speech-based interfaces in the developing world, where cultural differences need to be taken into account in successfully deploying speech-based interfaces.

To summarize, we can say that there are certainly some approaches that are helping increase the performance of speech recognition systems when faced with non-native and accented speech. On the other hand, many questions are left unanswered, leaving plenty of room for future research.

Chapter 10

Speech-to-Speech Translation

Stephan Vogel, Tanja Schultz, Alex Waibel, and Seichii Yamamoto

10.1 Introduction

Speech-to-speech translation is the task of translating speech input in one language into speech output in another language. This process consists of three basic steps: speech recognition, translation, and speech generation. In this chapter, the main focus is on the translation component and the challenges posed by spoken language input. Discussions of multilingual speech recognition and speech generation aspects can be found in Chapters 4 to 7.

A variety of approaches to speech-to-speech translation have been developed, including interlingua-based, example-based, statistical, and transfer approaches. The first section of this chapter compares interlingua-based and statistical implementations within the framework of the NESPOLE! system developed at Carnegie Mellon University. It discusses decoding strategies allowing for flexible phrase reordering, spoken language specific problems, such as the removal of disfluencies, and coupling of the speech recognition and translation components. The second section of this chapter focuses on example-based and transfer approaches and describes their realization within the spoken language translation system developed at ATR in Japan.

Translation of *speech* (as opposed to text) is greatly complicated by the fact that spontaneously spoken speech is ill-formed and errorful. Speech translation is compounding three sources of errors: errors introduced by the speaker (disfluencies, hesitations), errors of the recognizer, and errors of the translation system. In addition to the need for robust individual components (recognition, translation, synthesis), a successful speech-to-speech machine translation system (SSMT) must also be error-tolerant in its architecture. It must allow for and represent near-miss alternatives at each level of processing, statistically score the likelihood of each hypothesis, and prune and reject unpromising candidates as new knowledge and information is applied at each processing stage.

In order to provide the translation component with multiple hypotheses, the recognition output string is replaced by an output graph of possible near-miss word hypotheses (the "word-lattice"), and the subsequent translation module selects the most likely sentence during its attempt to find a translation. The translation module should also allow for missing or wrong words and provide multiple interpretation results in its attempt to produce an output sentence. Two schools of thought diverge on how to achieve that: a direct approach (here, statistical machine translation) and an Interlingua approach. Both approaches handle ambiguity stochastically and by retaining and evaluating competing alternatives. Both also take advantage of semantic and domain knowledge to subsequently reduce ambiguity and extract useful information from an otherwise ill-formed sentence.

10.1.1 Speech Translation Strategies

In the following sections, we present several speech translation strategies that we have explored at our laboratories. We describe and evaluate each method according to the following criteria:

- First and foremost, spoken language input and translation performance have to be considered.
- Cost of adding a new language: As the number of languages grows, the number of language pairs increases quadratically. Translation based on a language independent intermediate representation (**Interlingua**) alleviates this problem by requiring translation

into the Interlingua only. However, it requires designing the Interlingua and developing analysis and generation modules. More flexible and automatically scalable approaches must be found.

- Cost of development: To achieve good performance, good coverage of the many possible translations from one language to the other has to be achieved. Therefore, if translation models are (semi-)manually designed, each new language and each domain require additional development effort.
- Cost of data collection: Performance is dramatically affected by the amount of available data. The cost of collecting large parallel or tagged corpora (tree banks) increases with the number of languages and/or domains.
- Explicit representation of meaning: It is useful to be able to represent meaning for a system to produce paraphrases and/or to link to other applications in addition to the translator.

10.1.2 Language Portability

Speech translation has made significant advances over the last years with several high-visibility projects focusing on diverse languages in restricted domains (e.g., C-Star, NESPOLE!, Babylon). While speech recognition emerged to be rapidly adaptable to new languages in large domains, translation still suffers from the lack of both hand crafted grammars for Interlingua-based approaches and large parallel corpora for statistical approaches. Both facts prevent the efficient portability of speech translation systems to new languages and domains. We believe that these limits of language- and domain-portable conversational speech translation systems can be overcome by relying more radically on learning approaches using easy/cheap-to-gather data and by applying multiple layers of reduction and transformation to extract the desired content in another language. Therefore, as shown in Figure 10.1 we cascade several stochastic source-channel models that extract an underlying message from a corrupt observed output. The three models effectively translate: (1) speech into word lattices (ASR), (2) ill-formed fragments of word strings into a compact well-formed sentence (Clean), and (3) sentences in one language into sentences in another (MT).

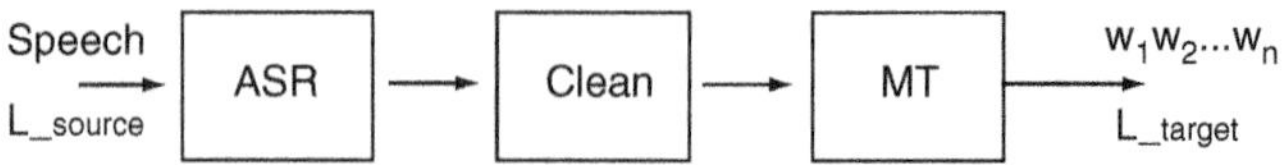

Figure 10.1: Stochastic source-channel speech translation system.

10.2 Statistical and Interlingua-Based Speech Translation Approaches

In response to these considerations, we begin by exploring direct and Interlingua-based translation approaches, as well as modifications on Interlingua-based approaches. While direct approaches can be trained on data, they require $O(N^2)$ parallel corpora if all of the N languages need to be connected. Interlingua-based approaches require fewer translation modules, but they typically require costly design and grammar development.

After a brief description of the Interlingua we introduce statistical machine translation methods. We then explore Interlingua-based statistical methods, interactively learning Interlingua and statistical translation via intermediary natural languages (pivot languages). Furthermore, we discuss the integration of speech recognition and translation modules into integrated speech translation systems. These integration methods are critical to scale speech translation systems to domain unlimited performance. Central aspects are cleaning disfluencies in speech and dealing with recognizer errors by translating not only the first best recognition result but all paths in the word lattice generated by the speech recognition system. Last but not least, we present a prototype of an integrated domain unlimited speech translation system.

10.2.1 Interlingua-Based Speech Translation

Interlingua-based MT analyzes a sentence into a language independent semantic representation using an analysis grammar, and generates the target sentence using a generation grammar. Therefore, building a domain-specific translation system first requires designing an Interlingua that is rich enough to capture the semantics of the domain. Second, analysis and generation grammars need to be developed.

Interlingua Design

The Interlingua, as for example used in the NESPOLE! system, is called **Interchange Format** (IF) (Levin et al., 1998). The IF defines a shallow semantic representation for task-oriented utterances that abstracts away from language-specific syntax and idiosyncrasies while capturing the meaning of the input. Each utterance is divided into semantic segments called semantic dialog units (SDUs), and an IF is assigned to each SDU. An IF representation consists of four parts: a speaker tag, a speech act, an optional sequence of concepts, and an optional set of arguments. The representation takes the following form:

speaker: speech act + concept* (argument*)

The speaker tag indicates the role of the speaker in the dialog, for example, "agent" and "client" in a hotel reservation dialog. The speech act captures the speaker's intention—for example, getting information or confirming. The concept sequence, which may contain zero or more concepts, captures the focus of an SDU—for example, getting information about the price of the hotel room. The speech act and concept sequence are collectively referred to as the domain action (DA). The arguments use a feature-value representation to encode specific information from the utterance (e.g., a double room, a nonsmoker). Argument values can be atomic or complex. Very simple and more complex examples of utterances with corresponding IFs are shown here:

On the twelfth we have a single and a double available.
a:give-information + availability + room
(time = (md12), room-type = (single + double))

Thank you very much.
c:thank

The NESPOLE! Interlingua is based on representing the speaker's intention rather than the literal meaning of the utterance. The design of an Interlingua has to balance expressive power and simplicity. The inventory of domain actions needs to be sufficiently expressive in order to capture speaker intention. The specification at the argument level attempts to distinguish between domain-dependent and domain-independent sets of arguments, to better support portability to new domains. The Interlingua also has to be simple and straightforward enough so that grammar developers can independently work on different languages at different sites without the need of constant information exchange.

Semantic Grammars

The advantage of semantic grammars is that the parse tree that results from analyzing an utterance is very close to its final semantic interpretation (as opposed to, say, the laborious step of transforming syntactic constituent structures into semantic functional structures). A disadvantage however, is that a new grammar has to be developed for each domain, although some low-level modules—such as those covering time expressions—can be reused across domains. Once an utterance comes in, it is run through a parser that, together with the grammar in question, produces an analysis in the form of a parse tree. A statistical parser such as SOUP (Gavalda, 2004) is especially suitable for handling spoken language because it represents semantic grammars as probabilistic recursive transition networks. Thus, the search for the optimal interpretation is easily implemented as a beam search, with a scoring function that maximizes the coverage (number of words parsed) and likelihood (sum of arc probabilities) of an interpretation, but minimizes its complexity (number of nonterminals in parse lattice) and fragmentation (number of parse trees per utterance). Moreover, SOUP allows for skipping of words (with a penalty), character-level parsing (morphology rules can be defined using the same formalism as rules operating at the word level), multiple-tree interpretations (interpretations can be a sequence of top-level parse trees covering nonoverlapping fragments of the utterance), and dynamic modifications (rules can be modified on the fly), which enables the interactive grammar learning described below.

Interactively Learning Semantic Grammars

One of the greatest difficulties in the successful deployment of natural language processing (NLP) systems lies in the inherent richness of human language: it is impossible to capture a priori all the different ways in which people may choose to express a particular idea. Therefore, NLP

> **do I have any mail**
I understand “do I have any mail”
> **arrange by recency**
I don’t understand right away what you mean but let me guess
“arrange by recency” is a way to express
1. count mail, e.g., “count”
2. list mail, e.g., “list”
3. sort mail, e.g., “sort”
0. None of the above
> **sort**
“recency” is a way to express
1. sort by size, e.g., “size”
2. sort by date, e.g., “date”
3. sort by sender, e.g., “sender”
0. None of the above
> **by date**
Thanks for teaching me the meaning of “arrange by recency”
I understand “arrange by recency”
> **Please arrange messages from Bob by recency**
I understand “please arrange messages from Bob by recency”

```
...
public <sortMail> = <_SORT>
   <_MAIL_ARGUMENT>  [<_SORT_MODE>]
   [<_SORT_BY>] [ <_SORT_BY>  <_SORT_MODE>]:
...
<_SORT> = [please] (sort | arrange)

<sortBY_date> = date | time | recency:
...
```

Figure 10.2: Dynamic grammar acquisition as by-product of clarification dialogs using the GSG system on an e-mail client application. The left panel shows the dialog maintained between the user (preceded by “>”) and the system; the right panel shows the dynamic extension of the underlying grammar, particularly the acquisition of the meaning of “arrange” and “recency.”

systems that are capable of learning the meaning of extra-grammatical utterances—that is, capable of dynamically extending the coverage of their underlying grammars—become of paramount importance. An example is GSG (Gavalda, 2000), a system that extends the semantic grammar of a particular domain simply by asking clarification questions to the nonexpert user of the application in question (see Figure 10.2).

This is accomplished by augmenting the initial grammar using external knowledge sources (such as a part-of-speech tagger or a shallow syntactic grammar) in combination with various machine learning strategies within a coherent, mixed-initiative conversation with the end user. The result is the acquisition of new grammar rules. Then, a sophisticated rule management scheme—which includes detection of rule subsumption and rule ambiguity, vertical generalization (bottom-up generalization along ontological IS-A links), and horizontal generalization (making certain right-hand-side constituents optional and/or repeatable)—is employed to add the new rules to the current grammar, seamlessly and without disrupting analyses that are already correct.

10.2.2 Statistical Direct Translation

Statistical machine translation (SMT) was proposed in the early 1990s by the IBM research group (Brown et al., 1993b) and has since grown into a very active research field. Its key advantage is that it can be automatically trained from large corpora. The approach is based on Bayes' decision rule: Given a source sentence $\mathbf{f} = f_1^J$ of length J, the translation $\mathbf{e} = e_1^I$ is given by:

$$\hat{\mathbf{e}} = \arg\max_{\mathbf{e}} p(\mathbf{e}|\mathbf{f}) = \arg\max_{\mathbf{e}} p(\mathbf{f}|\mathbf{e})p(\mathbf{e}). \tag{10.1}$$

Here, $p(\mathbf{e}) = p(e_1^I)$ is the language model of the target language, typically a trigram language model, and $p(f_1^J|e_1^I)$ is the **translation model**. The argmax operation denotes the search problem.

Word Alignment Models

A number of different translation models, also called alignment models, have been described (Brown et al., 1993b; Wu, 1995; Vogel et al., 1996; Och and Ney, 2000). The most simple **word alignment** models are based only on word co-occurrence statistics. In Brown et al. (1993b), this is the first in a sequence of models:

$$p(\mathbf{f}|\mathbf{e}) = \frac{1}{I^J} \prod_{j=1}^{J} \sum_{i=1}^{I} p(f_j|e_i). \tag{10.2}$$

This is the so-called **IBM1 model**. The **IBM2 model** also includes position-alignment probabilities $p(i|j,J,I)$, resulting in:

$$p(\mathbf{f}|\mathbf{e}) = \prod_{j=1}^{J} \sum_{i=1}^{I} p(f_j|e_i)p(i|j,J,I). \tag{10.3}$$

As an alternative to the IBM2 model, which is based on absolute positions, the so-called **HMM alignment model** has been formulated (Vogel et al., 1996), which uses relative positions to model word reordering between two languages:

$$p(\mathbf{f}|\mathbf{e}) = \sum_{a_1^J} \prod_{j=1}^{J} p(a_j|a_{j-1}, I) \cdot p(f_j|e_{a_j})$$

Here, a_j denotes the alignment to the j'th word in **f**.

Phrase Alignment Models

A simple approach to extract phrase translations from a bilingual corpus is to harvest the Viterbi path generated by a word alignment model. The phrase alignment is then essentially a post-processing step following the word alignment. For any word sequence in the source sentence, the aligned words in the target sentences are taken from the Viterbi alignment. The smallest and the largest index on the alignment side are then the boundaries for the entire target phrase aligned to the source phrase.

Many word alignment models are not symmetrical with respect to source and target language. To make up for the asymmetry of the word alignment models, training can be done in both directions: source to target, and target to source. This results in two Viterbi paths for each sentence pair. Different ways have been explored to combine the information given by those alignments. Och and Ney (2000) described experiments using intersection, union, and a combination along with some heuristic rules. Koehn et al. (2003) studied different combination schemes and concluded that using the right one has a bigger impact on the resulting performance of the translation system than the underlying word alignment model.

Some alternative phrase alignment approaches have been developed, which do not rely on the Viterbi word alignment. Both Marcu and Wong (2002) and Zhang et al. (2003b) consider a sentence pair as different realizations of a sequence of concepts. These alignment approaches segment the sentences into a sequence of phrases and align those phrases in an integrated way.

In Vogel et al. (2004) phrase alignment is described as sentence splitting, as shown in Figure 10.3. The goal is to find the boundaries for the

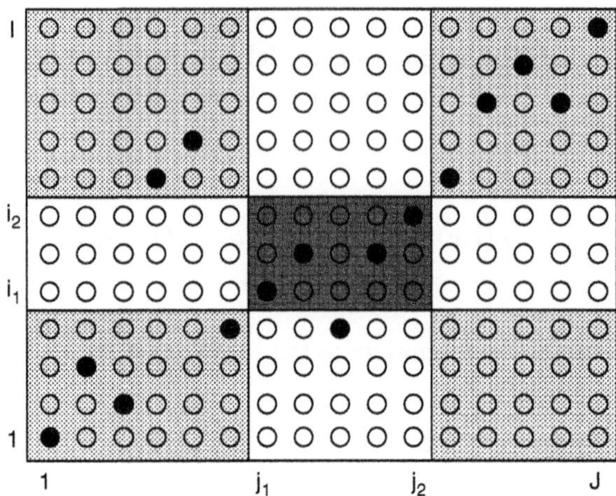

Figure 10.3: Phrase alignment as sentence splitting.

target phrase, given some source phrase. This can be done with a modified IBM1 alignment model:

- For words inside the source phrase, we sum only over the probabilities for words inside the target phrase candidate; for words outside of the source phrase, we sum only over the probabilities for the words outside the target phrase candidate.
- The position alignment probability—which for the standard IBM1 alignment is $1/I$, where I is the number of words in the target sentence—is modified to $1/(k)$ inside the source phrase and to $1/(I-k)$ outside the source phrase.

This leads to the following calculation:

$$p_{i_1,i_2}(\mathbf{f}|\mathbf{e}) = \prod_{j=1}^{j_1-1} \sum_{i \notin (i_1..i_2)} \frac{1}{I-k} p(f_j|e_i) \times \prod_{j=j_1}^{j_2} \sum_{i=i_1}^{i_2} \frac{1}{k} p(f_j|e_i) \times \prod_{j=j_2+1}^{J} \sum_{i \notin (i_1..i_2)} \frac{1}{I-k} p(f_j|e_i) \tag{10.4}$$

The target side boundaries i_1 and i_2, which give the highest alignment probability, are chosen by:

$$(i_1, i_2) = \arg\max_{i_1, i_2} \{p_{i_1,i_2}(\mathbf{f}|\mathbf{e})\}$$

Similar to $p_{i_1,i_2}(\mathbf{f}|\mathbf{e})$, we can calculate $p_{i_1,i_2}(\mathbf{e}|\mathbf{f})$—now summing over the source words and multiplying along the target words.

To find the optimal target phrase, we interpolate the log probabilities and take the pair (i_1, i_2) that gives the highest probability.

$$(i_1, i_2) = \arg\max_{i_1, i_2} \{(1 - c)\log(p_{(i_1,i_2)}(\mathbf{f}|\mathbf{e})) + c \cdot \log(p_{(i_1,i_2)}(\mathbf{e}|\mathbf{f}))\}$$

The Decoder

A decoding approach that is based on

$$\hat{e} = \arg\max_{e} p(\mathbf{e}|\mathbf{f}) = \arg\max_{e} p(\mathbf{f}|\mathbf{e})p(e) \tag{10.5}$$

requires that, for any given word, its preceding word history is known. This leads to a search organization that constructs the target sentence in a sequential way. However, to incorporate the different word order of various languages, the words in the source sentence have to be covered nonsequentially while the translation is being generated. The most general form is to allow for any permutation of the source words—that is, without restrictions on the possible reordering. Such a search organization, restricted to word-based translation, has been described in Nießen et al. (1998). However, this leads to a very large search space and high computational costs. Therefore, various restrictions on reordering have been proposed.

The approach described here allows for phrase-to-phrase translation. Decoding proceeds along the source sentence. At each step, however, the next word or phrase to be translated may be selected from all words or phrases starting within a given look-ahead window from the current position. The decoding process works in two stages: First, the word-to-word and phrase-to-phrase translations and, if available, other specific information like named-entity translation tables are used to generate a translation lattice.

Second, a standard N-gram language model is applied to find the best path in this lattice. It is during this search step that reordering has to be taken into account. Both steps will now be described in more detail.

We define a transducer as a set of translation pairs generated by the methods described above as well as alternative knowledge sources, such as manual dictionaries and named entity lists. Each translation pair is given as a quadruple:

label # source words # target words # probability

For decoding, the transducers are organized as trees over the source side, and the translations are attached to the final nodes. This allows for efficient processing, since a node in the transducer represents all source phrases that consist of the words along the path to this particular node and include all possible paths that lead to final nodes of this particular node's subtree.

The first step in the decoding process is to build a translation lattice by applying the transducers. We convert the sentence to be translated into a lattice structure, in which the nodes are the positions between the words, and the words are attached to the edges. The nodes ν are numbered from 0 to J, the length of the source sentence. We also use ν to simply denote the node number.

To search for matching phrases, we encode the relevant information in a hypothesis structure

$$h = (\nu_1, \nu_2, \sigma, h_p, \varepsilon),$$

which means that starting from node ν_1 and ending in node ν_2, a sequence of words has been found that corresponds to a transducer path from state σ_0 to state σ, and whereby in the last step, the hypothesis h_p has been expanded over edge ε.

The matching process between a path through a transducer and a segment of a sentence can start at all positions in the sentence. Therefore, an initial hypothesis $(\nu, \nu, \sigma = \sigma_0, h_p = \emptyset, \varepsilon = \emptyset)$ is set for each node except the final node in the lattice.

Expanding hypotheses is structured in the following way: Let ν be a node in the translation graph, and $E(\nu)$ be the set of incoming edges for this node. Let $\nu^s(\varepsilon)$ denote the start node of an edge ε. Then, for each incoming edge $\varepsilon \in E(\nu)$, all hypotheses h_p in $\nu^s(\varepsilon)$ are expanded with the word f attached to ε. That is to say, if σ_p is the transducer state of hypothesis h_p, then σ is the transducer state that can be reached from σ_p over the transition labeled f. If expansion is possible, then a new hypothesis h is generated:

$$h_p = (\nu_1, \nu^s(\varepsilon), \sigma_p, h', \varepsilon') \rightarrow h = (\nu_1, \nu, \sigma, h_p, \varepsilon).$$

If expanding a hypothesis leads into a final state of the transducer, a new edge is created and added to the translation lattice. The new edge is labeled with the category label taken from the transducer. The additional information stored with this edge is the translation and the sequence of edges traversed, which corresponds to the sequence of source words.

Once the complete translation lattice has been built, a one-best search through this lattice is performed. In addition to the translation probabilities, or rather translation costs, as we use the negative logarithms of the probabilities for numerical stability, the language model costs are added and the path that minimizes the combined cost is returned.

The search for the best translation hypothesis involves generating partial translations and expanding them until the entire source sentence has been accounted for. The information accumulated during search is stored in the following hypothesis structure:

$$h = (Q, C, \Lambda, i, h_p, \varepsilon),$$

where Q is the total cost; C, the coverage information; Λ, the language model state; i, the number of the words in the partial translation; and h_p and ε, the trace-back information. In the case of a trigram language model, the language model state comprises just the last two words of the partial translation—that is, the history in the next expansion step.

To allow for reordering, we organize the search in the following way. Assume we have a partial translation, which already covers c words of the source sentence, $n < c$ of which are the first words of the sentence. (In other words, the initial section of the sentence has already been completely translated, the remainder only partially.) To expand this partial translation, we have to extend it over one of the edges in the translation lattice that corresponds to one of the remaining untranslated source words. We allow for phrases—that is, longer edges in the translation lattice. It can be the case, therefore, that such an edge spans over some words that have already been covered. This causes a collision, and so an expansion over such an edge is not possible.

Reordering is now restricted to be within a window of given size. That is to say that the next word to be translated has to be taken from positions $n <= j <= n + d$, where d is the size of the reordering window. In terms of nodes: if ν_1 is the node with number n and ν_2 is the node with number $n + d$, then expansion is restricted to edges starting from nodes $\nu_1 <= \nu' <= \nu_2$. With $d = 0$, there is no reordering; therefore, decoding is monotone.

Expansion of hypotheses is organized according to overall coverage—that is, the number of words already translated. So we start with coverage zero and expand until we have reached coverage J, where J is the number of words in the source sentence. At the sentence end, the language model probability for the sentence end is applied. In addition, a sentence-length model can be used. The best hypothesis is then used to trace back and collect the actual words generated along this path.

To reconstruct the path taken through the translation lattice, we need to store additional back-pointer information. Traveling back using these pointers allows us to generate the actual sequence of words. The back-pointer information consists of the edge that was traversed during the last expansion and the pointer to the predecessor of the current hypothesis.

10.2.3 Statistical Translation Using a Formal Interlingua

Interlingua-based translation as described above requires, in addition to the design of the interlingua, the development of handwritten (or interactively learned) semantic grammars—analysis grammars for each input language and generation grammars for each output language. Here we describe a method to automatically train a semantic mapping between source text and the tree-structured interlingua, which replaces the analysis grammar. We show that this can be done given a corpus of semantically tagged data (from source language to IF).

A Language Model for Trees

In the usual situation, where $\mathbf{e} = (e_1, \ldots, e_l)$, language modeling is typically based on the decomposition

$$p(\mathbf{e}) = \prod_{i=1}^{l} p(e_i | e_1, \ldots, e_{i-1}),$$

where the conditional probability $p(e_i | e_1, \ldots, e_{i-1})$ is approximated by the relative frequencies of N-grams seen in the training corpus. While $\mathbf{e}$ may in this case be defined as some token e together with a subsequence $\mathbf{e}'$, a tree $\mathbf{e}$ may be defined as consisting of some token e together with a set of $a \geq 0$ subtrees $\mathbf{e}_1, \ldots, \mathbf{e}_a$ (a is the arity of the tree). This leads to the decomposition

$$p(\mathbf{e}) = p(e | \mathbf{e}_1, \ldots, \mathbf{e}_a) \cdot \prod_{i=1}^{a} p(\mathbf{e}_i | \mathbf{e}_1, \ldots, \mathbf{e}_{i-1}),$$

which corresponds to a bottom-up decoding in the order $\mathbf{e}_1, \ldots, \mathbf{e}_a, e$.

It is a special feature of the IF that the ordering of subtrees is unimportant for the semantics they cover—that is, the term $a(b, c)$ is semantically equivalent to $a(c, b)$. This justifies the assumption that the probabilities $p(\mathbf{e}_i | \mathbf{e}_1, \ldots, \mathbf{e}_{i-1})$ are independent of $\mathbf{e}_1, \ldots, \mathbf{e}_{i-1}$, given the recursive formula

$$p(\mathbf{e}) = p(e | \mathbf{e}_1, \ldots, \mathbf{e}_a) \cdot \prod_{i=1}^{a} p(\mathbf{e}_i),$$

in which $p(\mathbf{e}_i)$ is to be decomposed further in the same way as $p(\mathbf{e})$. To approximate $p(e | \mathbf{e}_1, \ldots, \mathbf{e}_a)$ with relative frequencies, "tree-N-grams" are used. As Figure 10.4 shows, these N-grams use only the roots of the subtrees.

Translation Models for Trees

As described previously, the standard translation models use the concept of word alignment: each word in the source sentence is aligned to a word

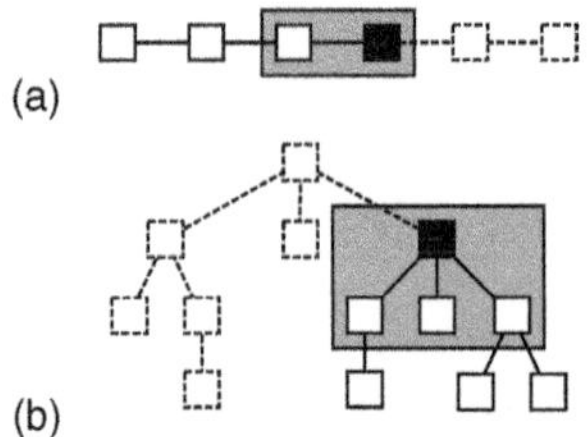

Figure 10.4: The concept of N-grams (a) in sequences (b) in trees.

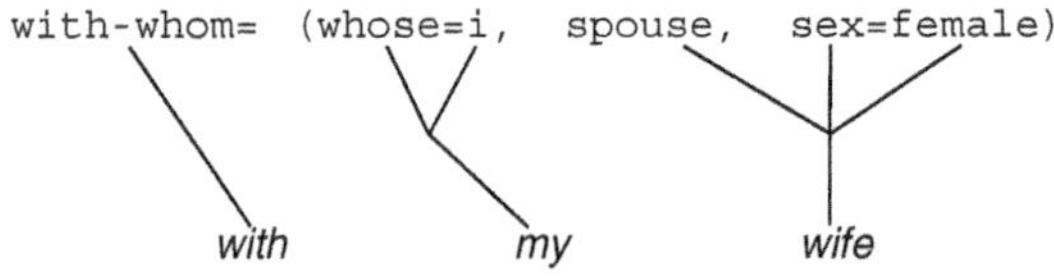

Figure 10.5: An alignment between an English phrase and its corresponding IF representation.

in the target language. Words that have no correspondence in the target sentence are aligned to the so-called empty word added to the sentence at position 0. This concept of alignment can also be used when translating into IF, as illustrated in Figure 10.5.

The IBM1 and IBM2 translation models proposed in Brown et al. (1993b) can be generalized to the case in which the **e**'s are trees. For the IBM1 translation model, this is straightforward because the model makes no assumptions that are specific to sequences. In the model's formula

$$p(\mathbf{f}|\mathbf{e}) = \frac{1}{(I+1)^J} \prod_{j=1}^{J} \sum_{i=0}^{I} p(f_j|e_i),$$

we just need to assign an index to each of the I nodes in the tree **e** in some arbitrary way.

The IBM2 model also includes alignment probabilities $p(i|j,I,J)$, resulting in the estimation formula

$$p(\mathbf{f}|\mathbf{e}) = \prod_{j=1}^{J} \sum_{i=0}^{I} p(f_j|e_i) p(i|j,J,I)$$

in the sequential case. The idea is that the jth token of the source sentence $\mathbf{f}$ is aligned to the ith token of the target sentence $\mathbf{e}$ with probability $p(i|j,J,I)$ provided that J is the length of $\mathbf{f}$ and I is the length of $\mathbf{e}$.

Now, indexing the nodes of a tree is no longer arbitrary because it affects the values of $p(i|j,J,I)$. Using again the fact that the IF is commutative, it seems appropriate to consider only the depth of a particular node. Its position within the level does not contribute any information. Taking d as the depth of $\mathbf{e}$ and $\#e_i$ as the number of nodes in level i, this leads to

$$p(\mathbf{f}|\mathbf{e}) = \prod_{j=1}^{J} \sum_{i=0}^{d} \left(\frac{p(i|j,J,d)}{\#e_i} \sum_{e_i} p(f_j|e_i) \right)$$

where the e_i in the right sum run over all nodes of level i.

Decoding Trees

For generating $\mathbf{e}$ from $\mathbf{f}$ given a language model $p(\mathbf{e})$ as well as a translation model $p(\mathbf{f}|\mathbf{e})$, a stack decoder similar to the one described in Wang and Waibel (1997) is used but adapted to generate trees rather than linear sequences. To cope with the huge search space, "bad" hypotheses are pruned after each iteration.

The algorithm starts with an empty hypothesis. In the case of linear sequences, new hypotheses are generated by iteratively appending new target words to existing hypotheses. The tree decoder takes a *set* of existing hypotheses $\mathbf{h}_1, \ldots, \mathbf{h}_n$ and forms a new tree, with the $\mathbf{h}_i$ as subtrees and an additional target token as its root. If the algorithm is restricted to choose only sets of size 1, it reduces to the sequential version.

Generating trees gives a much larger search space than generating linear sequences. In fact, while n hypotheses and k words to append lead to $n \cdot k$

new hypotheses for the next iteration of the sequence decoder, the tree decoder generates $2^n \cdot k$ new hypotheses because a set of size n has exactly 2^n subsets.

Three methods are applied to reduce the search space. First, hypotheses that are not legal terms according to the IF specification are not generated. Second, the branching factor of generated trees is restricted to three, the depth to four. Finally, standard pruning is used. In less than 5% of the test sentences used in our experiments, the decoder generated an IF that had a lower score than the reference IF, indicating that the number of search errors due to pruning was small.

10.2.4 Using English as Interlingua

Interlingua-based translation systems and statistical translation systems are both well-known approaches with inherent advantages. However, in most systems, only one of the approaches can be fully implemented, at the expense of the other. Using English as interlingua tries to combine the advantages of a system with an explicit Interlingua and the advantages of a pure data-driven system. The Error-Driven Translation Rule Learning (EDTRL) system overcomes existing limitations by (1) avoiding the need for an explicit handcrafted interlingua specification and (2) tackling the "Parallel Data Sparseness Problem" that statistical machine translation faces for unusual or low-resources language pairs.

While translation from and to English has made significant progress partly due to large parallel corpora, the situation drastically changes if non-English language pairs are supposed to be translated into each other. The amount of parallel text corpora is much smaller than the parallel text corpora from each of these languages paired with English. The intuitive solution to this problem is to cascade two translators using English as an intermediate language. However, the pure cascading of two machine translation systems using the output of the first as input to the second results in a multiplication of translation errors and therefore in a significantly higher error rate compared to each translation step. The reduction of this multiplication in translation errors should be achieved by introducing a suitable interlingua and appropriate training and decoding methods. This is the focus of the EDTRL approach.

Standardized and Simplified English as Interlingua

Using ordinary English makes it more difficult to take advantage of formal aspects of an interlingua. To take this into account and to improve the cascaded translation, the intermediate English is transformed to a standardized and simplified form. The standardization step maps alternative expressions with similar or equal meanings to the most commonly used alternative. Sometimes English utterances have some freedom in word order without changing the main meaning of the utterance. To obtain a consistent word order in such cases, a reordering step can be applied, as in the following:

> please give me ... → give me ... please

Furthermore, the sentence structure is simplified (SimplifiedEnglish, 2004), with more complex, rarely used tenses being replaced by easier ones:

> He had spoken. → He spoke.
> He would be speaking. → He would speak.

Although these kinds of simplifications do remove information, often such fine nuances are of little value to the quality of the translation given the current performance of MT systems. In most cases, the translation profits from the transformations through more reliable alignments and better utilization of the training data. Even humans can benefit from Simplified English in some technical domains (AECMA, 2004).

Linguistically Enhanced English as Interlingua

Besides standardizing and simplifying the intermediate English, adding further information to the structure and the semantic content of a sentence can be helpful for the second translation step. To be independent of the source and target language and to minimize manual work, additional knowledge sources should only use information that can be automatically obtained from parallel text or be derived from the intermediate representation based on English as interlingua.

We examined the incorporation of the following additional knowledge sources to provide additional information for the translation process:

- Morphological Analyzer: Starting from the WordNet ontology (Miller et al., 2005), we built a system to analyze English word forms and determine its base forms and derivation rules. The analyzer contains a set of common transformation rules and an even larger list of exceptions from these rules. In the current implementation, each word is analyzed without using its context or information from former sentences. The precision for finding the base class is 95%, while the determination of the derivation rules is not yet that good.
- Sense Guesser: The sense guesser tries to find the sense of a word. Many words have different meanings depending on the context in which they occur—for example, "table" can have the senses "desk" or "chart." Often the context of the word can be used for disambiguation. In our example, the context "in the" assigns "table" to the chart class, while "on the" assigns it to the desk class. We used the sense hierarchy from WordNet.
- Synonym Generator: WordNet also lists synonyms for words, all within a well-structured and linked hierarchy. Both the sense guesser and synonym generator only use open word classes like nouns, verbs, adjectives, and adverbs.
- Part-of-Speech Tagger: A statistical part-of-speech tagger is used to provide POS-tags. The tagger uses the tag set described in Brill (1995) and trained on the tagged Brown corpus.
- Named Entity Tagger: Handwritten rules are used to find named entities, which often need to be treated in a special way.

Further knowledge sources such as sentence type, active or passive voice, politeness, domain, and category could also be added.

Connecting the Translation Steps

The translation errors from intermediate English to the target language can be reduced if not only the best hypothesis but also additional information from the search is used. We have examined the following methods:

- N-best list of complete translations: The translation system produces up to N alternative translation hypotheses and passes them to the second translation step. The number of hypotheses has to be kept small to guarantee fast overall decoding, thereby allowing only for little variability. This approach did not improve the translation in our experiments.

- N-best word or phrase alternatives to the best hypothesis: This method selects the single best hypothesis from the first translation step, but augments it by adding alternative words or phrases that have high translation probabilities. This strategy results in a noticeable improvement in the translation performance.
- Full lattice: In order not to fix one translation hypothesis as the basis for constructing these alternatives, we can also pass on full translation lattices. This method has the highest potential because it keeps all promising alternatives. But without pruning, this approach increases the search space considerably. Using a lattice as input for the second translation step has been shown as the most profitable way to use translation alternatives to improve the translation quality.

Besides the information about alternatives and additional knowledge sources their probability or confidence measure can be part of the interlingua. Therefore words, phrases, and their alternatives carry probabilities as well as attributes and classes. All this information together with the formalization step forms the interlingua, which allows improvement of the cascaded translation.

10.2.5 Comparing the Translation Strategies

In this section, we evaluate the different translation strategies on two speech-to-speech translation tasks. In the first part, we compare the grammar-based system to the statistical interlingua-based system and to the direct statistical approach. (The experiments were carried out using travel planning dialogs from the NESPOLE! project [Lavie et al., 2001a].) In the second part, a comparison between the Error-Driven Translation Rule Learning (EDTRL) system and the direct statistical system on the BTEC (Kikui et al., 2003) corpus is presented.

Interlingua versus Direct Statistical Translation

Dialogs in the travel planning domain have been collected, transcribed, and annotated with IF representations (Lavie et al., 2001a). From this

Table 10.1 Corpus statistics for the NESPOLE! training and test set.

	Training		Test	
Language	Ger	Eng	Ger	Eng
Sentences	2,427	2,427	194	194
Tokens	11,236	11,729	889	955
Vocabulary	1,196	1,010	269	241
Singletons	566	429	152	123

Table 10.2 Scores of the translations generated by systems *IL*, *SIL*, and *SMT*; the values are percentages and averages of four independent graders.

	IL	*SIL*	*SMT*
Perfect	18.9	15.1	40.3
Okay	36.3	30.2	22.7
Bad	44.8	54.7	37.0
Acceptable	**55.2**	**45.3**	**63.0**

database, we extracted a trilingual corpus of about 2,500 triples German-English-IF as a training set. One hundred ninety-four German sentences were held out to be used as a test set. Detailed corpus statistics are given in Table 10.1.

For each German test sentence, three IF representations were generated using (1) the grammar-based system (*IL*); (2) the statistical system (*SIL*), with a model trained on German/IF; and (3) the direct statistical translation system (*SMT*). For the interlingua-based systems, the IF expressions were converted into English using the same IF-to-English generation grammar.

The translations from the different systems were then presented to six human evaluators. Each translation was assigned one of three grades: "perfect" (the translation is semantically complete and grammatically correct); "okay" (the main part of the original semantics is covered and expressed understandably); "bad" (otherwise). The "perfect" and "okay" translations form the class "acceptable."

The evaluation results are given in Table 10.2. The statistical IL system does not perform quite as well as the grammar-based system. Given the very small training corpus, with about 40% of all words seen only once during training, this is not surprising, on the contrary, the results show the potential of the proposed approach. However, the direct statistical system—not using any syntactic structure information beyond what is implicit in the phrase-to-phrase alignments—outperformed the grammar-based system, despite the small training corpus and the significant amount of effort that had been put into the development of the grammars.

EDTRL versus Direct Statistical Translation

In the following experiments, we first evaluate the concept of English as an interlingua, and then compare this to the direct statistical translation.

To evaluate the concept of English as an interlingua, we chose Chinese as the input language and Spanish as the output language, since, in spite of the widespread use of these languages, comparatively few direct Chinese-Spanish translations are available. We trained the EDTRL system for Chinese to English ($C \rightarrow E$), English to Spanish ($E \rightarrow S$), and Chinese to Spanish ($C \rightarrow S$). We then cascaded the $C \rightarrow E$ and $E \rightarrow S$ systems by simply feeding the output of the former into the latter. The translation was then done on the same test set using the full definition of an augmented, formalized version of English as an intermediate step. Additionally, we trained a statistical MT system on the same language pairs and cascaded the $C \rightarrow E$ and $E \rightarrow S$ translations to generate a $C \rightarrow E \rightarrow S$ translation in comparison to a direct $C \rightarrow S$ translation. To evaluate the translation quality, we used the NIST standardized tool for benchmark evaluations (MTeval) in version 10 (NIST, 2000). For comparison, we give also the results for Systran's publicly available online machine translation system.

The SMT system and the EDTRL system both use the same bilingual training corpus, while the EDTRL system uses additional dictionaries for initialization. An additional difference is the handling of punctuation. While EDTRL ignores punctuation marks, SMT treats them as normal words. In the reported experiments, the EDTRL system does not make use of the sense guesser, the named entity tagger, and full lattice. The data for these experiments were taken from the Basic Travel Expression Corpus (BTEC), a multilingual collection of conversational phrases in the travel domain (Kikui et al., 2003) as briefly introduced in Section 10.4.2. Table 10.3 shows the training and test material for Chinese, English, and Spanish phrases. Since only a subset of 6,027 phrases was available for

Table 10.3 Training (test in parentheses) corpora.

Train (Test)	Chinese	English	Spanish
sentences	162,316 (506)	162,316 (506)	6,027
-unique	96,074 (497)	97,500 (503)	5,934
-avg. length	7.0 (7.3)	7.5 (7.5)	9.8
words	1,134,417 (3681)	1,216,207 (3779)	58,834
vocabulary	13,793 (954)	16,224 (843)	4,651
-singletons	4,745 (590)	6,705 (523)	2,370
-unseen	(29)	(22)	

Spanish, the training data for the E $\rightarrow$ S and C $\rightarrow$ S systems was reduced to the corresponding parallel phrases. The scores were calculated based on 16 English and 3–4 Spanish reference translations.

The first four lines in Table 10.4 give NIST scores of direct translations from the source to the target language. For the direct translation, the EDTRL system does not use any interlingua. While the lower rows refer to an internal evaluation of a preliminary version of the systems (01/2004), both systems were lately compared among several systems in an official evaluation (IWSLT, Kyoto Japan, August 2004). Based on the NIST score of the unlimited Chinese-English track, the SMT system came in first and the EDTRL system came in second.

The second and third column of Table 10.4 show the comparison between EDTRL and direct statistical translation. As the fifth line shows, the cascaded EDTRL system outperforms the directly trained systems. Using augmented and formalized English as interlingua yields to further improvements of the EDTRL system (last line of Table 10.4).

Table 10.4 NIST scores for translation from Chinese to Spanish.

Translation tasks	EDTRL	SMT	Systran
C $\rightarrow$ E (IWSLT eval 08/2005)	7.5	9.56	-
C $\rightarrow$ E (internal eval 01/2004)	7.34	7.35	5.74
E $\rightarrow$ S	5.17	4.57	6.06
C $\rightarrow$ S	3.17	3.04	-
C $\rightarrow$ E $\rightarrow$ S	3.41	2.60	2.84
C $\rightarrow$ E_{IL} $\rightarrow$ S	3.69	-	-

10.3 Coupling Speech Recognition and Translation

Due to the peculiarities of spoken language, an effective solution to speech translation cannot be expected to be a mere sequential connection of automatic speech recognition (ASR) and machine translation components but rather a coupling between both. This coupling can be characterized by three orthogonal dimensions: (1) the complexity of the search algorithm, (2) the incrementality, and (3) the tightness, which describes how close ASR and MT interact while searching for a solution (Ringger, 1995). The benefits and drawbacks have been widely discussed along aspects such as modularity, scalability, and complexity of systems (Ringger, 1995; Harper et al., 1994). State-of-the-art translation systems use a variety of different coupling strategies. Examples of loosely coupled systems are IBM's MASTOR (Liu et al., 2003), ATR-MATRIX (Takezawa et al., 1998c), and NESPOLE! (Lavie et al., 2001a), which uses the interlingua-based JANUS system. Examples for tightly coupled systems are EuTrans (Pastor et al., 2001), developed at UPV, and AT&T's Transnizer (Mohri and Riley, 1997).

10.3.1 Removing Disfluencies

Spontaneous spoken speech usually contains disfluencies such as filler words, repairs, or restarts, which do not contribute to the meaning of the spoken utterance and cause sentences to be ill-formed, longer, and thus harder to process for translation. We developed a cleaning component based on a noisy-channel model that automatically removes these disfluencies (Honal and Schultz, 2003, 2005). Its development requires no linguistic knowledge but rather annotated texts and therefore has large potential for rapid deployment and adaptation to new languages.

In this approach, we assume that "clean" (i.e., fluent) speech gets passed through a noisy channel that adds "noise" to the clean speech, and thus outputs disfluent speech. Given a noisy string N, the goal is to recover the clean string C such that $p(C|N)$ becomes maximal. Using Bayes' rule, this problem can be expressed as:

$$\hat{C} = \arg\max_{C} P(C|N) = \arg\max_{C} P(N|C) \cdot P(C). \qquad (10.6)$$

We model the probability $P(C)$ with a trigram language model trained on fluent speech. To establish correspondences between the positions of the source and the target sentences, word-alignment models as described previously can be used. However, in the case of disfluency cleaning, only deletions of words needs to be considered. Assuming that each target sentence is generated from left to right, the alignment a_j defines whether the word n_j in the source sentence is deleted or appended to the target sentence. Let J be the length and n_j the words of the source sentence N, I the length, and c_i the words of the target sentence C; and m the number of deletions (of contiguous word sequences) that are made during generation of the target sentence. We can then introduce an alignment a_j for each word n_j and rewrite $P(N|C)$ as:

$$P(N|C) = P_{I,J}(m) \cdot \prod_{j=1}^{J} P_w(n_j). \tag{10.7}$$

The probability $P_{I,J}(m)$ models the number m of contiguous word sequences that can be deleted in N to obtain C. $P_w(n_j)$ is the probability that word n_j of the string N is disfluent.

Each of the probabilities $P_w(n_j)$ is finally composed of a weighted sum over the following six models: (M1) models the length of the deletion region of a disfluency; (M2) models the position of a disfluency; (M3) models the length of the deletion region of a disfluency with a word fragment at the end of the reparandum; (M4) models the context of a potentially disfluent word; (M5) uses information about the deletions of the last two words preceding a potentially disfluent word; and (M6) takes into account whether a potentially disfluent word is part of a repeated word sequence.

The system can be optimized on a development test set by training the scaling factors for the different models using a gradient descent approach.

The probability distributions for the models are obtained from the training data using relative frequencies. All experiments are conducted on spontaneously spoken dialogs in English from the Verbmobil corpus, and, in order to demonstrate the feasibility of rapid adaptation, on the spontaneous Mandarin Chinese CallHome corpus. The highest performance gain results from model (M4), which considers the context of a potentially disfluent word. This can be easily explained for filler words, since it allows discriminating between the deletion of the word "well" in the

Table 10.5 Results for automatic disfluency removal on the English Verbmobil (EVM) and the Chinese CallHome (CCH) corpora.

Corpus	Setup	Precision	Recall	F_1
EVM	Baseline	90.2	77.2	0.832
	Hand optimized	91.5	86.2	0.888
	Gradient descent	93.1	85.1	0.890
CCH	Baseline	76.8	49.4	0.601
	Hand optimized	77.8	53.4	0.634
	Gradient descent	79.0	53.0	0.634

context "Well done!" and "Alright, well, this is a good idea." The impact of (M1), (M2), and (M3) is a slight increase of the number of hits at the cost of a slight increase or decrease of the number of false positives. Model (M5) causes a large number of false positives and was therefore disregarded in the best system.

Overall, the baseline system shows a precision of 90.2% and a recall of 77.2% for English dialogs, as shown in Table 10.5. Almost no effort was required for the adaptation to Mandarin Chinese. The same algorithms and the same statistical models were used, achieving 76.8% precision and 49.4% recall on the Mandarin corpus (after retraining on the CallHome data). When adjusting the weighting parameters for the models, a small improvement was achieved.

10.3.2 Lattice Coupling

Research on spoken language translation must support scalable systems capable of handling complex translation tasks, but it must also allow for the improvement of individual components. Our own system is therefore structured in a loosely coupled, nonincremental way. Initially, the link between the ASR and MT components was established solely through the single best hypothesis generated by the speech recognizer. Since it is well known that the speech recognition word error rate can be dramatically decreased by generating many alternatives in the form of N-best lists or word lattices, we expect that it could also improve translation performance.

We conducted several studies to investigate whether lattice-based coupling between ASR and MT improves speech translation performance.

The studies were carried out on German-to-English travel-arrangement dialogs originally recorded as part of the NESPOLE! speech translation project (Lavie et al., 2001a). The applied version of the JANUS speech recognition toolkit (JRTk) (Metze et al., 2003) achieved 23.5% lattice word error rate in 1.3× real time, given a lattice density of 22. In JRTk, a word lattice is represented as a directed graph in which nodes are associated with words, and links represent the possible succession of words. The lattice density is defined as the number of words in the lattice divided by the number of words in the transliteration. JRTk allows for various lattice-related functions, such as filler word removal, as well as beam-width pruning to obtain cleaned lattices in the desired density.

The translation system used is the CMU statistical machine translation (SMT) system, as described above and in more detail in Vogel et al. (2003) and Vogel (2003). The decoder was extended to read entire word lattices. A word lattice is traversed from left to right, and word-to-word and phrase-to-phrase translations are added by extending the lattice with edges to which the translations are attached. The enlarged word lattice represents the search space in which the best path is found, with path scores accumulated over translation model scores, target language model scores, acoustic scores, and source language model scores.

The baseline system uses the single best hypothesis for coupling ASR and MT, and gives a BLEU score of 16.83. The BLEU score is a standard measure of translation quality (Papineni et al., 2002). It is measured as a weighted sum of N-gram precision counts up to $N = 4$, modified by a length penalty. Higher BLEU scores indicate better translation quality.

The first experiment applied lattice-based coupling without taking into account acoustic scores or language model scores of the paths in the JRTk lattice. This was to determine if the SMT system could benefit from those additional paths in the word lattice that have a significantly lower error rate than the first-best hypothesis. However, despite the advantage of the lattice topology with different densities, it did not outperform the baseline. Since the ASR lattice typically contains not only paths that are better than the first-best hypothesis but also many paths that are worse, the decoder can choose a path that is easy to translate—that is, a path that gives high probabilities for the translation model and target language model—but that can have many recognition errors. In other words, the resulting translation is not guaranteed to be a translation of what the speaker originally said. This result indicates that it is necessary to incorporate the ASR acoustic and source language model scores into the selection process of the MT system.

In the second experiment, weighted acoustic scores were added to the translation scores. The BLEU score was calculated for acoustic score weights ranging from 0.01 to 0.29 and for lattices of different densities. The best improvement for the BLEU score over the baseline was 7.3%, obtained with a lattice density of 3 and an acoustic score weight of 0.28. The BLEU score increased from that of the one-best hypothesis with the addition of acoustic scores; however, no smooth transition could be found with increasing densities, which indicates that source language model scores need to be included in the translation system.

A closer analysis showed that the length of an utterance has an impact on the improvements on the BLEU score. We therefore used a development test set to find the best BLEU score improvements given the utterance lengths, acoustic score weights, and lattice densities. Table 10.6 shows the optimal settings for different utterance lengths.

The optimized parameters were finally applied to the test set and resulted in a relative improvement over the baseline of 16.22%.

Finally, both ASR scores, the acoustic and the source language model scores were included in order to identify paths in the word lattice that have few recognition errors and, at the same time, represent good translation hypotheses.

Figure 10.6. demonstrates the effect of adding both ASR scores to the translation model and target language model scores.

For a lattice of density 4 and an ASR score weight of the 0.89, the improvement of the BLEU score over the baseline is 12.71% (on the test set). When adding source language model scores, the system was again tuned separately for sentences of different lengths using a development set.

Table 10.6 Optimal density and acoustic score weight based on utterance length.

	Words	Optimal Acoustic Weight	Optimal Lattice Density
Short	1–5	0.0	8
Medium	6–10	0.22	3
Long	11–23	0.01	2

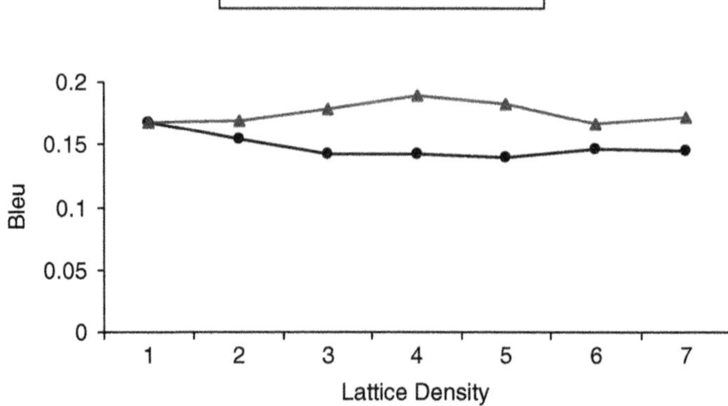

Figure 10.6: Translation quality with and without ASR scores (acoustic model and source language model scores).

The resulting optimal model scaling factors and lattice densities are shown in Table 10.7.

These optimal parameter settings are now much smoother than in the previous case, in which only the acoustic model scores were used, indicating that adding the source language model makes the overall system more robust and stable.

When applying those settings to unseen test data, an improvement of 21.78% in the BLEU score was achieved. Further tuning of the parameters on the test set resulted in a relative improvement of 26.9% over the baseline.

Table 10.8 summarizes the results of the experiments in tight coupling between speech recognition and translation.

Table 10.7 Optimal density and acoustic score weight based on utterance length when using acoustic and source language model scores.

	Words	Optimal Acoustic Weight	Optimal Lattice Density
Short	1–5	0.61	4
Medium	6–10	0.93	2
Long	11–23	0.89	3

Table 10.8 Summary of translation results for tight coupling between recognition and translation (D = lattice density).

	BLEU Score	Improvement
Baseline	16.8	–
Lattice (D = 3)	14.2	−15.5%
with Acoustic Model (D = 3)	18.0	7.3%
- Length specific	19.5	16.2%
with AC and Source LM (D = 4)	18.9	12.7%
- Length specific	20.5	21.8%

10.4 Portable Speech-to-Speech Translation: The ATR System

This section describes example-based and transfer-based approaches to speech-to-speech translation (S2ST) in greater detail and exemplifies their use in a complete speech-to-speech system—namely, the system developed at ATR (Advanced Telecommunications Research Institute International) in Japan. ATR was founded in 1986 as a basic research institute in cooperation with the Japanese government and the private sector, and initiated a research program on Japanese-English speech-2-speech translation (S2ST) soon afterward. This program has addressed not only S2ST approaches proper but also the speech recognition, speech synthesis, and integration components that are required for a complete end-to-end system. The first phase of the program focused on a feasibility study of S2ST, which only allowed a limited vocabulary and clear, read speech. In the second phase, the technology was extended to handle "natural" conversations in a limited domain. The target of the current third phase is to field the S2ST system in real environments. The intended domain of the system is dialog applications.

While earlier phases of the research program were characterized by hybrid rule-based and statistical approaches, the current technology is heavily corpus-based and uses primarily statistical techniques to extract information from linguistically annotated databases. The reason is that corpus-based methods greatly facilitate the development of systems for multiple languages and multiple domains and are capable of incorporating recent innovative technology trends for each component. Domain portability is particularly important, since S2ST systems are often used for applications in a specific situation, such as supporting a tourist's conversation in non-native languages. Therefore, the S2ST technique must include automatic or semiautomatic functions for adapting to specific situations/domains in speech recognition, machine translation, and speech synthesis (Lavie et al., 2001b).

10.4.1 A Corpus-Based MT System

Corpus-based machine translation (MT) technologies were proposed in order to handle the limitations of the rule-based systems that had formerly been the dominant paradigm in machine translation. Experience has shown that corpus-based approaches (1) can be applied to different domains; (2) are easy to adapt to multiple languages because knowledge can be automatically extracted from bilingual corpora using machine learning methods; and (3) can handle ungrammatical sentences, which are common in spoken language. Corpus-based approaches used at ATR include, for example, Transfer-Driven Machine Translation (TDMT) (Furuse and Iida, 1994; Sumita et al., 1999), which is an exampled-based MT system based on the syntactic transfer method. One current research theme is to develop example-based translation technologies that can be applied across a wide range of domains, and to develop stochastic translation technologies that can be applied to language pairs with completely different structures, such as English and Japanese. Example-based methods and stochastic methods each have different advantages and disadvantages and can be combined into a single, more powerful system.

Our overall speech-to-speech translation system is shown in Figure 10.7. The system consists of three major modules: a multilingual

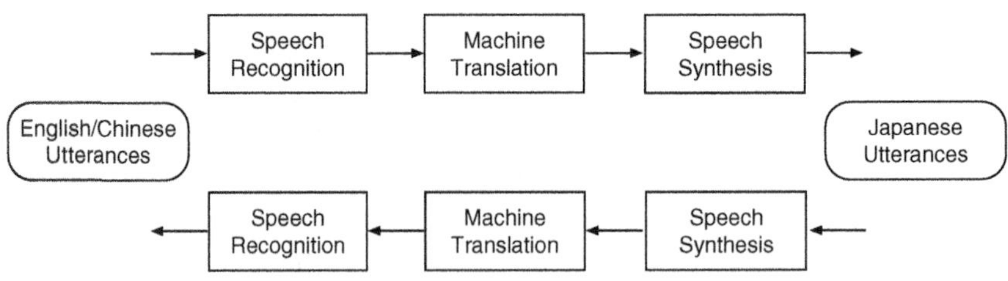

Figure 10.7: Block diagram of the ATR S2ST system.

speech recognition module, a multilingual machine translation module, and a multilingual speech synthesis module. These modules are designed to process Japanese, English, and Chinese using corpus-based methods. Each module is described in more detail below.

Multilingual Speech Recognition

The speech recognition component uses an HMM-based approach with context-dependent acoustic models. In order to efficiently capture contextual and temporal variations in the input while constraining the number of parameters, the system uses the successive state splitting (SSS) algorithm (Takami and Sagayama, 1992) in combination with a minimum description length criterion (Jitsuhiro et al., 2003). This algorithm constructs appropriate context-dependent model topologies by iteratively identifying an HMM state that should be split into two independent states. It then reestimates the parameters of the resulting HMMs based on the standard maximum-likelihood criterion. Two types of splitting are supported: contextual splitting and temporal splitting, as shown in Figure 10.8.

Language modeling in the multilingual speech recognizer is performed by statistical N-gram models with word classes. Word classes are typically established by considering a word's dependencies on its left-hand and right-hand context. Usually, only words having the same left-hand and right-hand context dependence belong to the same word class. However, this

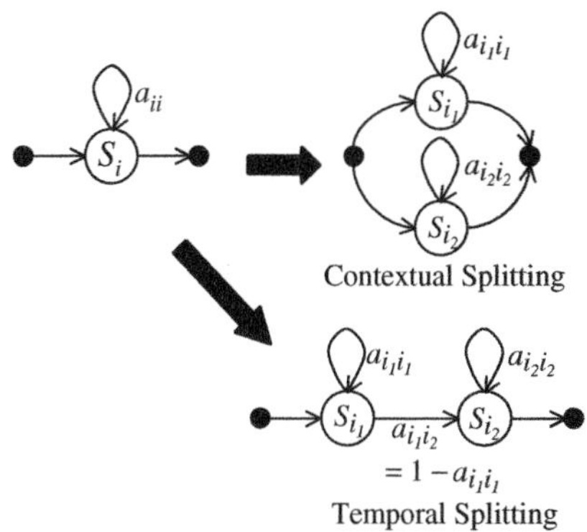

Figure 10.8: Contextual splitting and temporal splitting.

word class definition is not adequate for representing the distribution of words that have the same left-hand or right-hand context but not both, as exemplified by the words "a" and "an." The left-hand context of "a" and "an" is almost equivalent; however, the right-hand context is significantly different. Such differences are more common in languages with inflection, such as French and Japanese. For example, the Japanese inflection form has an influence only on the right-hand context, while the left-hand context can be shared between the same words with different inflection forms.

We therefore use multidimensional word classes to represent left- and right-hand context dependence separately (Yamamoto et al., 2003). Multidimensional word classes can assign the same word class to "a" and "an" to represent the left-context Markovian dependence (left-context class), and assign them to different word classes to represent the right-context Markovian dependence (right-context class). Each multidimensional word class is automatically extracted from the corpus using statistical information, rather than grammatical information such as part of speech. Formally, this is defined as follows:

$$P(w_i|w_{i-N+1}\ldots w_{i-1}) = P(C^l(w_i)|C^{rN-1}(w_{i-N+1})\ldots C^{r2}(w_{i-2}C^{r1}(w_{i-1}))$$
$$P(w_i|C^l(w_i)), \tag{10.8}$$

where the suffix for class C is used to represent position-dependent (left- and right-context) Markovian dependence. Here, $C^l(w)$ represents the left-context class to which the word w belongs, and $C^{ri}(w)$ represents the right-context class to which the ith word w belongs. Hereafter, we refer to these class N-grams based on multidimensional classes as multiclass N-grams.

Multilingual Machine Translation

The translation engine (named C-cube, for "corpus-centered computation") relies heavily on corpus-based technology. Translation knowledge is extracted from corpora, translation quality is gauged and optimized by reference to corpora, and the corpora themselves are paraphrased or filtered by automated processes. Figure 10.9 shows an overview of the machine-translation system developed in the C-cube project.

There are two main approaches to corpus-based machine translation: (1) Example-Based Machine Translation (EBMT) (Nagao, 1984; Somers,

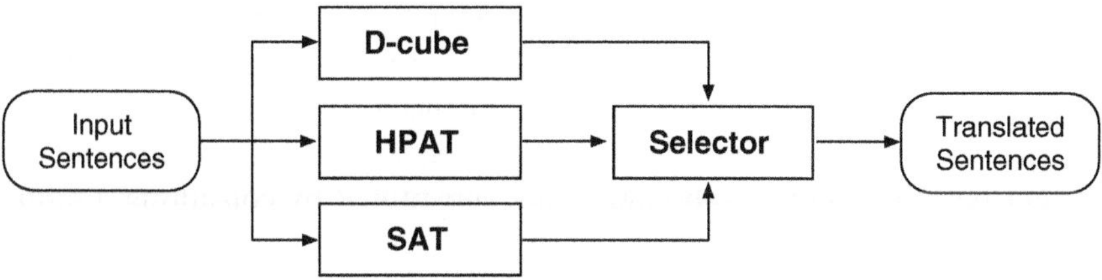

Figure 10.9: An overview of the machine translation system developed in the C-cube project.

1999); and (2) Statistical Machine Translation (SMT) (Brown et al., 1993; Knight, 1997; Ney, 2001; Alshawi et al., 2000; Wang and Waibel, 1998; Och et al., 1999; Venugopal et al., 2003), which was already described above. C-cube develops both technologies in parallel and blends them into a single system. Three different machine-translation engines have been developed in the course of this project: D-cube, HPAT, and SAT. D-cube is a sentence-based EBMT engine. It retrieves the most similar example of the input from example sentences using dynamic programming based matching, and adjusts the gap between the input and the retrieved example by using dictionary information (Sumita, 2001). HPAT is a phrase-based EBMT engine. Based on phrase-aligned bilingual trees, transfer patterns are generated. According to these patterns, the source phrase structure is obtained and converted to generate target sentences (Imamura, 2002). SAT is the Statistical ATR Translator. It translates between Japanese and English and builds on a word-based SMT framework. SAT is a developing series of SMT models, which includes phrase-based translation (T. Watanae and Sumita, 2002), chunk-based translation (Y. Akiba and Sumita, 2002), and sentence-based greedy decoding (Watanae and Sumita, 2003). These three systems have obtained good performance on Japanese and English translation, and have recently been successfully applied to Japanese and Chinese.

Sentence-Based EBMT

The ATR sentence-based EBMT approach relies on D3—a dynamic programming transducer that exploits matching between word sequences

(Sumita, 2001). Suppose we are translating a Japanese sentence into English. The Japanese input sentence (1-j) is translated into the English sentence (1-e) by utilizing the English sentence (2-e), whose source sentence (2-j) is similar to (1-j). The common parts are unchanged, and the different portions, shown in boldface, are substituted by consulting a bilingual dictionary.

;;; A Japanese input
(1-j) **iro**/ga/ki/ni/iri/masen
;;; the most similar example in corpus
(2-j) **dezain**/ga/ki/ni/iri/masen
(2-e) I do not like the **design**.
;;; the English output
(1-e) I do not like the **color**.

We retrieve the most similar source sentence example from a bilingual corpus. For this, we use *DP-matching*, which computes the *edit distance* between word sequences while simultaneously identifying the matched portions between the input and the example. The edit distance is calculated by summing the count of the inserted words, the count of the deleted words, and the semantic distance value of the substituted words, normalized by the sum of the lengths of the input and the source part of the translation example. The semantic distance between two substituted words is calculated by using the hierarchy of a thesaurus (Sumita and Iida, 1991a). Thus, the language resources required in addition to a bilingual corpus are a bilingual dictionary, which is used for generating target sentences, and thesauri of both languages, which are used for incorporating the semantic distance between words into the distance between word sequences. Furthermore, lexical resources are used for word alignment.

Phrase-Based EBMT

The second EBMT system is different from the first in that it parses bitexts of a parallel corpus with grammars for both source and target languages. It incorporates a new phrase alignment approach called Hierarchical Phrase Alignment (HPA), which was proposed by Imamura (2001). HPA retrieves equivalent phrases that satisfy two conditions: (1) words in the pair correspond with no deficiency and no excess; and (2) the phrases are of the same syntactic category. This was subsequently extended to HPA-based translation (HPAT) (Imamura, 2001). HPAed bilingual trees include all information necessary to automatically generate transfer patterns. Translation is done according to transfer patterns using a Transfer-Driven MT (TDMT) engine (Furuse and Iida, 1996). Finally, a *feedback cleaning* method (Imamura, 2003) is applied that utilizes automatic evaluation to remove incorrect/redundant translation rules. The standard BLEU method was utilized to measure translation quality for the feedback process, and a hill-climbing algorithm was applied to search for the combinatorial optimization. Finally, incorrect/redundant rules were removed from the set of all rules initially acquired from the training corpus, which improved the translation quality considerably.

Phrase alignment refers to the extraction of equivalent partial word sequences between bilingual sentences. The term "phrase alignment" is used since these word sequences include not only words but also noun phrases, verb phrases, relative clauses, and so on.

For example, when the following bilingual sentence is given,

English: *I have just arrived in New York.*
Japanese: *Nyuyooku ni tsui ta bakari desu.*

the phrase alignment should extract the following word sequence pairs.

- *in New York* ↔ *Nyuyooku ni*
- *arrived in New York* ↔ *Nyuyooku ni tsui*
- *have just arrived in New York* ↔ *Nyuyooku ni tsui ta bakari desu*

We call these *equivalent phrases* and define this task as extracting phrases that satisfy the following two conditions.

Condition 1 (Semantic constraint):
Words in the phrase pair correspond to no deficiency and no excess.
Condition 2 (Syntactic constraint):
The phrases are of the same syntactic category.

In order to extract phrases that satisfy the two conditions, corresponding words—called *word links*, represented as $WL(word_e, word_j)$—are first extracted by word alignment. Next, the sentence pair is parsed and phrases and their syntactic categories are acquired. Those phrases—which include some word links, exclude other links, and are of the same syntactic categories—are regarded as equivalent.

For example, in the case of Figure 10.10(a), `NP(1)` and `VMP(2)` are regarded as equivalent because they only include *WL(New York, Nyuyooku)* and are of the same syntactic category. In the case of *WL(arrived, tsui)*, `VP(3)` is regarded as equivalent, and in the case of both word links, `VP(4)`, `AUXVP(5)`, and `S(6)` are regarded as equivalent. Consequently, six equivalent phrases are extracted hierarchically.

Even though word links are available, the parts-of-speech (POS) of the words are sometimes different in different languages, as shown in the second example in Figure 10.10(b). In this case, the phrases that contain only *WL(fully, ippai)* or only *WL(booked, yoyaku)* are not regarded as equivalent because of the syntactic constraint, and `VP(2)` nodes are extracted first. Thus, few unnatural short phrases are extracted as equivalent.

The problem besetting the method described above is that the result of the phrase alignment directly depends on the parsing result. This problem can be solved by using the following features and techniques.

Disambiguation Using Structural Similarity: As Kaji et al. (1992) and Matsumoto et al. (1993) showed, some parsing ambiguities can be resolved when the two languages are made to correspond. This disambiguation utilizes structural similarity. For example, a PP attachment in English is ambiguous as to whether it modifies a noun or a verb, but this is nearly always definite in Japanese. Hence, when the attachment is assumed to modify the same word, the ambiguity is resolved. Accordingly, the structures between the two languages become similar.

We employ a *phrase correspondence score* to measure structural similarity. This measure is calculated by counting the phrases that satisfy the above two conditions. The parsing candidate that has the maximal score is selected.

Combination of Partial Trees: Partial parsing is an effective way to avoid a lack of grammar or to parse ungrammatical sentences. It is used to combine partial candidates in the parser. Therefore, a criterion as to whether the part is valid or not is necessary for the combining process. We utilize the phrase correspondence score as the criterion, and a partial tree sequence that maximizes the sum of the phrase correspondence scores is searched for.

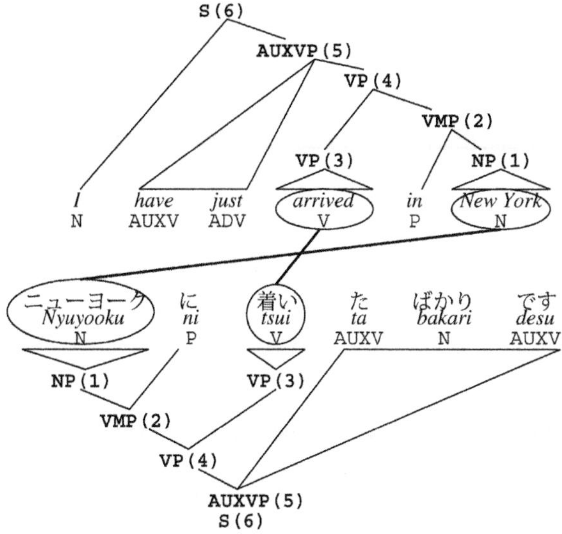

(a) Example of Simple Translation

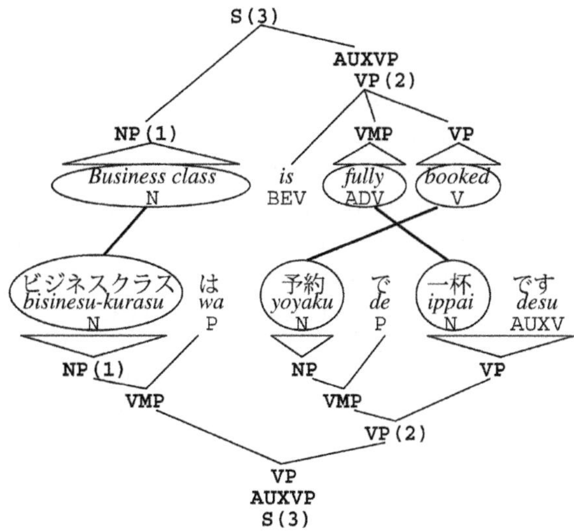

(b) Example of Different POS Translation

Figure 10.10: Examples of Hierarchical Phrase Alignment (top and bottom trees denote English and Japanese, respectively; lines between languages denote word links).

Syn. Cat.	Source Pattern		Target Pattern	Source Example
VP	X_{VP} *at* Y_{NP}	$\Rightarrow$	Y' *de* X'	*(present, conference)*
			Y' *ni* X'	*(stay, hotel), (arrive, p.m.)*
			Y' *wo* X'	*(look, it)*
NP	X_{VP} *at* Y_{NP}	$\Rightarrow$	Y' *no* X'	*(man, front desk)*

Figure 10.11: Examples of transfer rules in which the constituent boundary is "at."

The forward DP backward A^* search algorithm (Nagata, 1994) is employed to speed up the combination.

Transfer Driven Machine Translation (TDMT)

The Transfer Driven Machine Translation system, or TDMT (Furuse and Iida, 1994; Sumita et al., 1999), used here is an example-based MT system (Nagao, 1984) based on the syntactic transfer method (called transfer-based MT).

Transfer rules of TDMT represent the correspondence between source and target language expressions. These are the most important kinds of knowledge in TDMT. Examples are shown in Figure 10.11 that include the preposition "at." In this rule, source language information is constructed by a source pattern and its syntactic category. The source pattern is a sequence of variables and constituent boundaries (functional words or part-of-speech bigram markers). Each variable is restricted by a syntactic category using daughter rules. Namely, source language information is equivalent to a context-free-grammar such that the right side of each rewrite rule absolutely contains at least one terminal symbol.

Target patterns are similarly constructed with variables and constituent boundaries, but they do not have POS bigram markers. In addition, each rule has source examples, which are instances of variables. The source examples are head-words acquired from training sentences. For instance, the first rule of Figure 10.11 means that the English phrase "present at (the) conference" was translated into the Japanese phrase "*kaigi* (conference) *de happyo-suru* (present)."

At the time of translation, the source sentence is parsed using source patterns. Then, the target structure, which is mapped by target patterns, is generated (Figure 10.12). However, as shown in Figure 10.11, one transfer

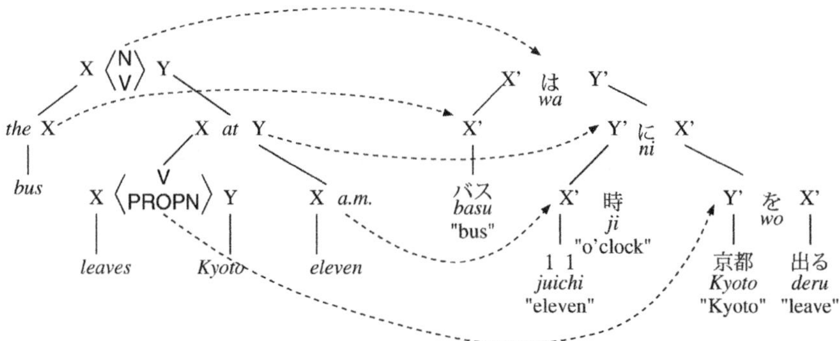

Figure 10.12: Example of TDMT transfer process.

rule has multiple target patterns. In order to select the appropriate target pattern, semantic distances (node distances on the thesaurus; refer to Sumita and Iida (1991b)) are calculated between the source examples and the daughter head-words of the input sentence, and the target pattern corresponding to the best matching example is selected. Therefore, each rule also has head information.

For example, when the input sentence "The bus leaves Kyoto at eleven a.m." is given, the source pattern (X *at* Y) is used. Then, the head-word of the variable X is "*leave*" and Y is "*a.m.*" According to the semantic distance calculation, the source example (*arrive*, *p.m.*) is the closest match. Therefore, the target pattern (Y' *ni* X') is selected. The semantic distance is also applied to parsing disambiguation.

Application of HPA Results for TDMT

The transfer rules described previously are constructed by source patterns that include their syntactic category, target patterns, source examples, head information, and local dictionaries. They are generated as follows from the HPA results (Figure 10.13).

1. First, the result of HPA is transformed into a structure that can construct transfer rules.

    - If an input word sequence includes continuous content words, insert a bigram marker in the intermediate of the content words.

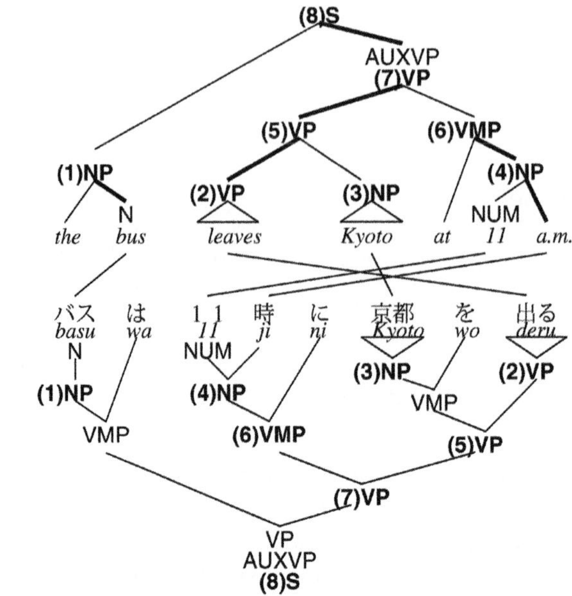

(1) Result of HPA (Bold lines denote head information)

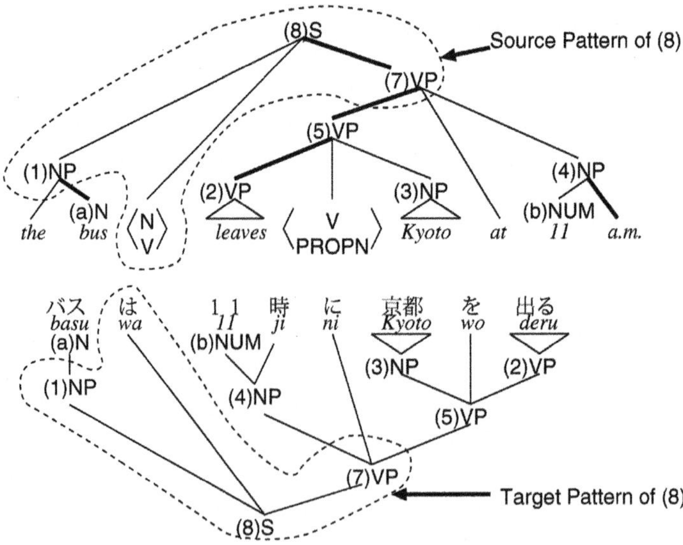

(2) Tree structure after transformation

Figure 10.13: Example of transfer rule generation.

The bigram marker is an artificial word, which works as a functional word when the translator parses the input sentence.

- If the edges of a word link are content words and are of the same POS types, a new word-level correspondence is added. The function of this correspondence is to translate unseen words by referring to the translation dictionary.

  In Figure 10.13, the correspondences `(a)N` and `(b)NUM` are supplied from the word links *WL*(*bus*/N, *basu*/N) and *WL*(*11*/NUM, *11*/NUM).
- When the source pattern is generated, the correspondence is removed if variables are continuous because TDMT does not accept the series of variables. In Figure 10.13, `(6)VMP` is removed.
- All nodes that do not have correspondences are removed except for the top node.

2. Next, source patterns, target patterns, source examples, head information, and local dictionaries are created as follows.

   - Source patterns and target patterns are generated from the correspondences. The patterns are generalized by regarding daughter corresponding nodes as variables.
   - Head information is acquired from grammar, and source examples are identified by tracing the parsing tree to the head branch.
   - Local dictionaries are created by word links and by extracting leaf equivalent phrases in which the source phrase contains only a word.

In addition, because the inputs of phrase alignment are aligned sentences, sentence correspondences are added to the phrase alignment results as equivalent when the top nodes of the trees do not have a correspondence.

When the result of HPA is given (as shown in Figure 10.13), five rules are generated (as shown in Figure 10.14). Note that rules are not generated from the correspondences `(2)VP`, `(3)NP`, `(a)N`, and `(b)NUM` because they are output to the local dictionaries.

Syn. Cat.	Source Pattern		Target Pattern	Source Example
(8) S	X_{NP}<N-V>Y_{VP}	⇒	X′ *wa* Y′	(*bus, leave*)
(7) VP	X_{VP} *at* Y_{NP}	⇒	Y′ *ni* X′	(*leave, a.m.*)
(5) VP	X_{VP}<V-PROPN>Y_{NP}	⇒	Y′ *wo* X′	(*leave, Kyoto*)
(1) NP	*the* X_N	⇒	X′	(*bus*)
(4) NP	X_{NUM} *a.m.*	⇒	X′*ji*	(*11*)

Figure 10.14: Example of generated rules from the sentence "The bus leaves Kyoto at 11 a.m."

SAT (Statistical ATR Translator)

As described above the framework of statistical machine translation formulates the problem of translating a sentence of language J into another language E as the maximization problem of the conditional probability $\hat{E} = \text{argmax}_E P(E|J)$ (Brown et al., 1993b). The application of the Bayes rule resulted in $\hat{E} = \text{argmax}_E P(E)P(J|E)$. The former term $P(E)$ is called a language model, representing the likelihood of E. The latter term $P(J|E)$ is called a translation model, representing the generation probability from E into J.

Under this concept, Brown et al. (1993b) presented a translation model in which a source sentence is mapped to a target sentence with the notion of word alignment.[1] Although it has been successfully applied to similar language pairs, such as French-English and German-English, little success has been achieved for drastically different language pairs, such as Japanese-English. The problem lies in the huge search space resulting from frequently occurring insertion/deletion, the larger numbers of fertility for each word, and the complicated word alignments. Due to its complexity, a beam search decoding algorithm often leads to suboptimal solutions.

Word alignment based statistical translation represents bilingual correspondence by the notion of word alignment A, allowing one-to-many generations from each source word. A is an array for target words describing the indices to the source words. For instance, Figure 10.15 illustrates an example of English and Japanese sentences, E and J, with sample word alignments A. In this example, the "show$_1$" has generated two words,

[1]The source/target sentences are the channel model's source/target that correspond to the translation system's output/input.

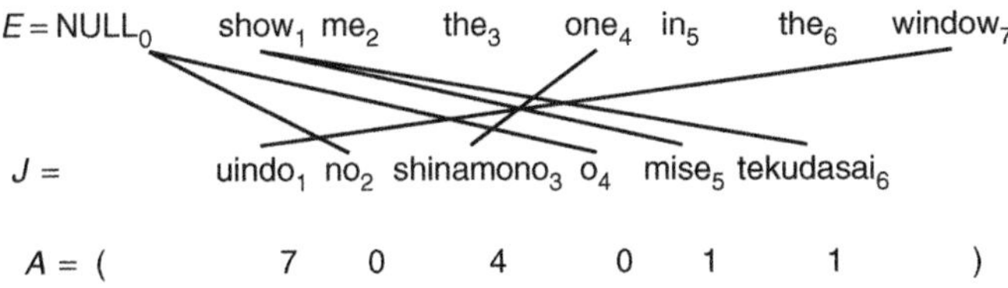

Figure 10.15: Example of word alignment.

"mise_5" and "tekudasai_6." The word alignment assumption, the translation model $P(J|E)$, can be further decomposed without approximation:

$$P(J|E) = \sum_A P(J,A|E)$$

The word alignment based statistical translation model was originally intended for similar language pairs, such as French and English. When applied to Japanese and English, for instance, the resulting word alignments are very complicated, as seen in Figure 10.15. The complexity is directly reflected by the structural differences—that is, English takes an SVO structure while Japanese usually takes the form of SOV. In addition, insertion and deletion occur very frequently as seen in the example.

Example-Based Decoder

Instead of decoding word-by-word and generating an output string word-by-word, as seen in beam search strategies, Watanabe and Sumita (2003) proposed an alternative strategy taken after the framework of example-based machine translation: Retrieve a translation example from a parallel corpus whose source part is similar to the input sentence, then slightly modify the target part of the example so that the resulting part becomes the actual translation (see to Figure 10.16).

Retrieval of Translation Examples

Given an input sentence J_0, the retrieval process looks up a collection of translation examples $\{(J_1, E_1), (J_2, E_2), \ldots\}$, where J_k is similar to J_0 using

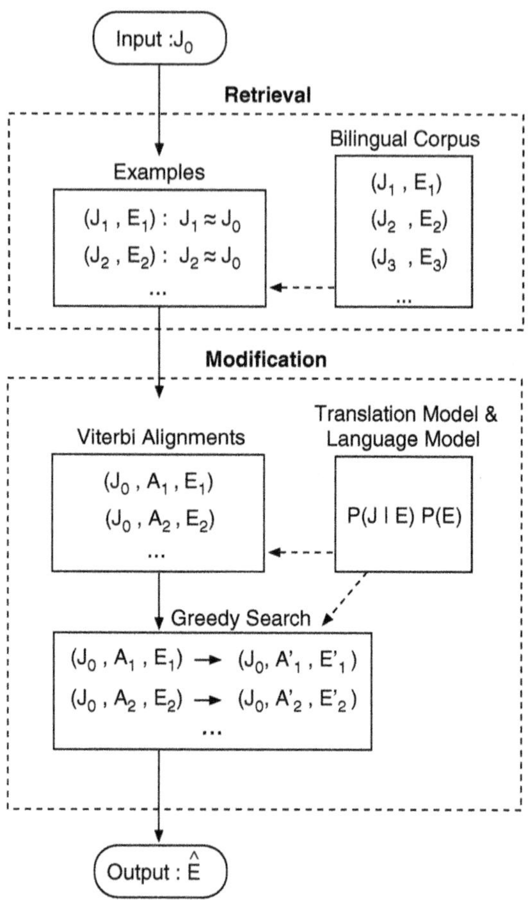

Figure 10.16: Example-based decoding.

the edit distance criteria, penalizing an insertion/deletion/substitution with one editing step. A simple implementation of the multiple alignment problem resulted in an NP-hard problem in which Dynamic Programming (DP) algorithms should be applied to all the examples in a bilingual corpus. An additional problem results from the fact that the DP matching criteria does not reflect the closeness of two sentences. For instance, the sentence "I'm a computer system engineer" can match many examples—such as "I'm a graduate student" or "I'm an engineer"—with the same edit distance of 3.

In order to overcome those problems, a tf/idf criterion was introduced to search for the relevant examples by treating each example as a document. Particularly, when given an input J_0, the decoder first retrieves $N_r(\leq N)$ relevant translation samples $\{(J_1, E_1), (J_2, E_2), \ldots\}$ using the tf/idf criterion, as seen in the information retrieval framework (Manning and Schütze, 1999):

$$P_{tf/idf}(J_k, J_0) = \sum_{i:J_{0,i} \in J_k} \frac{\log(N/df(J_{0,i}))/\log N}{|J_0|},$$

where $J_{0,i}$ is the ith word of J_0, $df(J_{0,i})$ is the document frequency for the word $J_{0,i}$, and N is the total number of examples in a bilingual corpus. Note that the frequency is set to 1 if the word exists in J_k, otherwise 0; and tf/idf scores are summed and normalized by the input sentence length.

Then, for each sample (J_k, E_k), DP matching is performed against J_0 to compute the edit distance:

$$dis(J_k, J_0) = I(J_k, J_0) + D(J_k, J_0) + S(J_k, J_0),$$

where $k \leq N_r$ and $I(J_k, J_0)$, $D(J_k, J_0)$, and $S(J_k, J_0)$ are the number of insertions, deletions, and substitutions, respectively. All samples are scored by the following criteria:

$$score = \begin{cases} (1.0 - \alpha)(1.0 - \frac{dis(J_k, J_0)}{|J_0|}) & dis(J_k, J_0) > 0 \\ \quad +\alpha P_{tf/idf}(J_k, J_0) & \\ 1.0 & \text{otherwise} \end{cases}$$

In this scoring, $dis(J_k, J_0)$ is transformed into the word error rate by normalization with the input length $|J_0|$, then subtracted from 1 to derive the correction rate. The correction rate is linearly interpolated with the normalized tf/idf score with a tuning parameter α that was set to 0.2 in the experiments. Note that when the distance of the input sentence and the source part of an example is zero, the example is treated as an exact match and is scored as one.

Modification of Translation Examples

After the retrieval of similar examples $\{(J_1, E_1), (J_2, E_2), \ldots\}$, the modification step tunes the sample translations according to a statistical translation model. In this step, the greedy algorithm was applied, originated in Germann et al. (2001). However, it differs in that the search starts from the retrieved translation example, not from a guessed translation.

For each translation example (J_k, E_k),

1. Compute the Viterbi alignment A_k for the pair (J_0, E_k)
2. Perform the hill-climbing algorithm for (J_0, A_k, E_k) to obtain (J_0, A'_k, E'_k) by modifying A_k and E_k

A_k is computed through hill climbing by rearranging particular word alignments as proposed by Brown et al. (1993b). When the retrieved samples contain an exact match scored as one, the search terminates and returns the retrieved examples with the highest probability together with the Viterbi alignment.

When the samples are not an exact match, the decoder performs hill climbing, modifying the output and alignment for a given example (J_0, A, E), where A is the word alignment initially assigned by the Viterbi alignment and E is the target part of the example. In this greedy strategy, the operators applied to each hill-climbing step are:

- Translate words: Modify the output word E_{A_j} to e aligned from J_{0j}. If $e = \text{NULL}$ then J_{0j} is aligned to NULL and $A_j = 0$. When the fertility of E_{A_j} becomes zero, the word E_{A_j} is removed. e is selected from among the translation candidates, computed from the inverse of the lexicon model (Germann et al., 2001).
- Translate and insert words: Perform the translation of a word, and insert a sequence of zero fertility words at appropriate positions. The candidate sequence of zero fertility words is selected from the Viterbi alignment of the training corpus (Watanabe and Sumita, 2002).
- Translate and align words: Move the alignment of A_j to i and modify the output word from E_i to e.
- Move alignments: This operator does not alter the output word sequence, but modifies the alignment A through moving/swapping (Brown et al., 1993b).

- Swap segments: Swap nonoverlapping subsets of E by swapping a segment from i_0 to i_1 and from i_2 to i_3. Note that $i_1 < i_2$.
- Remove words: Remove a sequence of zero fertility words from E.
- Join words: Join words of E_i and $E_{i'}$ when the fertility of both of the words is more than or equal to one.

For each hill-climbing step, the decoder tries all the possible operators, then selects the best step for the next iteration. The hill-climbing operators were taken from Germann et al. (2001), but two new operators were added: the "translate-and-align-words," and the "move-alignment." If at the first step of computing the Viterbi alignment an input word is found for which no translation exists among the retrieved samples, this word will either be aligned to NULL or to an irrelevant word, raising the fertility. Therefore, the translate-and-align operator can force it to move the alignment to another word and to choose the right word-for-word translation using the lexicon model. Similarly, the move-alignment operator can resolve the problem by simply alternating the existing word alignments.

Selector Approach

Today, no single system can translate everything. The translation quality changes from sentence to sentence and system to system. Since each system has its own strengths for translating particular kinds of sentences, the differences in quality can be substantial. The translations provided by multiple engines are often complementary. Thus, we could obtain a large increase in accuracy if it were possible to select the best out of different translations for each input sentence.

We adopted a selector approach, as shown in Figure 10.9, by using both the language and translation models of SMT (Akiba et al., 2002). The selector outperformed conventional selectors using only the language model (the target N-gram in the experiment with the three previously mentioned machine translation systems).

Although it is likely that the selector based on SMT models always favors SMT, that was not true for this experiment. This suggests that although an SMT engine fails to produce good translation under time and space restrictions, SMT models can be used as a good measure for comparing the quality of multiple translations. The second suggestion is amplified to a general framework in that a translation system can be divided into two parts: (1) a generator of translation candidates and (2) a selector of the best one by using automatic quality evaluation. This contrasts with a conventional system, which often determines a single translation in a deterministic fashion. The emphasis here lies on focus on the independence of quality evaluation as an automated process, and the fact that an SMT model works well as an evaluator.

The selector scheme is an easy-to-implement method of improving overall performance, since there is no need to investigate the complex relationships between the resources and processes of the component MT systems by hand. This simplifies the development process because each element MT system can be improved without any consideration of the other components, and this loosely coupled system automatically boosts the performance of the whole system by exploiting the elemental improvements.

Multilingual Text-to-Speech

The synthesis component of the S2ST system is based on the concatenative synthesis approach (see Chapter 7). As one of the pioneers in corpus-based speech synthesis technology, ATR has made contributions to the progress of the technology through various studies, which have led to the development of three text-to-speech (TTS) systems, namely ν-talk (Sagisaka et al., 1992b), CHATR (Black and Taylor, 1994), and XIMERA, which is currently under development. The framework of XIMERA is essentially the same as that of CHATR, ν-talk, and other corpus-based TTS systems. The general aim is to achieve a substantial improvement in naturalness through the optimization of the component technologies of TTS.

A block diagram of XIMERA is shown in Figure 10.17. Similar to most concatenative TTS systems, XIMERA is composed of four major modules, namely a text processing module, a prosodic parameter generation module, a segment selection module, and a waveform generation module.

The target languages of XIMERA are Japanese and Chinese. Although the framework of corpus-based synthesis is language independent, most modules, in reality, must be developed or tuned toward the target language. Language dependent modules include the text processing module, acoustic models for prosodic parameter generation, speech corpora, and the cost

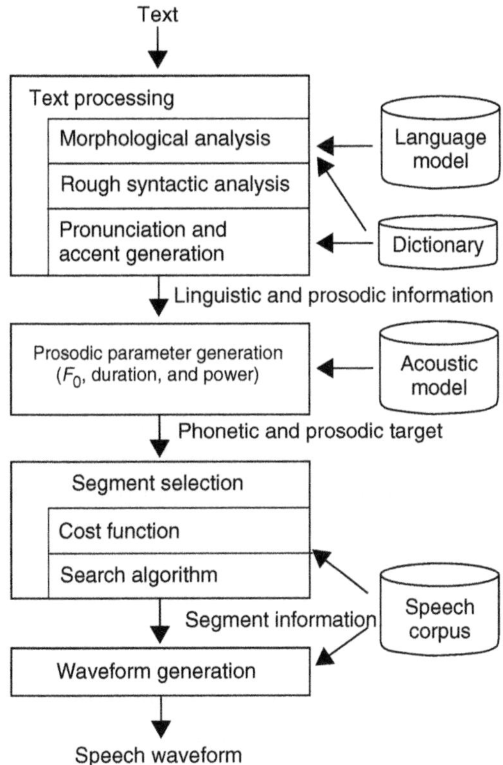

Figure 10.17: A block diagram of a TTS module.

function for segment selection. The search algorithm for segment selection is also related to the target language via the cost function.

Although emotion and speaking style variations are indispensable for achieving a speech synthesizer that can be used in place of humans, XIMERA is currently focused on a normal reading speech style suitable for news reading and emotionless dialog between man and machine.

The prominent features of XIMERA are: (1) its large corpora (a 110-hour corpus of a Japanese male, a 60-hour corpus of a Japanese female, and a 20-hour corpus of a Chinese female); (2) HMM-based generation of prosodic parameters (Tokuda et al., 2000); and (3) a cost function for segment selection, optimized on perceptual experiments (Toda et al., 2004).

Text Processing and Speech Corpora

The text processing module consists of three submodules for morphological analysis, rough syntactic analysis, and pronunciation and accent generation. The morphological analysis is conducted based on a bigram language model and a morpheme dictionary consisting of 239,591 Japanese and 195,959 Chinese entries. The rough syntactic analysis determines (1) a dependency between adjacent words, which is mainly used for F_0 generation, and (2) clause boundaries, which are mainly used for pause insertion. The pronunciation generation determines the readings of homographs and euphonic changes of unvoiced to voiced sounds. The accent generation determines the accent type of an accentual phrase based on accent types and the accent concatenation features of the constituent morphemes. In terms of the dichotomy of corpus-based versus noncorpus based, the first two subprocesses are corpus-based technologies while the latter one is not.

Three large-scale single-speaker speech corpora were collected. One corpus of Japanese male of 111-hour length, one corpus of Japanese female of 60-hour length, and one corpus of Chinese female of 20-hour length. The contents of the Japanese corpora include news, novels, and travel conversations. The travel conversations were uttered in a reading style. The given corpus size in hours include speech pauses within utterances but not between utterances. The speakers were professional narrators. The recordings took 181 days over a span of 973 days (Japanese male corpus), 95 days over a span of 307 days (Japanese female corpus), and 32 days over a span of 63 days (Chinese female corpus). The speech was recorded in a soundproof room. To obtain high signal-to-noise ratio (SNR), a large diaphragm condenser microphone with a cardioid directional pattern was used. The speech data were digitized at a sampling frequency of 48 kHz with 24-bit precision, recorded onto a hard disk. After reading errors were removed by human inspection, speech data were separated into utterances, high-pass filtered at 70 Hz, precision-converted down to 16 bits after amplitude adjustment, and stored in files. Phonemic transcriptions in katakana characters were also inspected and corrected by humans. Phone segmentation was conducted automatically by using speaker-dependent monophone HMMs. The results were not corrected manually.

Generation of Prosodic Parameters

Prosodic parameters, namely F_0, phone duration, and power, are generated applying the HMM-based speech synthesis algorithm (Tokuda et al., 1995, 2000). Japanese speech is modeled with context-dependent HMMs of 42 phonemes consisting of five states each, while Chinese speech is modeled with context-dependent HMMs of 60 initials and finals. The generated prosodic parameters are sent to the succeeding module to be used as targets for segment selection.

Cost Function

The cost function of a sentence for segment selection is given by

$$C_g = \frac{1}{N} \sum_{i=1}^{N} C_l(u_i, t_i)^p, \tag{10.9}$$

where N denotes the number of targets in the sentence, C_l denotes a local cost at the target t_i, and u_i and t_i, respectively, denote the ith target and segment candidate. Power p was determined to be 2 as a result of perceptual experiments (Toda et al., 2003). The local cost is given by

$$\begin{aligned} C_l(u_i, t_i) = {} & w_{F0} \cdot C_{F0}(u_i, t_i) + w_{dur} \cdot C_{dur}(u_i, t_i) \\ & + w_{cen} \cdot C_{cen}(u_i, t_i) + w_{F0c} \cdot C_{F0c}(u_i, t_i) \\ & + w_{env} \cdot C_{env}(u_i, t_i) + w_{spg} \cdot C_{spg}(u_i, t_i), \end{aligned} \tag{10.10}$$

where $C_{F0}(u_i, t_i)$, $C_{dur}(u_i, t_i)$, and $C_{cen}(u_i, t_i)$, respectively, denote errors in F_0, segment duration, and spectral centroid between the target and a segment candidate (target costs). On the other hand, $C_{F0c}(u_i, t_i)$, $C_{env}(u_i, t_i)$, and $C_{spg}(u_i, t_i)$, respectively, denote discontinuities of F_0, phonetic environment, and spectrum between adjacent segments (concatenation costs). w_{F0}, w_{dur}, w_{cen}, w_{F0c}, w_{env}, and w_{spg} are corresponding weights for the local costs. Mappings from acoustic measures into the above local costs and weights were optimized based on perceptual experiments (Toda et al., 2004).

10.4.2 Multilingual Corpora

ATR has been constructing three different types of corpora in the travel domain in addition to speech databases for training acoustic models: (1) a large-scale multilingual collection of basic sentences that covers many topics in travel conversations, called BTEC (Basic Travel Expression Corpus) (Takezawa et al., 2002); (2) a small-scale bilingual collection of spoken sentences that reflects the characteristics of the spoken dialogs, called SLDB (Spoken Language Database) (Morimoto et al., 1994); and (3) a small-scale corpus of the MT-assisted dialog, called MAD (Kikui et al., 2003). The first one is used to train the multilingual translation component; the second one is used to link spoken sentences to basic sentences, and the third one is used mainly for evaluation of S2ST.

Basic Travel Expression Corpus (BTEC)

The Basic Travel Expression Corpus (BTEC) was designed to cover utterances for all potential topics in travel conversations, together with their translations (Kikui et al., 2003). Since it is almost infeasible to collect them through transcribing actual conversations or simulated dialogs, sentences from the memories of bilingual travel experts are used. We started by investigating phrase books that contain bilingual (in our case Japanese-English) sentence pairs that experts consider useful for tourists traveling abroad. We collected these sentence pairs and rewrote them to make translations as context independent as possible and to comply with our speech transcription style. Sentences outside of the travel domain or those containing very special meanings were removed.

Table 10.9 shows the basic statistics of the BTEC collections, called BTEC1, 2, and 3. Each collection was created with the same procedure in a different time period. We used a morpheme as the atomic linguistic unit for Japanese (instead of a word), since the morpheme unit is more stable than the word unit. This table shows that the BTEC sentences are relatively short and contain duplications.

Since the BTEC sentences did not come from actual speech conversation, we were able to efficiently create a broad coverage corpus; however, it has two potential problems. First, the corpus may lack utterances that appear in real conversation. For example, when people ask the way to a

Table 10.9 Size of BTEC and SLDB.

	BTEC1	BTEC2	BTEC3	Total (BTEC)	SLDB
Utterances (E)					
# of tokens	175,512	46,288	198,290	420,070	16,110
# of types	102,869	37,791	141,504	254,766	14,266
Words (E)					
# of tokens	1,089,570	311,537	1,316,188	2,717,295	199,951
# of types	14,725	11,796	22,925	27,998	4,544
Words per Utterance (E)	6.21	6.73	6.64	6.47	12.75
Utterances (J)					
# of tokens	172,468	46,268	198,290	417,026	16,110
# of types	108,612	36,869	143,454	264,401	14,259
Morphemes (J)					
# of tokens	1,181,763	341,353	1,434,175	2,957,291	204,894
# of types	20,363	15,081	31,155	39,316	5,765
Morphemes per Utterance (J)	6.85	7.38	7.25	7.09	12.75

bus stop, they often use a sentence like (1). However, BTEC1 contains (2) instead of (1).

(1) I'd like to go downtown. Where can I catch a bus?
(2) Where is a bus stop (to go downtown)?

The second problem is that the frequency distribution of the corpus may be different from the "actual" one. In this corpus, the frequency of an utterance (indirectly) corresponds to how many travel experts came up with this sentence and in how many situations they think the sentence will appear. Therefore, it is possible to think of this frequency distribution as a first approximation of reality, but this is an open question.

BTEC was used as test sentences for evaluating various S2ST systems at IWSLT (International Workshop for Spoken Language Translation), held at ATR in October 2004.

ATR is expanding the volume of BTEC by about one million English-Japanese sentence pairs and creating Chinese-Japanese sentence pairs by translating from English/Japanese to Chinese.

Spoken Language Database (SLDB)

For a comparison of the features of expressions in BTEC and utterances in real conversation, we employed an existing corpus, SLDB. SLDB consists of simulated dialogs between a hotel clerk and a tourist. We used a Japanese speaker and an English speaker to achieve some dialog tasks (e.g., to book a hotel room) through conversation mediated by two professional interpreters: one translating from Japanese to English, the other from English to Japanese. Thus, this corpus can be seen as a collection of bilingual dialogs using ideal S2ST systems.

The statistics of SLDB corpora are also seen in Table 10.9. The number of utterance tokens are the same for each language since the utterances were translated simultaneously.

The travel conversation task was selected for the SLDB corpus because of its familiarity to people and its expected use in future speech translation systems. The interpreters speak English and Japanese in all of the conversations and take the role of the speech translation system. One remarkable characteristic of the database is the integration of speech and linguistic data. Each conversation comprises recorded speech, transcribed utterances, and their correspondences.

In creating SLDB, we have tried to collect conversations consisting of speech and sentences that are acceptable for a speech translation system in the near future. For that reason, we have imposed the following constraints on utterances and turns.

1. A speaker can only speak when it is his/her turn. When a speaker wants to stop speaking and listen to a response, he/she actively transfers the speaking permission to the other speaker.
2. Each utterance must be concluded in ten seconds or less, and is then sent to an interpreter.
3. The interpreter translates the speaker's utterance and conveys it to the other speaker. The speaker's utterance is translated consecutively rather than simultaneously, so that the length of the interpreter's utterance is almost the same as the speaker's original utterance. This is also done to resemble the output of speech translation systems in the near future.
4. When a speaker cannot finish what he/she wants to say within the ten seconds, he/she must use several utterances of ten seconds or less.

Each utterance (or translation) continues until the speaker transfers his/her speaking permission (or alike).
5. No interruption of a speaker (or interpreter) is allowed.

With the above constraints, we can avoid extremely long sentences and overlapping of utterances. All human interpreters were professionals resulting in a high translation quality that never caused mistranslations during the conversations.

According to our transcribed bilingual text, all of the expressions in both English and Japanese are acceptable if we consider that they are spoken languages.

As previously noted, SLDB can be seen as a collection of bilingual dialogs using ideal S2ST systems. However, utterances in SLDB have different characteristics from utterances collected from conversation in a real environment. This can be observed when comparing SLDB to a Japanese monolingual travel conversation database, "The Travel Arrangement Task (TRA)," with conversations collected using similar scenario settings and the same recording conditions for robust speech recognition research (Nakamura et al., 1996; Takezawa et al., 1998a). Table 10.10 shows the characteristics of the bilingual and monolingual travel conversation databases. The frequency of filled pauses and the frequency of self-repairs of the bilingual travel conversation database are less than those of the monolingual travel conversation database. Several constraints may

Table 10.10 Characteristics of bilingual and monolingual travel conversation databases.

Conversation style	Bilingual (J–E)	Monolingual (J–J)
Number of collected conversations	618	892
Speaker participants	71	499
Interpreter participants	23	0
Total number of utterances	16,107	22,874
Total number of Japanese words	301,961	491,159
Utterances including one filled pause or more	24%	42%
Utterances including one self-repair or more	3%	6%
Average unit length (per utterance)	30 morae	35 morae

cause these significant differences. The situations in the Japanese monolingual conversation database involve direct communications between two Japanese speakers, but the bilingual situations involve indirect communications. When collecting conversations, the turn-around time is much longer for the bilingual case than for the monolingual case, such that the speakers in the bilingual case may have more than enough time to consider what to say and how to say their next utterances.

MT-Assisted Dialogs (MAD)

The last approach is intended to collect representative utterances that people will input to S2ST systems. For this purpose, we carried out simulated (i.e., role-play) dialogs between two native speakers of different mother tongues using a Japanese-English bidirectional S2ST system instead of human interpreters. We replaced the speech recognition modules with human typists for most parts of the dialogs in order to concentrate on the effects of MT by circumventing communication problems caused by speech recognition errors (Takezawa and Kikui, 2003). In this case, the resulting system is considered equivalent to using an S2ST system whose speech recognition part is almost perfect. The environment with human typists is somewhere between the "Wizard of Oz" approach used in Verbmobil (Jekat and Hahn, 2000) and in creating the SLDB, which replaced the entire S2ST process with humans, and an approach that relies only on an S2ST system (Costantini et al., 2002). In order to investigate the effects of including speech recognizers, we substituted real speech recognizers for human interpreters in some dialogs. An overview of the data collection environment is shown in Figure 10.18.

Translation quality generally depends on the linguistic and acoustic properties of the training and test corpora. We have carried out five sets of simulated dialogs so far, changing the complexity of the task that the speakers needed to carry out and the instructions given to the speakers.

We divided the dialog tasks into three classes: simple, medium, and complex. A task in the simple class consisted of a single request to which a simple answer was expected, such as "asking an unknown foreigner where a bus stop to downtown is." The medium class tasks included two or three negotiations, each requiring two or three utterance turns, such as ordering meals at a restaurant by asking today's special. A task in the complex

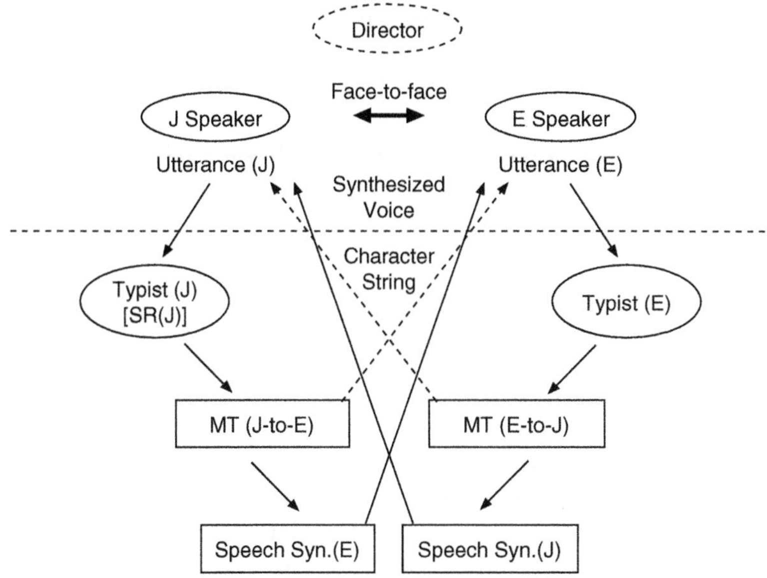

Figure 10.18: Data-collection environment of MAD.

class consisted of mutually dependent sequences of information exchange. A typical example is planning a guided tour through a conversation with a travel agent.

The MAD corpus consists of five sets of dialogs. The first set of dialogs (MAD1) was collected to see whether conversation through a machine translation system is feasible. The second set (MAD2) focused on task achievement by assigning complex tasks to participants. The third set (MAD3) contained carefully recorded speech data using medium complexity. The fourth set (MAD4) aimed at investigating the relations between instructions and utterances. For this set, we gave different types of instructions before starting the dialog. The least restrictive instruction was something like "Speak loudly and clearly," whereas a more restrictive one was "Try to divide utterances into sentences that include one topic in one sentence." For the final set (MAD5), we controlled extra information given to speakers, such as the translated text displayed on a PDA, to see whether the information affected the dialogs.

Table 10.11 shows a summary of the five experiments, MAD1–MAD5. In this table, the number of utterances includes both Japanese and English.

Table 10.11 Statistics of MAD corpora.

Subset ID	MAD1	MAD2	MAD3	MAD4	MAD5
# of utterances	3022	1696	2180	1872	1437
# of morphs per utterance	10.0	12.6	11.1	9.82	8.47
# of utterances per dialog	7.8	49.3	18.8	22.0	27.0
Task Complexity	Low	High	Medium	Medium	Medium

Average numbers depend on experimental conditions

10.4.3 Component Evaluation

Japanese Speech Recognition

The ML-SSS algorithm and the MDL-SSS algorithm were compared to create Japanese context-dependent HMMs. As described in the previous sections, the resulting acoustic models can capture both contextual and temporal variations, while decision-tree clustering can capture contextual variations only. Furthermore, the MDL-SSS algorithm can estimate the best model almost automatically, while the ML-SSS algorithm needs to find the best parameters via experiments.

For the acoustic training set, we used the Japanese travel dialogs in "The Travel Arrangement Task (TRA)." We also used 503 phonetically balanced sentences (BLA) read by the same 407 speakers of the TRA. TRA includes about 5 hours of speech, and BLA includes about 25 hours of speech. The TRA corpus includes many expressions that are similar to those of the BTEC, and the BLA corpus is helpful for creating Japanese standard phoneme models.

The analysis conditions were as follows: The frame length was 20 ms and the frame shift was 10 ms; 12-order MFCC, 12-order ΔMFCC, and Δ log power were used as feature parameters. The cepstrum mean subtraction was applied to each utterance. We used 26 Japanese phonemes and silence. Table 10.12 shows the phoneme units for the Japanese ASR.

Table 10.12 Phoneme units for Japanese ASR.

Vowels	a, i, u, e, o
Consonants	b, ch, d, g, f, h, j, k, m, n, ng, p, q, r, s, sh, t, ts, w, z, zh

The silence model with three states was built separately from the phoneme models. Three states were used as the initial model for each phoneme. The scaling factors $C_c = 2, C_t = 20$ were used for the MDL-SSS. After a topology was obtained by each topology training method, mixture components were increased, and a five Gaussian mixture model was created.

The transcriptions of the BTEC and SLDB corpus were used to create language models. The training databases included 6.2 million words. A word bigram model, a word trigram model, and a Multi-Class Composite (MCC) bigram model were created. The MCC bigram model included 4,000 classes for each direction. The size of the lexicon was 54,000 words, and the number of extracted composite words was 24,000 words. For recognition, the gender-dependent acoustic model and the MCC bigram model were used in the first pass, and the word trigram model was used to rescore word lattices in the second pass.

For the test set, the Japanese test set 01 of the BTEC was used, which contains 510 sentences. As a speech database, we collected utterances by 20 males and 20 females. Each speaker uttered 102 utterances included in the test set 01.

As Table 10.13 shows, the perplexity of the MCC bigram model lies between the word bigram and the word trigram model.

Table 10.14 shows the recognition performance represented by word accuracy (WA) rates for the Japanese BTEC test set 01. The model with 2,086 states, created by the MDL-SSS, using scaling factors $C_c = 2$,

Table 10.13 Perplexity for Japanese BTEC test set 01.

word 2-gram	word 3-gram	MCC 2-gram
30.64	17.45	24.81

Table 10.14 Recognition performance for Japanese BTEC test set 01.

	#states	WA[%]
ML-SSS	2,100 (max state length = 4)	94.51
MDL-SSS	2,086 ($C_c = 2, C_t = 20$)	94.41

Table 10.15 Word accuracy rates [%] for two different language model combinations (the MDL-SSS acoustic model).

Model	Word 2-gram		MCC 2-gram	
3-gram Rescore	No	Yes	No	Yes
Acc.(%)	91.98	93.84	93.37	94.41

$C_t = 20$, obtained almost the same performance as that with 2,100 states created by the ML-SSS.

Table 10.15 shows the word accuracy rates by two combinations of language models. For the first-pass search, one used the word bigram model and the other used the MCC bigram model. Furthermore, both of them used the word trigram model for rescoring in the second pass search. The acoustic model was the same as the MDL-SSS model with scaling factors $C_c = 2, C_t = 20$ in Table 10.14. The MCC bigram model obtained a 17.3% error reduction rate compared to the word bigram model, and the combination of the MCC bigram model and the word trigram model obtained a 9.25% error reduction rate compared to the combination of the word bigram model and the word trigram model.

English Speech Recognition

In contrast to the Japanese speech recognition system, in-domain acoustic training data were not available at the first stage of developing the acoustic model of the English speech recognition system. However, as Lefevre et al. (2001) demonstrated, out-of-domain speech training data do not cause significant degradation of system performance. In fact, it was found to be more sensitive to the language model domain mismatch. Thus, we chose the *Wall Street Journal* (WSJ) corpus for acoustic model training, since we needed a speech database that was large enough and that contained clean speech from many speakers. About 37,500 utterances recommended for speaker-independent training (WSJ-284) were selected as the training set for our acoustic model. The total number of speakers was 284 (143 male and 141 female). Feature extraction parameters were the same as for the Japanese language system: 25 dimensional vectors (12 MFCC + 12 Delta MFCC + Delta pow) extracted from 20 ms long windows with a 10 ms shift. First, we trained a model of the same size and topology with the same training method as the Japanese baseline—that is, 1,400 states with five mixture components per state and the ML-SSS algorithm. This was rather small compared to the other models that have been built on the same data (IWSLT workshop 1994), so it was not expected to have high performance. Nevertheless, we regarded it as a starting point for further model development and optimization. Next, we trained several models using the MDL-SSS algorithm, in which the temporal splitting constant C_t was set to 20 and the contextual splitting constant C_c took values from 2 to 10. In this way, we obtained models with state numbers ranging from about 1,500 to about 7,000. Initially, they all had five mixture components per state. The preliminary tests showed that the model with 2,009 states was the best and was therefore selected for further experiments. Two more versions of this model—with 10 and 15 mixture components per state—were trained as well.

The language model training data consisted of 600,000 English sentences of BTEC and SLDB and about 3.4 million words. Standard bigram and trigram models were trained as well as one MCC word bigram model. The number of classes was 8,000, while the number of composite words was about 4,000.

Although the BTEC task domain is quite broad, there are many travel-oriented words that are not included in publicly available pronunciation dictionaries. Also, there are many specific proper names of sightseeing places, restaurants, travel-related companies, and brand names. A large portion of the task word list represents Japanese words, including Japanese first and family names. In total, there were about 2,500 such words ($\approx$10% of the 27,000-word dictionary), and to develop good pronunciation variants for them was quite a challenge. For Japanese words, because there is no principled way to predict how a native English speaker would pronounce them, the pronunciation will depend heavily on the speaker's proficiency in Japanese, ranging from being fluent to speaking just a couple of widely known words. Therefore, we decided to mimic the first extreme by taking one pronunciation variant from the Japanese dictionary and converting the Japanese to the English phone set, and to mimic the second extreme by generating another pronunciation variant by following the English grapheme-to-phoneme rules. The latter was done by using the TTS software "Festival" followed by a manual correction of some of the pronunciations judged as "making no sense."

The English phoneme set consisted of 44 phonemes, including silence. They were the same as those used in the WSJ corpus official evaluations

Table 10.16 Acoustic model performance comparison.

<table>
<tr><td>Model</td><td>ML-SSS</td><td colspan="4">MDL-SSS</td></tr>
<tr><td>State #</td><td>1,400</td><td>1,578</td><td colspan="2">2,009</td><td>3,028</td></tr>
<tr><td>Mix. #</td><td>5</td><td>5</td><td>5</td><td>15</td><td>5</td></tr>
<tr><td>Acc.(%)</td><td>87.5</td><td>88.1</td><td>88.5</td><td>89.4</td><td>88.2</td></tr>
</table>

Table 10.17 Language model performance comparison.

<table>
<tr><td>Model</td><td colspan="2">Word 2-gram</td><td colspan="2">MCC 2-gram</td></tr>
<tr><td>3-gram Rescore</td><td>No</td><td>Yes</td><td>No</td><td>Yes</td></tr>
<tr><td>Acc.(%)</td><td>89.21</td><td>92.35</td><td>89.63</td><td>93.29</td></tr>
</table>

because in this way, we could use its dictionary as a source of pronunciation base forms. In addition, we could run the WSJ task tests with our model to compare performance.

In the first series of experiments, we evaluated the performance of the several acoustic models we have trained. The test data comprised 1,200 sentences from 35 speakers. Small conventional bigram and trigram language models covering about 25% of all text training data were used to speed up the evaluation. The recognition results in terms of word accuracy are given in Table 10.16. As can be seen, the MDL-SSS model with 2,009 states and 15 mixture components was the best one, thus it was used for the next experiments involving different types of language models.

Next, we evaluated the language model's performance. In these experiments, we used 204 utterances taken randomly from the larger BTEC test set. The results are summarized in Table 10.17.

Chinese Speech Recognition

The basic subword units used for the Chinese speech recognition front end were the traditional 21 initials and 37 finals, as illustrated in Table 10.18.

The acoustic model was developed using a well-designed speech database: the ATR Putonghua (ATRPTH) speech database of 2003 (Zhang et al., 2003). This database has a rich coverage of the triplet initial/finals

Table 10.18 Subword units for Chinese ASR system.

Unit	Types
Initials	b, p, m, f, d, t, n, l, g, k, h, j, q, x, z, c, s, zh, ch, sh, r
Finals	a, ai, an, ang, ao, e, ei, en, eng, er, i1, i2, i3, ia, ian iang, iao, ie, ing, in, iu, iong, o, ou, u, ua, uai, uang uan, ui, un, uo, ong, v, van, ve, vn

Table 10.19 Token coverage rates of different subword units.

Unit	792 Set	Newspaper	Token Coverage
A	974	1,306	98.81%
B	402	408	99.99%
C	10,906	48,392	70.15%
D	4,653	4,598	99.42%

phonetic context, and sufficient samples for each triplet with respect to balanced speaker factors, including gender and age.

The phonetically rich sentence set of ATRPTH has 792 sentences. An investigation on the *token coverage rates* has been carried out on a one-month volume of daily newspapers for different types of phonetic units. Table 10.19 shows the results, where

- Unit *A*: represents the tonal syllable
- Unit *B*: represents the base syllable without tone discrimination
- Unit *C*: represents the normal initial/final triplets
- Unit *D*: represents the context-tying initial/final triplets, which are tied based on phonetically articulatory configurations, and are assumed to cover the major variants of each triplet phonetic context (Zhang et al., 2001).

The speakers were chosen to have a balanced coverage of different genders and ages. Each unique triplet has at least 46 tokens in the speech database, guaranteeing a sufficient estimation for each triplet HMM.

During the model estimation, accurate pause segmentation and context dependent modeling were done iteratively to guarantee the model's accuracy and robustness (Zhang et al., 2003a). The HMM structure was derived

Table 10.20 Chinese character based recognition performance.

Group	Character Corr.	Character Acc.
Male	96.1%	95.7%
Female	95.2%	94.4%
Total	95.7%	95.1%

through a phonetic decision tree based maximum likelihood state splitting algorithm. The acoustic feature vector consisted of 25 dimensions: 12 dimensional MFCCs, their first order deltas, and the delta of frame power. The baseline gender-dependent HMM had 1,200 states, with 5 Gaussian mixtures in each state.

The language model for Chinese ASR also used the composite MCC N-gram model. The basic lexicon had 19,191 words, while the BTEC Chinese corpus contained 200,000 sentences for LM training. After they were segmented and POS tagged, word clustering was investigated based on the right- and left-context Markovian dependencies. A normal word based bigram model showed a perplexity of 38.4 for the test set with 1,500 sentences. With a clustering of 12,000 word classes, the MCC bigram model showed a perplexity of 34.8 for the same test data. The bigram language model was used to generate a word lattice in the first pass, and a trigram language model with a perplexity of 15.7 was used to rescore the word lattice.

The evaluation data were the BTEC Chinese language-parallel test data, which included 11.59 hours of speech by 20 females and 20 males. The ages of the speakers ranged from 18 to 55 years. All the speakers spoke Chinese Putonghua, with some accent.

Table 10.20 shows the gender-dependent, Chinese character based recognition performances. The total performance was 95.1% for Chinese character accuracy with a real-time factor of 26. The performance degraded to 93.4% when the search beam was narrowed to obtain a real-time factor of 6, emphasizing the need for algorithms to increase the search speed without sacrificing performance.

Text-to-Speech Evaluation

A perception experiment was conducted in which the naturalness of synthetic speech for XIMERA for Japanese and 10 commercial TTS systems

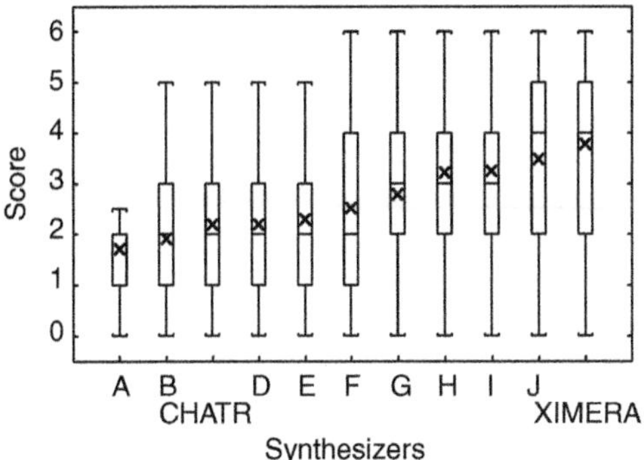

Figure 10.19: The result of an evaluation experiment for naturalness between several TTS products. The horizontal bars at the top, middle, and bottom of the boxes indicate 75%, 50%, and 25% quartiles. The vertical lines extend to the closer point of either the maximum/minimum value or 1.5 times the interquartile distance. Mean values are indicated by "x" marks.

was evaluated. A set of 100 Japanese sentences that were evenly taken from 10 genres was processed by the 11 TTS systems to form a set of stimuli comprising 1,100 synthetic speech samples. The stimuli were randomized and presented to 40 listeners through headphones in a quiet meeting room. The listeners rated the naturalness of each stimulus on a 7-point scale from 0 (very bad) to 6 (very good).

Figure 10.19 shows the result, in which XIMERA outperforms the other systems. However, the advantage over the second-best system, which is not a corpus-based system, is not substantial, although it is statistically significant.

A perception experiment for naturalness of synthetic speech for Chinese XIMERA is undergoing planning, and we have no evaluation results comparing other Chinese TTS systems.

10.4.4 Machine Translation Evaluation

Rank Evaluation Method

The rank evaluation methods are the simplest and the most common among test-set-based evaluations. The training corpora used for the machine

Table 10.21 Translation quality of four systems for BTEC.

	SAT	HPAT	D-cube	SELECTOR
A	67.2549	42.5490	63.7255	68.2353
AB	74.7059	63.7255	72.1569	75.8824
ABC	82.5490	79.0196	78.8235	83.5294

translation systems were the BTEC, described in Section 10.4.2. The test set consisted of 510 pairs randomly selected from BTEC and kept unused for training. The target part of the test set consisted of the paraphrasing of up to sixteen multiple reference translations for each source sentence, which are utilized for automatic evaluation programs.

Translations by four machine translation systems—SAT, HPAT, D-cube, and SELECTOR—were shown simultaneously to several native English professional interpreters. The evaluation was done according to ATR's evaluation standard of four grades: (A) Perfect: no problems in either information or grammar; (B) Good: easy to understand, with either some unimportant information missing or flawed grammar; (C) Fair: broken but understandable with effort; and (D) Nonsense or No-output. Each translation was finally assigned to the median grade from among its grades from multiple evaluators.

Table 10.21 shows the translation quality of Japanese to English translations for BTEC. The figures are accumulative percentages for the quality grade. It is fairly high even for the difficult language pair of Japanese to English. In addition, we can see in every grade, A, A+B, A+B+C, that the SELECTOR outperforms every single element machine translation.

Translation Paired Comparison Method

When considering the users' viewpoint, it is ideal to assemble people with various levels of skills in foreign languages and to ask them to evaluate their satisfaction with the achievement of the dialog through the system. This is important because they are thought to be influenced by their target language skill. However, experimental costs are so prohibitive that dialog experiments have never been carried out except evaluation tests for the final stage of development. To address this problem, Sugaya et al. (2000) proposed a translation paired comparison method that can precisely evaluate

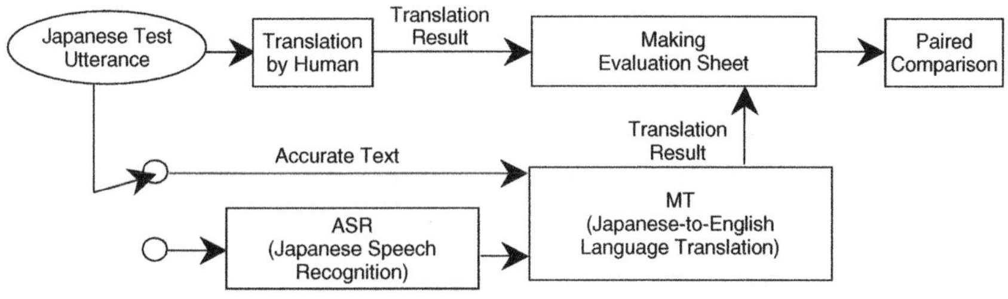

Figure 10.20: Diagram of translation paired comparison method.

systems' performance. For this method, the evaluation results are given as the systems' TOEIC scores. TOEIC is an acronym for **Test of English for International Communication**, which is a test for measuring the English proficiency of non-native speakers, such as Japanese (TOEIC, 2005). The total score ranges from 10 (lowest) to 990 (highest).

Figure 10.20 shows a diagram of the translation paired comparison method in Japanese-to-English translation. Here, Japanese native-speaking examinees were asked to listen to spoken Japanese text and then write an English translation. The Japanese utterance was presented twice within one minute, with a pause between the presentations. To measure the English capability of the examinees, their TOEIC scores were used. The examinees were asked to present their official certificates showing the TOEIC score they had earned on the test within the past six months.

In the translation paired comparison method, translations by the examinees and the output of the system were printed together in rows with the original Japanese text to form evaluation sheets for comparison by a bilingual evaluator. Each transcribed utterance on the evaluation sheets was represented by the Japanese test text and the two translation results.

The evaluator followed the procedure depicted in Figure 10.21. Ranks in the figure were defined as follows: (A) Perfect: no problem in both information and grammar; (B) Fair: easy-to-understand with some unimportant information missing or flawed grammar; (C) Acceptable: broken but understandable with effort; (D) Nonsense: important information has been translated incorrectly.

In the evaluation process, the human evaluator ignored misspellings because the capability being measured was not English writing but speech translation. From the scores based on these rankings, either the examinee or

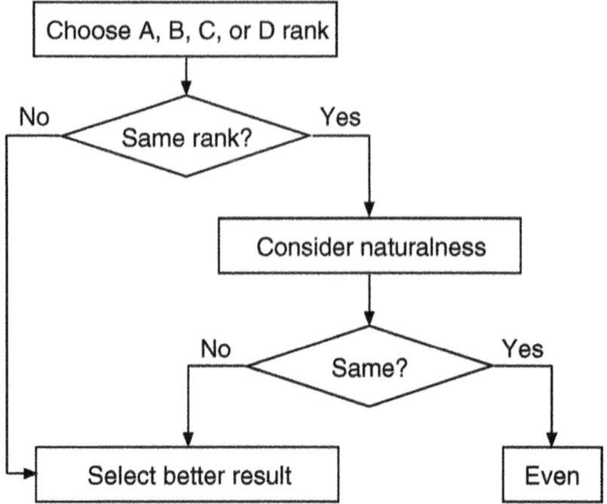

Figure 10.21: Procedure of comparison by bilingual evaluator.

the system was considered the "winner" for each utterance. If the ranking and the naturalness were the same for an utterance, the competition was considered "even."

To prepare for regression analysis, the number of "even" utterances were divided in half and equally assigned as system-won utterances and human-won utterances. Accordingly, we define the human winning rate W_H by the following equation:

$$W_H = (N_{human} - 0.5 \times N_{even})/N_{total}, \qquad (10.11)$$

where N_{total} denotes the total number of utterances in a test set, N_{human} represents the number of human-won utterances, and N_{even} indicates the number of even (nonwinner) utterances—that is, no quality difference between the results of the system and those of humans.

In the regression analysis, we regarded the examinees' TOEIC score as an independent variable and W_H as a dependent variable. In other words, W_H (Y_i) and the TOEIC scores for the examinees (X_i) are assumed to satisfy the population regression equation:

$$Y_i = \beta_1 + \beta_2 X_i + \epsilon_i \qquad (i = 1, 2, \ldots, m), \qquad (10.12)$$

where β_1 and β_2 are population regression coefficients and m is the number of examinees.

"System's TOEIC score" defines the capability-balanced point between the system and the examinees, and it can be determined as the point at which the regression line crosses half the total number of test utterances, that is, W_H of 0.5.

Figure 10.22 shows an example of evaluation results. In this case, the total number of examinees was 29, and their TOEIC scores ranged from 300 to 800. Every hundred-point range had five examinees, except for the 600s, which had four examinees. The test set consisted of 330 utterances from the SLDB corpus. In this figure, the straight line indicates the regression line. This system's TOEIC score is 705, the cross point of the regression line with $W_H = 0.5$. Consequently, the translation capability of the language translation system equals that of an examinee with a score of around 700 points on the TOEIC scale.

Automatic Evaluation Scheme

The translation paired comparison method can precisely evaluate the system's performance, but still leaves the problem of evaluation cost. Therefore, we developed an automatic evaluation scheme that can estimate the system's TOEIC score with an objective method. Basically, the method substitutes the human evaluation process with an objective evaluation method.

The first point to explain is the automation of the translation paired comparison method. There are two levels for applying an objective evaluation method to the automation of the translation paired comparison method: the utterance level and the test set level. However, both BLEU and NIST are inadequate for determining the relative merits at the utterance level. Consequently, we employed evaluation at the test set level, which consists of several hundred utterances. In a sense, this evaluation follows a different procedure from the original translation paired comparison method because the original method compared pairs at the utterance level.

The flow of the evaluation is shown in Figure 10.23. First, we apply the objective evaluation at the test set level to the evaluation of all the examinees' and the system's translations. Then, we carry out a regression analysis using the objective scores. In the regression analysis, we use the objective scores instead of W_H. More specifically, scores by the objective evaluation method (Y_{auto_i}) and the TOEIC scores for the examinees (X_i)

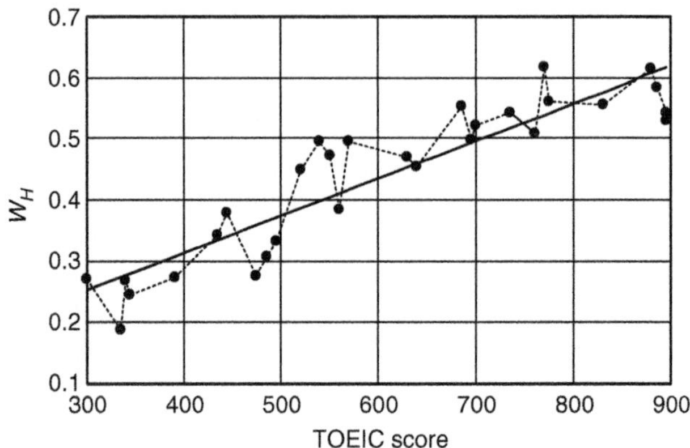

Figure 10.22: Evaluation result with the translation paired comparison method.

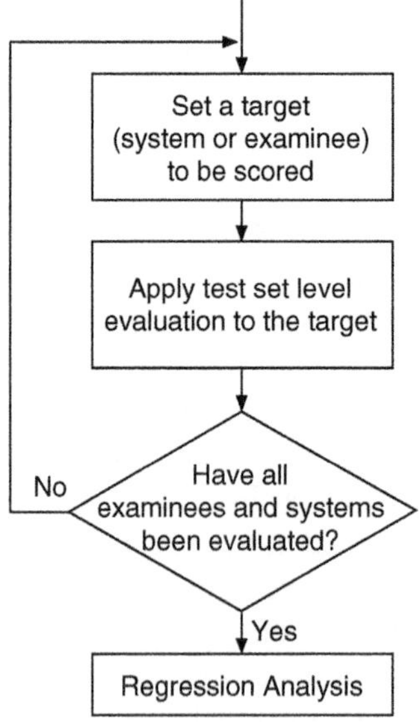

Figure 10.23: Procedure of the automatic evaluation method.

are assumed to satisfy the population regression equation:

$$Y_{auto_i} = \beta_{auto_1} + \beta_{auto_2} X_i + \epsilon_{auto_i} \qquad (i = 1, 2, \ldots, m), \qquad (10.13)$$

where β_{auto_1} and β_{auto_2} are population regression coefficients and m is the number of examinees.

In the automatic evaluation method, the system's TOEIC score is determined to be the point at which the regression line crosses the objective score evaluating the target system.

10.4.5 Overall Evaluation

Overall performance evaluation of the S2ST system was conducted with two modes: a laboratory test conducted under various controlled conditions, and a conversation test in real environments.

Laboratory Test

Translation performance was evaluated on two sets, BTEC1 and MAD4. Evaluation measures were subjective ranking, BLEU (Papineni et al., 2002), mWER (word error rate using multiple references), and the system's TOEIC score with the automatic evaluation scheme. Table 10.22 shows some of the evaluation results. In this table, subjective evaluation sections show the percentage of A, A+B, and A+B+C relative to the size of the test set.

Since BTEC originated from texts, we produced speech by reading them. In this sense, the resulting speech was artificial. In the case of MAD, we applied our speech recognizer to the recorded speech after the data collection. "MT only" for MAD4 shows the result of MT applied to correct transcriptions created manually. Speech recognition results show that we have relatively good results (i.e., just below 90% accuracy) for MAD because people spoke in a system-friendly way as described earlier. When we focus on translation performance, it is clear that BTEC is much easier than MAD. This is partly because the test sets share the same mother corpora with the training corpora. As shown in Table 10.9, BTEC contains many duplicates (e.g., 37% in Japanese) among sentences in the source and

Table 10.22 Translation performance.

Evaluation Set	BTEC1	MAD4
Speech Recognition (SR)		
Word Accuracy (%)	(94.8)	89.5
Sentence Perfect (%)	(81.4)	54.2
MT Only		
Subj (A/AB/ABC)	66.2/77.1/84.7	33.5/50.4/65.7
BLEU	0.59	0.48
mWER	0.31	0.45
SR+MT		
Subj (A/AB/ABC)	-	30.3/45.4/61.8
BLEU	-	0.37
mWER	-	0.54

target languages. Thus, some test utterances happen to be the same as training sentences even though we randomly separated the original BTEC into test and training sets. These "duplicate" utterances do not mean "closed" data since reference translations for the test-set utterances were created independent of original translations in the corpus. In order to know this effect, we divided our test data into those that were included in the training corpus (MATCHED) and the remaining data (UNMATCHED). These two subsets were applied to our MT system.

The results are shown in Table 10.23. We see that the subjective scores of the nonduplicate part are still better than the average scores of MAD. We conclude that BTEC is easier than MAD for the MT module based on the subjective evaluation scores.

Table 10.23 Translation performance for BTEC with/without duplications.

	UNMATCHED	MATCHED
Subj (A/AB/ABC)	43.9/60.4/73.7	88.6/93.7/95.7
BLEU	0.51	0.73
mWER	0.42	0.19
Perplexity	36.3	14.1

Table 10.24 Translation performance of test set without duplications.

	Low	Middle	High
Subj (A/AB/ABC)	62.3/78.4/89.8	23.4/40.7/62.9	14.9/32.1/44.6
BLEU	0.60	0.49	0.4
mWER	0.31	0.45	0.57
Perplexity	9.8	29.9	97.0
Length*	9.6	12.0	12.4

**Length = # of words per utterance*

To have a closer look at MAD, we divided the test utterances into three subsets with the same size in terms of their perplexity[2] and calculated scores for each subset, as shown in Table 10.24. The table shows that the perplexity of each subset is clearly different and that the translation performance negatively correlates with the perplexity. Consequently, we need to improve the translation performance of mainly high-perplexity utterances.

The system's TOEIC scores with the automatic evaluation scheme were also obtained. The system achieved on the BTEC and MAD subsets of low and middle perplexity a TOEIC score of more than 700. This is 50 points higher than the average score of a Japanese businessperson in the overseas departments of many Japanese corporations. On the contrary, the system's TOEIC score for all the test sets including the BTEC and every subset of MAD decreases to about 500 due to low TOEIC score for the subset of MAD with high perplexity, which is mainly due to low recognition accuracy, and a loss of fluency in the translated expressions.

These results support the bootstrapping style approach. Another promising approach is to create corpora that align high-perplexity utterances with their simplified expressions (i.e., paraphrases) and develop a corpus-based paraphrasing system (Shimohata et al., 2004). For this purpose, simplified utterances in MAD can be used as communication oriented paraphrases for S2ST.

Evaluation Tests in Real Environments

Real environments have a complex and often uncontrollable impact on S2ST systems. There is a limit to carrying out such evaluation tests with

[2]Perplexity was calculated in the source language.

regard to the cost of conversation experiments in real environments, however, evaluation tests in real environments are important for fully evaluating the availability of the system. Therefore, we conducted an evaluation test of S2ST systems from the viewpoint of user satisfaction in locations where S2ST systems may be necessary, such as international airports.

Prototype S2ST System

ATR has developed two kinds of prototype S2ST systems for conducting evaluation tests in real environments—one between English and Japanese, and the other between Chinese and Japanese. These prototype systems consist of two sets of handheld PDAs (Personal Digital Assistants) connected by means of wireless LAN to server modules of speech recognition, machine translation, and speech synthesis in the network. One PDA is used as an input device for Japanese and an output device for Japanese translated from English or Chinese. The other is used as an input device for English (Chinese) and an output device for English (Chinese) translated from Japanese. In this configuration, two sets of PDAs are used only as speech input and output devices and for some additional functions, such as speech detection, filtering, and noise reduction. We employed this configuration because we believe that (1) an input/output device should be portable and as light as possible in order to be taken everywhere, and (2) S2ST systems should accept expressions that are as diverse as possible to support communications in various situations. Since the ambient noise at the evaluation location (airport) is very high and nonstationary, we used head-mounted close-talking microphones.

Evaluation Test Procedure

The evaluation test sites were installed near the tourist information desk in the Kansai International Airport and downtown in Osaka city. The speakers of Japanese were staff members of the tourist information desks, who frequently speak English but not Chinese. The speakers of English and Chinese were tourists walking in the airport or downtown who volunteered to participate in the experiment.

After a short introduction on how to use the PDA, they were asked to do three kinds of tasks. In one, they were asked to present some predetermined simple questions, such as "Where is the taxi stand?" to the staff members. In the second, they were asked to make conversation consisting of two or three turns, such as how to go to downtown. In the third, the speakers were requested to ask some questions about their travel, which is the most difficult task for S2ST systems. After conversations using the S2ST systems, they were requested to fill in a questionnaire. We conducted the evaluation tests for three days at each site, with about fifty subjects each for English and Chinese.

Evaluation Results

Figure 10.24 shows the evaluation results based on two questionnaires: one about the user's impression on the success of communication, and one about response time. As can be seen, more than 70% of English subjects think that a conversation could be carried out (almost) successfully, while only 60% of the Chinese subjects believe this. The performance degradation of the Chinese S2ST may be due to (1) the smaller volume of the Chinese-Japanese BTEC corpus, which degrades the performance of the machine translation module between Chinese and Japanese, and (2) the lower accuracy of the Chinese speech recognition module, which also results from the smaller volume of the Chinese speech database.

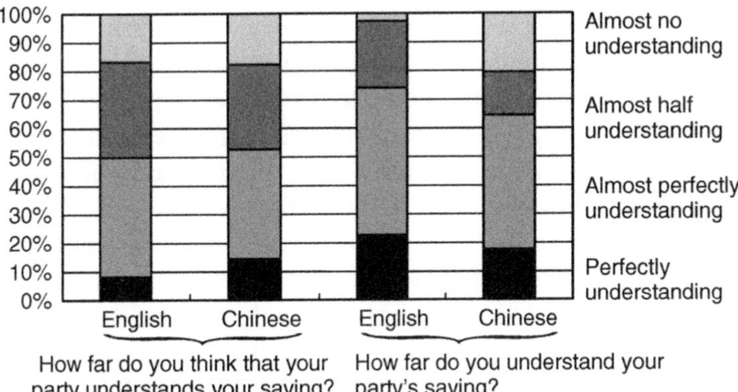

Figure 10.24: Evaluation results in real environments.

The evaluation results indicate that state-of-the-art S2ST systems evolved into a stage that allows communication across language boundaries, however, accuracy and response time (not shown in Figure 10.24) need further improvement.

10.5 Conclusion

10.5.1 Speech Translation Strategies

We investigated different translation strategies: Interlingua-based translation, statistical mapping into an interlingua representation, direct statistical translation, and using English as a pivot language. Designing an interlingua and writing the analysis and generation grammar is time consuming and requires highly trained linguists. We further introduced a statistical interlingua-based approach that applies techniques that have been initially developed for mapping word sequences into tree structures in direct statistical MT and have been extended for this purpose. This still requires the design of the interlingua and the annotation of sufficient data with the interlingua, but replaces the manual writing of grammars by automatically learning the mapping of the source sentence into the interlingua representation. In our experiments, this statistical approach to interlingua-based translation did not perform as well as the manually crafted grammars. To some extent this is due to data-sparseness problems. However, the manual grammars include some phrasal translations, which improve translation quality. Such a translation memory mechanism, mapping entire sentences to the appropriate interlingua representation, could be added to the statistical IL system as well to get further improvements.

We also investigated the performance of a direct statistical translation system, which is based on word-to-word and phrase-to-phrase alignments trained from the same data. Despite the general belief that statistical machine-translation systems can only work when large bilingual corpora are available, the direct statistical system outperformed the grammar-based system. As the statistical system used only the translations, this development cost is significantly lower than that for the interlingua system. And the statistical system is flexible in that additional data, like available dictionaries or additional monolingual data to train the language model, can be easily added to improve the performance.

The comparison of translation approaches suggest that MT systems can be successfully constructed for any language pair by cascading multiple MT systems via English. Moreover, end-to-end performance can be improved if the interlingua language is enriched with additional linguistic information that can be derived automatically and monolingually in a data-driven fashion.

When dealing with speech translation, we are faced with disfluencies and with errors from the speech recognizer. To handle disfluencies, we developed a consolidation module, which detects and removes these disfluencies. The approach is based on a noisy channel approach, borrowing essentially from the statistical machine-translation techniques.

Finally, we investigated better ways to couple speech recognition and translation to improve translation quality by optimizing the overall system. Translating all the paths in the word lattice generated by the speech recognition system and using the acoustic scores in addition to the translation and language model scores resulted in an improvement over translating only the first-best recognizer output.

Much remains to be done to bring robust speech translation to practical day-to-day use in the many languages of the world. Our experiments indicate that data-driven approaches—automatically learning from bilingual corpora—is the most competitive approach to rapid building of speech translation systems. So far, systems have been demonstrated for limited domain speech translation tasks, and these will remain important in the future. However, steps should and, we believe, can be taken now toward domain-unlimited speech translation.

10.5.2 Portable Speech-to-Speech Translation

We have conducted research on corpus-based technologies because we believe that corpus-based technologies are suitable for S2ST, taking into consideration the points of (1) multilanguage systems, (2) domain portability, and (3) the technology trend of each component technology for S2ST. In order to develop S2ST technologies, we have created various speech databases for speech recognition and speech synthesis, and also three different types of corpora in the travel domain: (1) a large-scale multilingual collection of basic sentences called BTEC, (2) a small-scale bilingual collection of spoken sentences called SLDB, and (3) a small-scale corpus of the MT-assisted dialogs called MAD.

We have developed two kinds of corpus-based S2ST systems based on various machine-learning algorithms for each component, a huge speech database, and multilingual corpora. One is the S2ST system between English and Japanese; the other, between Chinese and Japanese. We have conducted various evaluation tests of the S2ST systems using a subset of the BTEC and MAD. Thanks to many various technologies in each component module, and many large speech and text corpora, the S2ST systems provide good translation quality for utterances of the BTEC and also for utterances of the MAD with low and middle perplexity. We have also conducted evaluations of the systems from the viewpoint of user satisfaction in real environments, such as an international airport. The evaluation results show that the performance of the S2ST systems is not fully satisfactory, however, 60–70% of users think that conversation could be carried on successfully or almost successfully using the S2ST systems. One of the most important results is that about 60% of the subjects thought that the S2ST systems could be applicable to communication between people whose mother tongues are not understandable to each other.

However, the state-of-the-art S2ST technology still has a lot room for improvement. One of the problems to be resolved is to shorten the response time, especially for long utterances. It is desirable from the users' viewpoint to provide confidence measures for translated sentences. As is shown in the section on automatic evaluation schemes, we have an objective evaluation measure for estimating the quality of translated sentences in the test-set level, which shows the average performance for collection of utterances; however, we do not have adequate measure for estimating the quality in the utterance level. The other point is how to exchange information among each module. We have employed a simple cascade configuration of the speech recognizer and machine translation, in which the single best output of the former module is fed into the latter. The N-best lists of the former output contain the better recognition result, which may produce better translation results. In order to improve the performance of the S2ST system, it is important to investigate ways of improving the performance of each module but also to achieve more collaborative functions between the speech recognition module and the machine-translation module. One way is to develop larger multilingual corpora, which cover more domains and diverse expressions. A promising approach is to create corpora that align high-perplexity utterances with their simplified expressions (i.e., paraphrases) and develop a corpus-based paraphrasing system (Shimohata et al., 2004). For this purpose, simplified utterances in MAD can be used as communication-oriented paraphrases for S2ST.

Chapter 11

Multilingual Spoken Dialog Systems

Helen Meng and Devon Li

11.1 Introduction

Spoken dialog systems (SDS) are becoming increasingly pervasive in our everyday lives for information access. Efforts devoted to the design and development of SDS aim to bring the right information to the right people at the right time for a diversity of application domains, such as finance, air travel, train schedules, and weather. These domains typically involve dynamic information of recurring interests to the user. SDS encompasses a suite of speech and language technologies to offer a conversational interface to dynamic information, including speech recognition, natural language understanding, dialog modeling, and speech synthesis. Hence, the user can present queries to the system by speaking naturally, and the SDS can respond in real time in synthetic speech. Numerous commercial SDS have been deployed for multiple languages. For example, the SpeechWorks[1] voice-activated stock-trading system enables callers to get real-time stock

[1] Now Scansoft Inc. (www.scansoft.com)

and market information, place orders, and trade stocks at any time. The entire human-computer interaction is conducted over a fixed-line/mobile telephone channel—that is, in a screenless setting. An example is provided in Table 11.1 to illustrate the directed nature of the dialog interactions, in which the system elicits a series of information attributes from the user in a scripted order. Interactions of this style are often known as *directed dialogs* or ***system-initiative dialogs***. The system is always in control to guide the dialog at each step and constrain possible user input to a small set of options. In this way, the system is able to attain a decent level of robustness in recognizing and interpreting user input in order to achieve a high task-completion rate for informational transactions. However, the ease of use of these human-computer spoken-language interfaces is partially sacrificed due to constraints in interactivity. Recent research efforts in the field strive to relax the constraints in system-initiative dialogs.

Mixed-initiative dialog interactions allow both human and computer to influence the dialog flow and thus offer greater flexibility than system-initiative dialogs. If users could assume complete control of the

Table 11.1 An example dialog in the stocks domain illustrating the capabilities of a state-of-the-art spoken dialog system (source: www.speechworks.com).

System1	Welcome to Lim & Tan Securities voice-activated stock trading system. <tone>Stock trade menu. Please say the name of the stock you'd like to place an order for.
User1	*Singapore Airlines*
System2	Buy or sell?
User2	*Buy*
System3	How many shares would you like to buy?
User3	*Two thousand*
System4	Singapore Airlines Corporation. At what price would you <interrupted>. . . ?
User4	<barge in>*eighteen eighty*
System5	Would you like to use cash or CPF?
User5	*Cash*
System6	OK you want to buy 2,000 shares of Singapore Airlines Corporation at eighteen eighty using cash. Your order has been queued. Returning to main menu.

human-computer interactions and if the system could respond sufficiently to all user requests, we would have ***user-initiative dialogs***. However, this remains an elusive goal for existing technologies and systems. Recent efforts are devoted toward the development of mixed-initiative dialog models, in which the computer takes initiative at appropriate times to constrain the user's request in order to progress toward an intended/revised task goal (Horvitz, 1999; Levin, et al., 2000). Prominent mixed-initiative systems (Seneff et al., 1999; Rudnicky et al., 1999; Rosset et al., 1999) involve application domains such as air travel, train schedules, hotel, and traffic information.

A number of SDS have been extended to cover multiple languages. Examples include MIT's systems for English, Chinese, German, Japanese, and Spanish, and for a variety of domains, such as weather, flight scheduling, city navigation, and secondary language acquisition (Seneff et al., 2004; Zue et al., 2000). Other examples include the EU's Esprit MASK and ARISE projects that have developed spoken dialog systems in French, Dutch, and Italian for railway information inquiries (Mariani and Lamel, 1998). This chapter describes the design considerations in the development of an SDS for multilingual input and gives an overview of the approaches to this problem, followed by a detailed description of one system implementation—the ISIS system for the stocks domain (Meng et al., 2000, 2001, 2002). ISIS is a trilingual system that supports English, Cantonese, and Putonghua, the prime languages used in the area/region of Hong Kong. Putonghua (also known as Mandarin) is the official Chinese dialect, and Cantonese is a major Chinese dialect predominant in Hong Kong, Macau, and South China. ISIS resembles a virtual stockbroker who can provide the user with real-time stock market information, manage simulated personal portfolios, and handle simulated financial transactions via mixed-initiative conversational interactions. Within the context of ISIS, we also attempted an initial exploration along three possible directions of evolution related to multilingual conversational interfaces. The first is *adaptivity*, referring to the automatic expansion of the application's knowledge base to keep the dialog system up-to-date with new changes related to the domain in all supported languages. The second is *offline delegation*, which enables users to delegate tasks for the computer to execute offline—for example, in information monitoring. This is an instance of asynchronous human-computer interaction in which the two interlocutors do not take alternate turns. Instead, the dialog continues along a thread that may differ from the delegated task, and the response corresponding to the delegation may be generated at some indeterminate time in the future. This brings us to the third possible direction of evolution, *multithreaded dialogs*, in which an online interaction (i.e., an ongoing dialog thread) is *interrupted* by an alert message related to a previously delegated task. Such interruption calls for proper handling of the multiple discourse contexts as the interaction switches from one dialog thread to another. We believe that while support for offline delegation and multithreaded dialogs should be language independent, their entries into the main dialog flow should avoid code-switching.

While the aforementioned evolutionary developments are explored in concert in the context of ISIS, this work is also anticipatory of universal accessibility to the Web content at anytime, from anywhere, by anyone, and with any device. Web browsing by the Internet population is no longer restricted to desktop personal computers, but is expanding to mobile client devices of diverse form factors, such as personal digital assistants (PDAs), WAP phones, smart phones, as well as the conventional (displayless) telephones. We recently devoted some effort to developing a platform that supports *displayless* voice browsing of Web content through the telephone channel for both English and Chinese. The content has been transcoded from HTML to XML and transformed into VXML documents. These can be accessed by a voice gateway that connects to the Internet and the public switch telephone network. The voice browser is software that runs on the voice gateway and can interpret VoiceXML (VXML) documents. A VXML document specifies a human-computer dialog—human speech input is supported by speech recognition; and computer speech output is supported by speech synthesis/digitized audio. These core speech engines (for recognition and synthesis) are invoked by the voice browser. In addition, the voice browser can also handle touch-tone (telephone keypad) input. Existing voice browsers generally support a single, primary language, mostly English, for voice browsing. This limits displayless, telephone-based voice browsing to *English* Web content and English speaking users only. In this chapter, we present an attempt that extends displayless voice browsing to *both* English and Chinese Web content by developing a bilingual (English and Cantonese) voice browser for the VXML platform.

The rest of the chapter is organized as follows: Section 11.2 reviews the multilingual issues related to the development of SDS as well as related research areas in dialog modeling. Section 11.3 presents an overview of the

ISIS system. Explorations in adaptivity, offline delegation, interruptions, and multithreaded dialogs are described in Sections 11.4, 11.5, and 11.6. Section 11.7 presents some empirical observations of user interactions with ISIS. Section 11.8 takes a slight divergence to describe an example implementation of a simple multilingual SDS based on the VXML platform. Section 11.9 concludes and presents possible future directions.

11.2 Previous Work

This section presents a brief overview of multilingual SDS, implementation frameworks for dialog models, handling new vocabularies in spoken language input, and dialog model extensions to support delegation and interruption.

A generic SDS consists of:

- A speech recognizer that transcribes input speech into text
- A natural language understanding component that transforms the recognition output into a semantic representation (typically via parsing)
- A discourse and dialog manager that handles the inheritance of discourse history, content retrieval (often via database access), and dialog turn taking between the human and the computer
- A spoken response generator that verbalizes the retrieved content
- A text-to-speech synthesizer to generate a spoken presentation of the verbalized content

Development of multilingual SDS sees a drastic increase in system complexity for every additional language supported. Very often, multilingual SDS involves multiple speech recognizers, one for each supported language. This naturally creates the need for language identification as a preprocess unless the selected language is explicitly stated by the user. Alternatively, one may attempt to develop a single recognizer that supports vocabularies from multiple languages (see Chapter 5). This renders language identification an implicit process and can easily support code-switching within an utterance. However, experience has shown that the need to cover significantly expanded acoustic, phonetic, and lexical spaces often leads to large training data requirements across languages, difficulties in language modeling due to code-switching, and noticeable degradations in recognition performances. Multiple grammars must also be developed for parsing the different input languages, and their respective parse-tree nodes need to be mapped to a common set of attributes for database retrieval. This procedure, as in dialog management, should be language independent. Parallel generation grammars will also need to be developed for verbalizing responses in the various output languages, and responses in each language are passed to the appropriate text-to-speech synthesizer for generating a spoken presentation. As such, we also need multiple synthesizers—one for each supported output language. Recent efforts have been devoted toward the development of a single text-to-speech synthesizer that can support multiple languages, but this area is still relatively new. The rapid increase in complexity as we migrate from monolingual to multilingual SDS creates dire needs and challenges to identify the language-dependent and language-independent components in a multilingual SDS, as well as attempts to unify different components across languages to achieve a modular and compact implementation.

Regarding dialog management, previous approaches can largely be categorized into ***state-based dialog management*** and ***frame-based dialog management***. The state-based framework is often used for implementing system-initiative dialogs (see Table 11.1). Each dialog turn is modeled as a transition through a state sequence. The entire sequence prompts for the set of information attributes in stepwise order. Expert users may attempt to expedite transactions by providing multiple information attributes in a single dialog turn (e.g., in a natural language query/request). Hence, the state-based framework may allow certain states to be bypassed while maintaining the order of information entry. An illustration from the foreign exchange hotline[2] (Meng et al., 2000) is provided in Table 11.2. While the scripted order for information entry helps achieve a satisfactory level of robustness in task completion, it also constrains interactivity in the dialog. This constraint is relaxed in the *frame-based framework*, which is often used for implementing mixed-initiative dialogs. This framework resembles filling information slots in an electronic form (E-form), and the entries can be input in *any* order. At each dialog turn, the system may check for unfilled slots and prompt the user for associated information attributes. The user

[2]This is the CU FOREX system. A demonstration of the direct dialog is available at +852-2603-7884, and a demonstration of the natural language shortcut is available at +852-2603-7830.

Table 11.2 An example dialog from the CU FOREX hotline (Meng et al., 2000).

System1	Welcome to CU FOREX, a trilingual hotline for real-time foreign exchange inquiries developed at The Chinese University of Hong Kong. Which language would you prefer? Please say Cantonese, Putonghua or English ... (same prompt repeated in the other languages)
User1	*English*
System2	This system can handle a full question for your inquiry. <tone>Main menu. What kind of currency information are you interested in?
User2	*I'd like to know the three-month interest rate of the Eurodollar.*
System3	The quotes we provide are for reference only. Interest rate. Eurodollar for three months is two point four five. <tone>Main menu. What kind of currency information are you interested in?
User3	*How about the exchange rate between the British pound and the Hong Kong dollar?*
System4	The quotes we provide are for reference only. British pound to Hong Kong dollar. The buying rate is twelve point six zero seven. The selling rate is twelve point six zero one.

may choose to be either cooperative and answer according to the prompt, or uncooperative and provide an alternative information item. In this way, both the system and the user can influence dialog progression, which is characteristic of *mixed-initiative* interactions (Goddeau et al., 1996; Aust et al., 1995). Attempts are underway to achieve even greater flexibility in dialogs to support mixed-initiative negotiations and collaborative problem solving. Such dialogs are the least structured and involve planning and reasoning with knowledge and logic (Sadek and Mori, 1997). A comprehensive review can be found in McTear (2002).

There is a critical need for spoken dialog systems to keep up-to-date with expanding knowledge scopes in the application domains, which leads to new vocabularies in one or more of the supported languages. However, the recognizer(s) in an SDS are typically developed for predetermined and fixed vocabulary(ies). The occurrence of new words, also known as out-of-vocabulary (OOV) words, will inevitably lead to errors. Previous work has studied the OOV problem mainly in the context of speech recognition and language modeling (see Chapters 5 and 6). Here, a generic word model that permits arbitrary phone sequences is used as the OOV word model (Monas and Zue, 1997; Bazzi and Glass, 2000) to detect new English words. This may be used in conjunction with a confidence model to predict recognition errors due to OOV words that are misrecognized as in-vocabulary words (Hazen and Bazzi, 2001). (Issar 1996) studied the use of class-based language models that have expandable classes to include OOV words deemed appropriate for the class. This approach is extensible across different languages which makes handling of code-switching easier. Sublexical linguistic modeling has been applied to tackle the OOV problem (Lau and Seneff, 1998; Seneff et al., 1996). The sublexical model captures the relationships among morphs, graphemes, phonemes, and phonological rules to be applied to new word detection, letter-to-sound/sound-to-letter generation, and eventually new word acquisition. For the Chinese dialects, the inventories of written characters (some 10,000 in active daily use) and spoken syllables (roughly 1,500) as well as the character-to-syllable mappings are relatively stable. This simplifies the problem of pronunciation generation. However, Chinese does not have an explicit word delimiter, and a new word often incorporates a new sequence of one or more characters. Therefore, new word detection for Chinese will need strong language models and linguistic constraints. Automatic acquisition of new information items and nomenclatures, their new vocabularies, spellings, characters, and pronunciations across languages is an important direction of further development. This helps achieve *adaptivity* of multilingual conversational interfaces to application domains with growing knowledge scopes.

Thus far, we have seen many examples of using spoken dialog systems for information access. However, it is conceivable that the systems' utility may be extended to offline information monitoring. More specifically, if the users need to closely monitor changes in a piece of dynamic information, they will have to talk to the spoken dialog system rather frequently. The cognitive load of the users may be significantly reduced if they can *delegate* the task of information access/retrieval to software agents that can run continuously. This is exemplified by the multidomain ORION system developed at MIT (Seneff et al., 2000). Users can call ORION to enroll with their contact information. They can also call ORION to define a task, such as to alert the user an hour prior to touchdown of a flight. The system will alert the user at a designated time to deliver the requested information. Such alert messages may at times *interrupt* an online dialog between the human and the computer. A cost-benefit analysis of disrupting the online dialog may be analyzed prior to deciding on an interruption. Previous work by Horvitz et al. (2003) developed probabilistic user models that incorporate utility values in the deliberation of interrupting the user upon receiving alert messages. Related psychological studies investigate the effects of interruption at different phases of the primary, ongoing task (Cutrell et al., 2001) and propose visualization designs that enhance awareness of multiple, prioritized interruptions/alerts (Dantzich et al., 2002). Support for interruptions and multithreaded dialogs in conversational interfaces calls for a computational theory of discourse that lays out the structures necessary for proper treatment. Such theory is proposed by Grosz and Sidner (1985) for task-oriented dialogs. The aforementioned studies in alerts and interruptions lay the groundwork for developing intelligent systems that can automatically reason whether, when, and how to execute an interruption. The implementation to support delegations and interruptions in a multilingual SDS should be common and consistent across languages.

The following presents the ISIS system, a test bed in which we have implemented a mixed-initiative dialog model and explored adaptivity, delegation, and interruption within the stocks domain.

11.3 Overview of the ISIS System

11.3.1 The ISIS Domain

The ISIS application domain subsumes real-time stock information inquiries as well as transaction requests. Many subjects were recruited in a survey to provide the various types of queries they may wish to present to a financial information system. A conscious effort was made to achieve a somewhat balanced coverage among the languages and query types. The survey generated approximately two thousand textual queries in Chinese and English, respectively. These queries were referenced as we defined ten domain-specific task goals to determine the scope of the ISIS domain. The task goals are: QUOTE (asking for a real-time stock quote), NEWS (asking for the news about a listed company), TREND (asking about the movements of a stock's price), CHART (asking for a graphic display showing recent stock-price fluctuations), BUY (seeking to purchase shares of a stock), SELL (seeking to sell shares of a stock), ACCOUNT (asking for the user's portfolio/account information), NOTIFY (setting up the information profile for an alert service), AMEND (amending a previous order of transaction), and CANCEL (canceling a previous order of transaction).

ISIS integrates an array of core technologies for speech and language processing, which are described in the following.

11.3.2 Trilingual Speech Recognition (SR)

ISIS aims to handle three languages—English, Cantonese, and Putonghua. It integrates an off-the-shelf English speech recognizer as well as two home-grown HMM-based speech recognizers for the two Chinese dialects. Putonghua is the official spoken language used across China and is spoken in Taiwan and many overseas Chinese communities. Cantonese is a major Chinese dialect, predominant in South China, Hong Kong, Macau, and some overseas Chinese communities, spoken by about 60 million people around the world. Both Putonghua and Cantonese are based on the same writing system consisting of Chinese characters. In their spoken form, they are both monosyllabic and tonal. Each syllable can be decomposed into a syllable initial, a syllable final, and a tone. However, the two dialects also differ significantly, to the extent that a speaker knowing only one of the dialects may not be able to communicate with a speaker knowing only the other dialect. The differences between Putonghua and Cantonese reside in phonetics, syntax, and vocabulary selection. Putonghua has 24 initials and 37 finals, constituting 410 distinct base syllables. The dialect also has five lexical tones, giving a total of 1,400 tonal syllables. Cantonese has 20 initials and 53 finals, constituting 660 base syllables. It also has six lexical tones, giving a total of 1,800 distinct tonal syllables. The inventory of syllable initials and finals differs between the two dialects. We developed separate recognizers for Putonghua and Cantonese. The acoustic models for both recognizers are based on syllable initials (I) and finals (F). Recognition involves a two-pass search—the first creates a syllable lattice, and the second traverses the lattice with a language model to produce the output word sequence (Choi et al., 2000; Lo et al., 2000). In the ISIS system, the users are requested to specify their language of choice by mouse-clicking on radio buttons. This renders the system simpler, without the need for a language identification component.

11.3.3 Natural Language Understanding (NLU)

The NLU component accepts textual queries derived from typed input, recognized speech,[3] or recognized handwriting. NLU begins with parsing the user's query with a semantic grammar. To aid the task of domain definition and grammar development, we collected some sample queries in both English and Chinese. We requested that our subjects compose questions that they would ask of a stockbroker, such as questions on real-time stock quotes, or simulated investor accounts. In this manner, we collected 1,407 Chinese queries and 1,604 English queries. Examples include:

"Amend my purchase order for HSBC from three to six lots please."
如果和記黃埔升了兩塊錢, 請給我賣出兩千股
(Translation: If Hutchison Whampoa rises another two dollars, please sell two thousand shares.)

We hand-designed a set of semantic tags (or concept tags) based on the English sentences. Some syntactic tags are included as well. Had there been more data collected, we believe we could have applied a semiautomatic procedure for acquiring such structures from unannotated corpora (Siu and Meng, 1999). The set of concept tags forms the preterminal categories of the English grammar, and the tags are designed and processed to match the attribute labels for subsequent database access. A real-time data-capture component continuously updates a relational database based on a dedicated Reuters satellite feed. For example, consider the stock "HSBC" with concept tag STOCK_NAME. According to invocations specified in our grammar, this tag automatically invokes a procedure that converts it into 0005.HK (which signifies that it is the stock 0005 from the Stock Exchange of Hong Kong). The new tag 0005.HK matches the RIC4 code and can be

[3]Presently, the NLU component processes only the top-ranking speech recognition hypothesis. More sophisticated integration techniques between SR and NLU will be pursued as a next step.

used directly for database access. Examples of other concept tags include SHARE_PRICE, LOT_NUMBER, and SHARE_NUMBER.

Input queries in Chinese are first tokenized into words by means of a greedy algorithm together with a 1,100-word lexicon. The lexicon currently covers the 33 constituent stocks in the Hang Seng Index. We maximized the reuse of English concept tags (hand-designed with reference to the English queries) as preterminal categories for processing the Chinese sentences we had collected. At this initial stage, we are using a single Chinese grammar for both Putonghua and Cantonese queries. For example:

QUERY: 請問你有沒有長實的成交量
(Translation: Do you have the trading volume of Cheung Kong)
TOKENIZED WORDS: 請問你有沒有長實的成交量
TAGS: <QUES_TYPE><STOCK_NAME><DUMMY><TRADING_VOL>

The English and Chinese grammars have 169 and 174 preterminal tags, respectively. Of these, 143 are common between the two grammars, achieving about 82% sharing thus far. Examples of language-specific tags include: PREP (for English prepositions); TEENS (for English numbers like "eleven" and "twelve"); TWENTY and THIRTY (for the Chinese numbers "廿" and "卅"); and "號" (a number marker for Chinese numbers).

Verbalized numeric expressions abound in the ISIS domain, corresponding to stock codes, stock prices, number of lots, or number of shares. This is illustrated by the example query, "I want three thousand shares of Cheung Kong at one hundred and ten." In order to disambiguate among the possible semantic concepts that may correspond to a numeric expression, NLU in ISIS applies a set of *transformation rules*. For example, the transformation rule:

<numeric_exp><share_price>prev_bigram <stock_name><at>states that a parsed numeric expression (<numeric_exp>) should be transformed into a share price (<share_price>) if it is preceded by the concept bigram <stock_name><at>. This rule is applicable to the previous example query, which contains the numeric expression "one hundred and ten" and helps label it with the semantic concept of <share_price>. The format of the transformation rules resembles the rule templates in Brill (1995), so we can apply Brill's transformation-based tagger for part-of-speech (POS) tagging. We have adapted Brill's tagger for transforming semantic tags for semantic disambiguation in handling numeric expressions and

out-of-vocabulary words, as we found the approach to be directly applicable to tackling the problem of semantic disambiguation for both English and Chinese.

Parsing identifies a set of semantic concepts in the query and these are fed into a suite of Belief Networks (BN) for task goal inference (Meng et al., 1999). There are ten BNs in all for each language, where each BN corresponds to a single task goal. Each BN makes a binary decision regarding whether the input query relates to the task goal by generating an a posteriori probability that is compared with a threshold value. If all ten BNs vote negative for a given query, it is rejected as out of domain. Alternatively, a query may correspond to a single task goal or multiple task goals, as in the following.

QUERY: "Please show me the daily chart of HSBC and the closing price of Hang Lung."
GOAL(S): chart, quote
QUERY: 請告訴我匯豐的現價及給我在一百塊的時候入二千股
(Translation: Please give me the latest share price of HSBC and buy two thousand shares when it hits a hundred dollars per share.)
GOAL(S): quote, purchase_order

In order to acquire a disjoint test set, we further collected 484 Chinese queries and 532 English queries, all of which were annotated with communicative goals. While out-of-domain (OOD) queries are not used in training, a few are included in the testing corpora for investigating the capability of rejection. Examples of the OOD queries include:

QUERY: 請告訴我星加坡元兌港元的匯率
QUERY: "Give me the exchange rate for the Singapore Dollar against the Hong Kong Dollar." (Translation of the above)

Task goal identification accuracy of the English BNs for the English test is 92.5%. All the OOD queries were correctly rejected. However, only three of the six multiple-goal queries were correctly identified. Task goal identification accuracy of the Chinese BNs for the Chinese test is 93.7%. 11 of the 12 OOD queries were correctly rejected, and again only three of the six multiple-goal queries were correctly identified.

We also conducted a side experiment in which the BNs trained on Chinese were used for goal identification on the English queries. Both English training and testing sets were used, and goal identification accuracy was 82.9%. Similarly, the BNs trained on English queries were also used for goal identification of the Chinese training and testing sets. Goal identification accuracy was 80.5%. The main degradation here was concentrated on a few goals and was caused by language-specific elements as well as some degree of overfitting of the models to the small training set.

We consider this to be an interesting side experiment, as it illustrates that the BN framework can largely apply across languages. The identified goal together with the concepts and their values form the semantic frame that represents the meaning of the query. This is used to formulate an SQL expression for database access.

11.3.4 Discourse and Dialog

Discourse inheritance in ISIS uses an electronic form (E-form) for bookkeeping (Papineni et al., 1998; Goddeau et al., 1996; Meng et al., 1996). The information slots in the E-form are derived from the semantic concepts in the NLU grammars. The value of an information attribute obtained from the current user's query overrides that from the previous query (or queries) in discourse inheritance. A mixed-initiative dialog model is implemented using a turn-management table. For example, the turn-management table specifies that a query whose task goal is identified to be QUOTE will trigger a response frame labeled QUOTE_RESPONSE in order to use the appropriate template to generate the response text. Additionally, the turn-management table specifies the set of information slots that need to be filled before performing each inferred task goal. Should there be necessary but unfilled slots, the dialog model will prompt for missing information. Furthermore, queries inferred to be important transactions, such as BUY and SELL requests, will trigger a subdialog to confirm all details regarding the transaction prior to order execution. ISIS also has a list of meta-commands—HELP, UNDO, REFRESH,[4] GOOD-BYE, and so on—to enable the user to navigate freely in the dialog space. The dialog manager (DM) maintains the general flow control through the sequential

[4]The meta-command undo deletes the discourse inherited from the immediate past spoken query and is especially useful if the query has speech recognition errors. The command refresh deletes the entire inherited discourse to let the dialog start afresh.

process of speech recognition, natural language understanding; database access (for real-time data captured from a direct Reuters satellite downlink); and response generation in synthetic speech, text, and graphics. Speaker authentication is invoked at dialog turns that involve access to account information or execution of a transaction—that is, when the inferred task goals are BUY, SELL, ACCOUNT, AMEND, or CANCEL. The design and implementation of the discourse and dialog module are independent of the language in use. Only the language knowledge sources such as the grammar, etc. are language dependent and loaded as components.

11.3.5 Spoken Response Generation

Spoken responses in ISIS need to be generated for three languages. Language generation utilizes a response frame. The frame specifies the language to be generated, the response type (e.g., QUOTE_RESPONSE), associated information attributes (e.g., stock_name, bid price, and ask price), and their values as obtained from NLU or database access. While the response semantic frame is language independent, response generation involves the invocation of the generation template that corresponds to the response type. Each response type maps to three parallel templates, one for each language. Each template encodes a set of generation grammar rules for proper verbalization, and the grammar terminals are obtained from a three term list that was specially compiled. This list stores the English and Chinese lexical entries for every grammar terminal. At present, the generation template does not support code-switching (language switching) within a single response.

The response text generated is sent to a text-to-speech (TTS) synthesizer. ISIS integrates the downloadable version of the FESTIVAL system (Taylor et al., 1998) for English synthesis. Putonghua synthesis uses Microsoft's SAPI engine[5] and Cantonese synthesis uses the homegrown CU VOCAL engine (Meng et al., 2002). The synthesis quality of CU VOCAL can be optimized for constrained application domains to generate highly natural and intelligible Cantonese speech. This process of domain optimization has been demonstrated to be portable across Chinese dialects (Fung and Meng, 2000).

[5] www.microsoft.com/speech/

11.3.6 Speaker Authentication (SA)

Speaker authentication aims to automatically verify the speaker's claimed identity by voice prior to enabling access to personal financial information and execution of transaction orders. The current SA component is a *text-dependent* system—depending on the language spoken by the user in the immediately preceding dialog turn, the system randomly generates an English or Chinese digit string for the speaker to utter. SA uses Gaussian Mixture Models (GMM) (Reynolds, 1992) with 16 mixture components, and the models are trained by pooling data across languages. Authentication involves computing the likelihood ratio between the claimed speaker's model and the "world" model. This likelihood ratio is compared with a preset threshold, and hypothesis testing is applied to accept or reject the claimant for the purpose of user authentication.

11.3.7 System Architecture—CORBA

ISIS is a distributed system with a client/server architecture implemented on CORBA (Common Object Request Broker Architecture). CORBA is a middleware with specifications produced by the Object Management Group (OMG) and aims to provide ease and flexibility in system integration. The core technologies described in the previous subsection, (e.g., SR, NLU, and SA) reside in different server objects (see Figure 11.1). In addition, there is a server object for tracking timeouts[6] and another server object encapsulating two software agents responsible for message alerts. The client object is illustrated in Figure 11.2. CORBA provides the Object Request Broker (ORB) that handles communication between objects, including object location, request routing, and result returning. As illustrated in Figure 11.1, some server objects are implemented in Java or C on UNIX, while others are in Visual C++ on Windows NT. These server objects extend the stubs/skeletons (i.e., the glue to the ORB from the client/server) to the core speech and language engines. CORBA offers *interoperability* by the Interface Definition Language (IDL) to communicate with the variety of programming languages running on different operating

[6]The timeout manager (TM) monitors the time between successive user inputs. If the time duration exceeds a preset threshold, TM sends a message to the dialog manager, which in turn invokes the response generator to produce system responses such as "Are you there?"

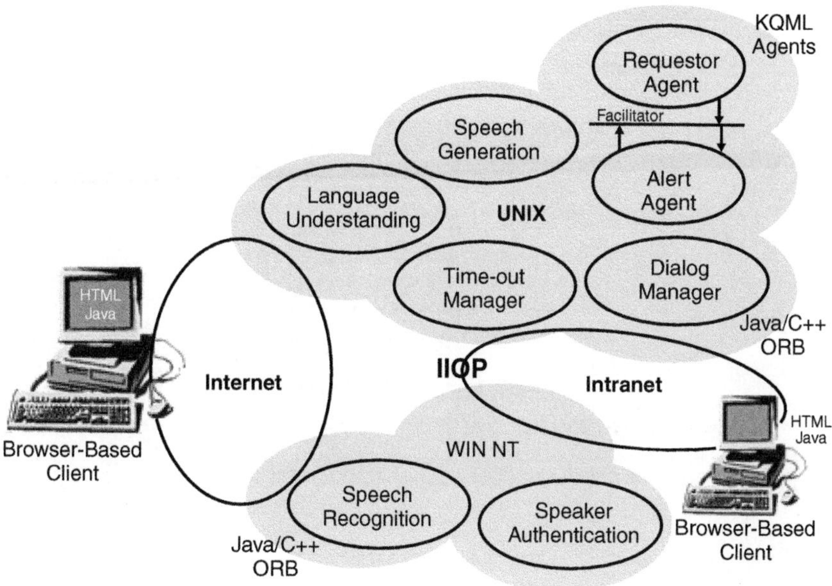

Figure 11.1: The ISIS architecture.

systems. CORBA also offers *location transparency* in that only the names of the server objects need to be known for two-way (sending/receiving) communication and no explicit host/port information is needed. A third advantage of CORBA is *scalability*—that is, a new object can be added to the system simply by adding its corresponding definition to the IDL, followed by recompilation.

Data are passed among the client and server objects in XML (Extensible Markup Language).[7] The data are labeled with descriptive tags characterizing the server class operation(s). For example, Figure 11.3 shows the XML output of the NLU component after processing a user request to purchase some stocks. Java applets in the client object can also capture mouse-clicks as events in order to support deictic expressions (see Figure 11.2). Client methods decode the semantic meaning of the mouse-clicks (e.g., the stock name of the table row that has been clicked) and send the information through CORBA to the NLU server object for subsequent processing.

[7] http://w3C.org/XML

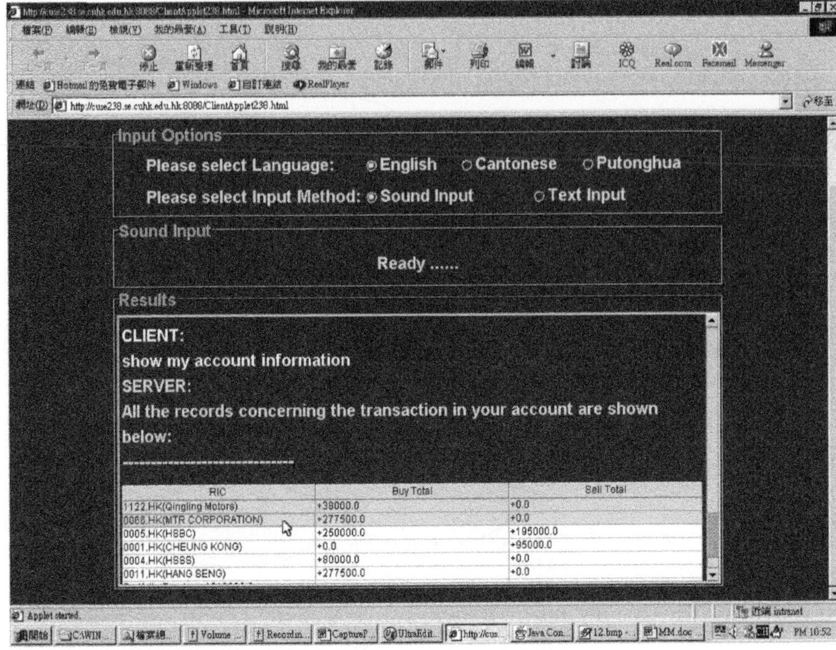

Figure 11.2: Screen shot of the client that provides options for language selection and input mode selection (text or voice). Should "Text Input" be selected, a text box will appear, and the user can input text by typing or stylus-writing. The client object can also capture click events to support deictic expressions. Clicked items in the table are highlighted.

```
<NLU>
<goal> buy </goal>
<ric> 0005.HK </ric>
<num_lots> 5 </num_lots>
<price> market </price>
</NLU>
```

Figure 11.3: Example of an XML message produced by the NLU server object. The user request was "Buy five lots of HSBC at the market price please." "ric" denotes Reuters Instrument Code.

Table 11.3 Example rejection dialog in the ISIS system.

User1	User logs into ISIS by typing in his account code.
ISIS1	Welcome to ISIS. Please input your query.
User2	I would like the exchange rate of the Singapore dollar please.
ISIS2	Sorry, I don't understand your request. I can provide only information about stocks such as real-time quotes, news, and charts. I can also provide information about a simulated personal portfolio, and handle buy and sell transactions. Description: ISIS rejects an out-of-domain query about foreign exchange.

A reference implementation based on the ISIS system, including software, documentation, and demo, is available online.[8] Other reference architectures are also available, such as the GALAXY system (Bayer et al., 2001) which has a hub interacting with several distributed information servers via a hub script that specifies server hosts, ports, and functionalities; and the Open Agent Architecture (OAA) (Cohen et al., 1994), which executes the user's request in a distributed manner via a blackboard server interacting with a community of cooperative software agents running on various machines.

11.3.8 An Example Dialog in ISIS

The initial prototype of the ISIS system has been developed and subjects have been invited to experiment with the system. Table 11.4 shows an example of a ISIS dialog and serves to illustrate some of the mechanisms that support the interaction.

11.4 Adaptivity to Knowledge Scope Expansion

As mentioned in the introduction, most conversational systems are designed with a *closed* knowledge scope. For *each* language in a multilingual system, the speech recognizer and language model typically have fixed and

[8] http://www.se.cuhk.edu.hk/ isis/download

Table 11.4 An example dialog from the ISIS system.

User1	I would like to buy three lots of Cheung Kong Holdings please.
ISIS1	Please provide the bid price. Description: Natural language understanding (NLU) infers that the task goal is BUY. Domain-specific constraints show that BUY requires three attributes: the stock name, the number of lots/shares, and a bid price. The last one is missing and hence the system prompts for it.
User2	我想看一看每日的走勢圖 Translation: Can you show me the daily chart first. Description: The user did not answer the system's prompt for a bid price. The discourse and dialog components offer this flexibility.
ISIS2	0001 長實請睇 Translation: *0001 (a stock code) Cheung Kong Holdings Ltd. Please check (daily chart shown).* Description: NLU infers the task goal is CHART. Domain-specific constraints in the turn-management table show that this requires a stock name and a chart type. The former is missing from the query but is inherited from discourse. The daily chart is shown on the graphical user interface.
User3	*Buy three lots at eighty-nine dollars, please.*
ISIS3	Please say the following digits in English to verify your identity: 1 3 7 6 4. Description: NLU infers that all the necessary attributes for a BUY request have been provided. The dialog manager thus invokes speaker authentication (SA) before executing the BUY request.
User4	*One, three, seven, six, four*
ISIS4	Please confirm the following action: buy three lots of Cheung Kong Holdings Limited at eighty-nine dollars. Please say yes to confirm or no to reject the action. (SA verified the user's identity and allowed the user to proceed with the BUY request. Transactions are always followed by a confirmation subdialog.)
User5	Yes.
ISIS5	The buy transaction is completed. Please input a new query.

(*continued*)

Table 11.4 Continued.

User6	*Show me the monthly chart of HSBC.*
ISIS6	The monthly chart of HSBC is shown below. (graphic shown)
User7	*Bye*
ISIS7	Goodbye and see you later. Description: Bye is a META-COMMAND that can be used to terminate the session at any time.

predetermined vocabularies; the NLU component typically has grammars with a fixed set of rules, categories, preterminals, and terminals; and the text-to-speech (TTS) synthesizer typically references pronunciation dictionaries with a fixed number of entries. Conversely, the knowledge scope of application domains tends to *grow* over time—for example, new companies continue to be listed at stock exchanges; new names of people and addresses are added to directory listings; and new products and services may become available. The widening knowledge gap between the conversational interface and the application domain creates an impending need for *adaptivity* in multilingual spoken dialog systems.

A direct impact of knowledge scope expansion is the OOV problem. New information that needs to be acquired by an adaptive interface inevitably involves new words that do not exist in the original vocabularies of the recognizers and their language models, the original set of grammar terminals in the NLU component, or the original pronunciation dictionaries of the TTS synthesizers. As mentioned earlier, a large body of previous work has addressed the OOV problem in the context of SR, language modeling, and letter-to-sound/sound-to-letter generation to augment existing vocabularies with spellings and pronunciations for recognition and synthesis. The current work takes on a slightly different focus in exploring adaptivity in the ISIS dialogs—that is, a newly listed stock (or OOV) detected in an input query will trigger a *subdialog* that seeks to elicit information about the new word from the user and automatically incorporates the word into the ISIS knowledge base. This strategy is largely adopted for both English and Chinese, with a minor enhancement for the latter, which will be described later. In the case of English, the new stock name may appear as a full name, an abbreviated name, or an acronym. It may also be input in spoken form via speech recognition, or in textual form via typing or stylus-writing on mobile PDAs or portable computers. The current work on OOV handling in ISIS is directly applicable to textual input and English acronyms with a spelling pronunciation (which can be recognized by the English recognizer). As mentioned in Chung et al. (2003), OOV handling involves acquisition of the spelling and pronunciation of the new word/name as well as its linguistic usage, such as semantic category. We focus on the latter aspect in this work—that is, handling primarily new stock names and bypassing temporarily the problem of OOV spelling/pronunciation acquisition in speech recognition.[9] However, in the case of Chinese, the method that we use in handling OOV *(N-gram grouping)* is conducive to speech recognition. This is by virtue of the syllabic nature of the Chinese language, in which every written character is pronounced as a spoken syllable.

Automatic incorporation of new stock names in ISIS involves two steps: (1) detecting new stock names and (2) invoking the subdialog for new stock acquisition. Details are provided in the next section.

11.4.1 Detecting New Stock Names

The detection process takes place in the NLU component. A new stock name is tagged with <oov>. For example, the Artel Group was listed during the time of development at the Stock Exchange of Hong Kong with a stock code of 0931. Some refer to the company as "Artel." Since the listing was new, the name did not exist in the original ISIS knowledge base. An input query such as "I'd like to check the news about Artel" will first undergo semantic parsing in the NLU component to yield the semantic sequence:

Semantic sequence from parser?

<dummy><check><news><about><oov>

The semantic label <dummy> is used to absorb an arbitrarily long text string while parsing the input query. Such a text string may be a grammatically ill-formed structure, a transcription of spoken disfluencies, or a series

[9] Automatic conversion of a spoken waveform with an unknown word to a grapheme sequence is a problem that merits a focused and dedicated effort. Several approaches to this problem have been developed in speech synthesis (see Chapter 7). Some studies have also addressed it in the context of speech recognition, for languages such as English (Chung et al., 2003), German (Schillo et al., 2000), and Dutch (Decadt et al., 2002).

of words that does not carry significant semantic content captured by the other semantic labels.

A set of transformation rules has been written for the purpose of determining whether an OOV word corresponds to a new stock name. The technique is similar to that used for disambiguation among multiple possible semantic categories that correspond to numeric expressions (see Section 11.3.3). The same transformation technique is applied here for both English and Chinese. The transformation rule that is applicable to the semantic sequence above is:

<oov><stock_name_oov>prev_bigram <news><about>

This rule states that the concept label <oov> should be transformed into <stock_name_oov> if it is preceded by the concept bigram <news><about>. Hence, the *transformed semantic sequence* becomes:

<dummy><check><news><about><stock_name_oov>

As a consequence, the OOV "Artel" in the example query is labeled as a new stock name—that is, <stock_name_oov>.

Application of this transformation technique to detection of new stock names in Chinese queries requires an additional procedure known as *N-gram grouping*. Consider another recently listed stock (China Unicom) that appeared in an input query:

Input Chinese Query: 我想知道中國聯通既股價
(Translation: I'd like to know China Unicom's stock price.)

Since this stock name does not exist in the original vocabulary, Chinese word segmentation fails to identify it as a word and tokenizes it into individual characters: Input Chinese Query after word tokenization: |我想知道|中|國|聯|通|既|股價|
Semantic Parser Output:

```
<ask>我想知道</ask>
          <singleton>中</singleton>
          <singleton>國</singleton>
          <singleton>聯</singleton>
          <singleton>通</singleton>
<particle>既</particle>
<price>股價</price>
```

An *N-gram grouping technique* checks for sequences of contiguous character unigrams and bigrams, which may be indicative of the occurrence of an OOV. These *N*-grams are then grouped together. To follow up with the example, we obtain:

N-gram Grouping Output:

<ask>我想知道</ask><oov>中國聯通</oov><particle>既</particle><price>股價</price>

Hence, the semantic sequence from parsing is: <ask><oov><price>
The transformation rule applicable to this semantic sequence is: <oov><stock_name_oov>prevtag <ask>nexttag <price>
The rule states that the concept label <oov> should be transformed into <stock_name_oov> if it is preceded by the tag <ask> and followed by the tag <price>. Hence, the transformed semantic sequence becomes:

<ask><oov><price>

Consequently the new stock name 中國聯通 (China Unicom) has been detected. It should be noted that the *N*-gram grouping technique used here is conducive to pronunciation acquisition of new stock names in Chinese speech recognition. This is because the Chinese language is syllabic in nature, and each written character is pronounced as a spoken syllable. Since the Chinese speech recognizer used in ISIS is syllable based, it is possible to map the character sequence tagged with <stock_name_oov> to the corresponding syllable sequence transcribed by the speech recognizer to derive a hypothesized pronunciation for the new Chinese stock name.

11.4.2 Invoking Subdialog for New Stock Incorporation

Detection of a new stock name causes the dialog manager to invoke a special subdialog that aims to incorporate the new stock into the ISIS knowledge base. The DM begins by triggering a substring match for all possible candidate listings that correspond to the new stock name. For example, the letter string in "Artel" can match "Artel Group" and "Hong Kong Artfield Group." Similarly, the abbreviation 中聯 (for China Unicom) can match 中國聯通, since the characters appear in the same order. The list of possible candidate listings is presented to the user onscreen for selection. Better

Table 11.5 Example illustrating the automatic acquisition of a new stock name in ISIS through a spoken dialog between the user and the system.

User1	*Do you have the real-time quotes of Artel?*
ISIS1	I do not know of the stock listed as Artel. This name is similar to the following list of stocks, together with their stock codes. Please speak the stock code for the correct choice. Otherwise, please ask for another stock. —— 0931.HK—Artel Group 1229.HK—HK-Artfield Group . . .
User2	*It should be zero nine three one.*
ISIS2	Here are the real-time quotes for Artel Group. <*quotes shown*> Do you generally refer to Artel Group as Artel? Please say yes or no.
User3	*Yes.*
ISIS3	I got it. How else may I help you?
User4	*Show me the daily chart of Artel, please.*
ISIS4	This is the daily chart of Artel Group. <*chart shown*>

matches (according to the edit distance) are ranked higher on the list (please see the dialog turn labeled ISIS1 in Table 11.5).

Spoken response generation can pronounce the new English stock names by means of letter-to-sound rules. Since the Chinese synthesizers have full coverage of single character pronunciations, new Chinese stock names can be spoken (character by character) in the ISIS system responses.

One may notice from ISIS1 in Table 11.5 that in order to select an item from a short list of new stock names, the user is constrained by the system to use either the stock code or an ordinal reference, such as "the first one" or "the next one." This is because new stock names contain OOV words that are problematic for speech recognition, but all digits in stock codes are in-vocabulary. However, constraining the user to utter only digits is cumbersome from a usability perspective.[10] This motivated

[10]This constraint may be relaxed in the future by invoking spelling-to-pronunciation generation using letter-to-sound rules and dynamically adding the listed stock names into the recognizer's vocabulary. We have recently been successful in using this method to develop speakable hyperlinks in Chinese speech-based Web browsing.

us to enhance the ISIS client object to capture mouse-clicks as events by using Java applets and Java classes. With this small addition of multimodal capability, deictic expressions that are spoken (e.g., "It should be this one") can be augmented by clicking. This is illustrated in Figure 11.4, in which the user has clicked on the selected entry. The mouse action is captured, and the contents of the table are retrieved; this stock information is packaged in an XML message in the NLU component and processed as usual.

Upon completion of this subdialog, the user has confirmed the correct listing for the new stock name. ISIS continues to elicit the preferred ways in which the user will refer to this new stock, including the full name and its abbreviations (e.g., "Artel Group" or "Artel"). Subsequently, two new

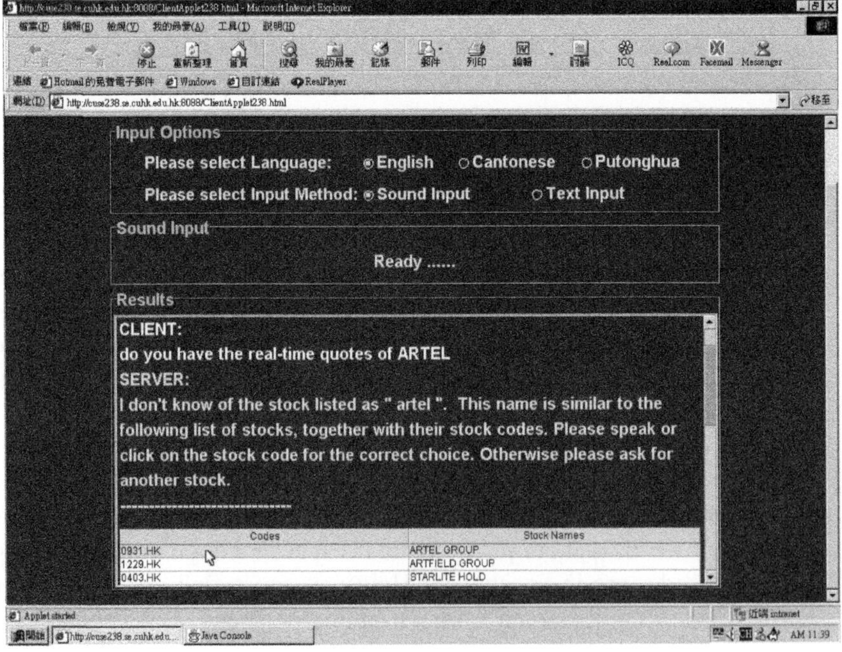

Figure 11.4: Screen shot of the ISIS client object presenting information in response to the user's query "Do you have the real-time quotes of Artel?" "Artel" is a new stock that did not exist in the original ISIS knowledge base.

grammar rules are automatically added to the NLU server:

stock_name –>Artel (Group)
stock_code –>0931.HK

Hereafter, the new stock is assimilated in the expanded ISIS knowledge base, and the user may refer to the newly listed company by its stock name or stock code.

11.5 Delegation to Software Agents

The financial domain has an abundance of time-critical dynamic information, and users of financial information systems often have to closely monitor for specific changes in various real-time information feeds, such as when a target bid/ask price is reached, or when the increment/decrement of a financial index exceeds a threshold. The cognitive load of such users can be greatly reduced if the task of monitoring information can be delegated to software agents. This work involves our initial effort in exploring *asynchronous* human-computer interaction by delegation to software agents within the context of ISIS. The interaction is asynchronous because the human and the computer no longer take alternate dialog turns in completing a task. Instead, the human verbally specifies the task constraints to software agents that will perform the task in the background (i.e., offline). ISIS supports two types of requests for alerts. The first kind is explicit, such as "Please notify me when HSBC rises above ninety dollars per share." When the prescribed condition is met at some indeterminate time in the future, the agents will return a notification message to the user. The second kind is implicit; if the user places a buy/sell order with a requested price that differs from the market price, ISIS will inform the user of the difference, offer to launch software agents to monitor the market price, and alert the user when the price reaches the requested level. Agent-based software engineering in ISIS uses the Knowledge Query and Manipulation Language (KQML), which provides a core set of speech acts (also known as *performatives*) for interagent communication. Alternative foundation technologies also exist for agent-based software engineering. Examples include the use of speech acts with Horn clause predicates for interagent communication in the Open

Agent Architecture (OAA)[11] mentioned previously, as well as the Belief, Desire, and Intention (BDI) paradigm (Rao and Georgeff, 1995) for structuring the content of the agent's messages in terms of the informational, motivational, and deliberative states of the agent. It should be noted that the implementation for delegation dialog is and should be independent of the language spoken by the user.

11.5.1 Multiagent Communication in ISIS

The ISIS implementation uses JKQML (i.e., Java-based KQML). There are three software agents in all: *requestor, facilitator, and alert* agents. If the user's requested transaction (e.g., "Buy three lots of HSBC at eighty-nine dollars, please") cannot go through due to a mismatch between the requested and market prices, ISIS will trigger the offline delegation procedures. First, a nonblocking XML message is sent from the DM server object to the requestor agent. This message encodes the information attribute-value pairs related to the requested transaction. The requestor agent receives this XML message, decodes it, and transmits a corresponding KQML message through the facilitator agent to the alert agent. The facilitator agent serves to locate the alert agent since the facilitator is an agent registry. The alert agent receives the KQML message, interprets it and inserts it by SQL into a relational database storing similar requests. The alert agent also keeps track of the alert conditions specified in the database and monitors the real-time data feed accordingly. The data feed is a direct satellite downlink from Reuters, received via a satellite dish mounted on the roof of our engineering building. If the previously specified condition is met (i.e., HSBC's market price hits 89 dollars per share), the alert agent will send a KQML message through the facilitator to alert the requestor agent. The performative *tell* is the expected response to *ask_all*. In the final step, the requester agent returns a KQML alert message to the dialog manager server object. This entire process of multiagent communication is depicted in Figure 11.5. Within this implementation, the client is able to receive alert messages as soon as the prespecified alert levels of various stock prices are met, without any noticeable delays. In order to maintain continuity in

[11] www.ai.sri.com

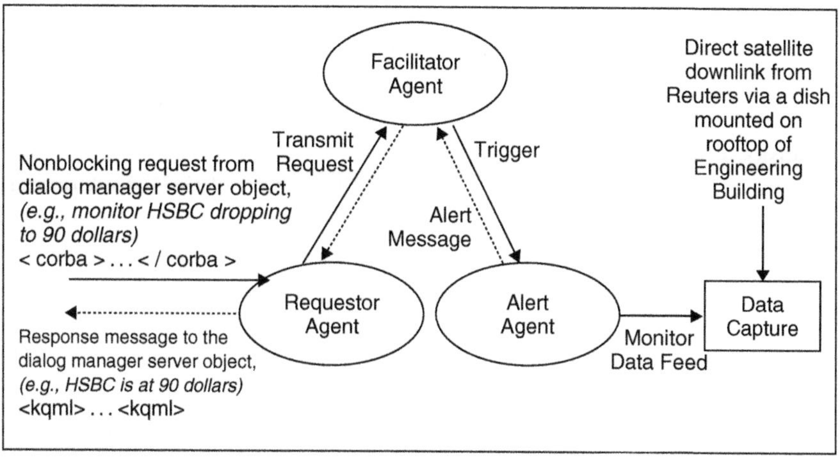

Figure 11.5: Multiagent communication in ISIS. The messages communicated between the agents and the dialog manager server object are in XML format and wrapped with indicator tags; the tag <corba> is used if the message originates from the DM, and the tag <kqml> is used if the message originates from the agents.

language use, the alert message also adopts the language spoken by the user in the immediately preceding dialog turn.

11.6 Interruptions and Multithreaded Dialogs

Delegational interactions involving software agents will generate alert messages that will subsequently interrupt an ongoing dialog (i.e., online interaction [OI]) between the user and the ISIS system. This brings the problem of maintaining multithreaded dialogs that correspond to: (1) the current (sub)task, which is in focus in the online interaction; (2) other concurrent (sub)tasks being pursued but temporarily not in focus; and (3) one or more pending notifications delivered by the software agents. The users may choose to stop their workflow in the online interaction, bring an alert message into focus to handle the pending task, and then return to the original interrupted workflow. This section reports on an initial investigation of interruptions and multithreaded dialogs in the context of ISIS. The scope of investigation is constrained to the task structure of the ISIS domain (as described in Section 11.3.1). The task structure is relatively simple, and an intended action may be easily decomposed into a sequence of simple tasks. For example, if the users are considering purchasing the stocks of a particular company, they may seek to obtain the real-time stock quotation, observe its fluctuation in a chart, check the news of competing companies in the same industry, and finally decide on placing a buy order. As can be seen, multiple tasks pertaining to the users' intent can be handled *sequentially* in the online interaction (OI) dialog in ISIS. Similarly, multiple alert messages resulting from offline interaction (OD) can also be handled one at a time. The situation is more complex in real applications—for example, users may be checking their appointment calendar while placing a purchase order with ISIS, which already involves concurrent tasks in the OI interaction dialog. Alternatively, multiple alert messages of equally high priority may arrive simultaneously and require immediate attention from the user. Such situations will require a complicated interruption framework. However, as a first step, this work aims to identify the necessary mechanisms involved in the simplified interruptions within the context of ISIS. In order to cater for language changes in the midst of a dialog, the alert messages generated by the system adopt the language used by the user in the preceding dialog turn. The same applies to the dialog turns in the interruption subdialogs as well.

In the ISIS implementation, a dedicated region on the screen display is reserved for icons that indicate the arrival of alert messages. The icon that corresponds to a target price notification will cause an icon labeled "Notify" to be placed on screen (see Figure 11.6). Similarly, the arrival of a buy/sell notification will place the icon "Buy/Sell" on screen (see Figure 11.7). It is possible for both icons to appear simultaneously. The user's eye gaze tends to scan from left to right in the box delineated as "Text Input" (i.e., the text box for typing or stylus-writing) or "Speech Input" (i.e., where the speech recognizer's transcription will appear). By placing the notification icons to the right of this box, the ISIS screen design intends to minimize disruptions to the ongoing workflow—that is, the users' eye gaze will reach the icon after inputting a request. The notification icons are also moderately close to the center of the screen in an attempt to provide sufficient awareness from the user. As will be elaborated on later, the users choose to handle an alert message by clicking on the notification icons. Hence, they have

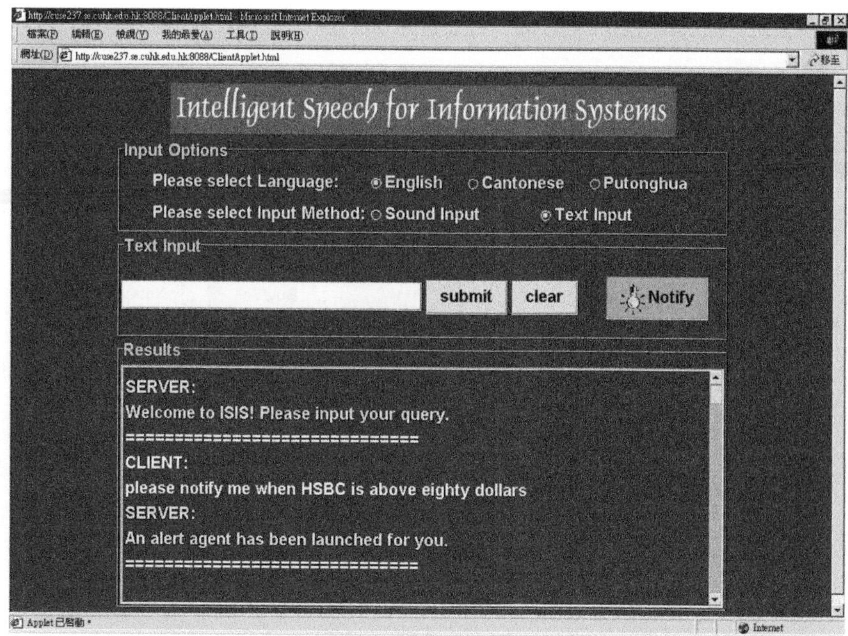

Figure 11.6: The "Notify" icon indicates the arrival of a notification message.

full control of when interruptions take place. Alert messages are queued chronologically as they arrive.

11.6.1 Interruptions of Online Interaction (OI) Dialogs by Offline Delegation (OD) Alerts

When icons appear to indicate the arrival of alert messages waiting in the queue, the user has the option of clicking on the icons at a convenient time to interrupt the online workflow and bring the alert message(s) into focus to handle the associated pending task(s). Alert messages relating to the icon "Notify" present less interruption to the user's workflow, since a click on the icon triggers delivery of the alert and thereafter the user can immediately resume the original workflow. Alert messages relating to the icon "Buy/Sell" are more complex, since they correspond to transactions.

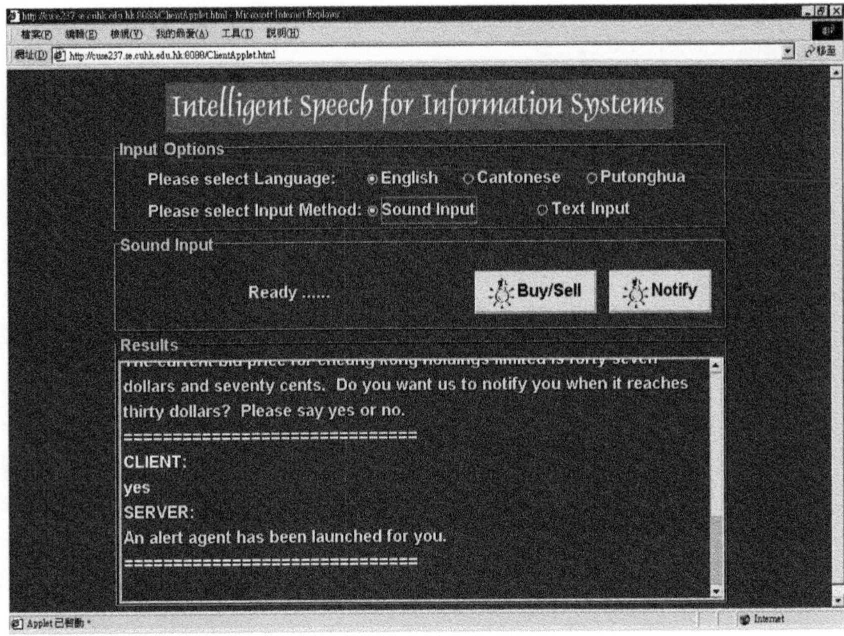

Figure 11.7: The presence of both the "Notify" and the "Buy/Sell" icons indicate that there are pending notification message(s) and buy/sell reminders in the queue.

Handling a buy/sell reminder involves bringing the transaction into the focus of attention, reinstating the transaction's information attributes, and values, allowing the user to make necessary adjustments to the attributes, and invoking the confirmation subdialog for transaction orders. These steps require a series of dialog turns before the user can revert back to the interrupted workflow. Consequently, the "Buy/Sell" reminders are more disruptive than the "Notify" alerts. This section describes the underlying structures and mechanisms in ISIS that support task interruption, task switching, and task resumption.

ISIS maintains two dialog threads that connect a chronological series of E-forms (or semantic frames), one for OI and the other for OD (see Figure 11.8). The OI thread maintains discourse of online interactions in which the user and the system take alternate dialog turns in mixed-initiative interactions. The task in focus is represented by the latest E-form in the

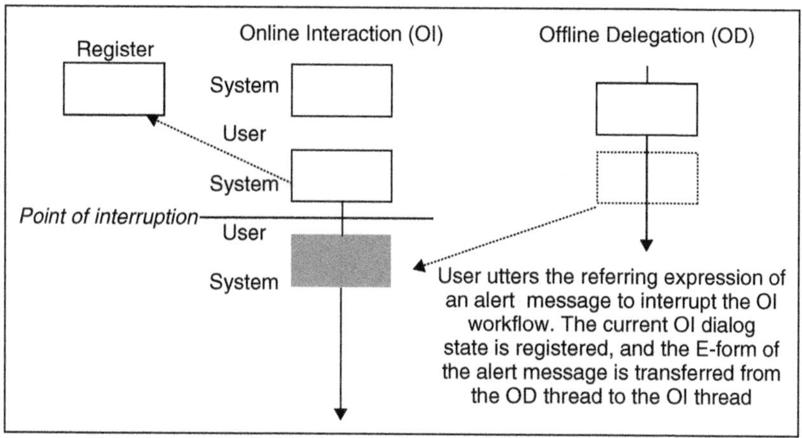

Figure 11.8: Structures and mechanisms supporting interruptions in ISIS. Online interaction (OI) and offline delegation (OD) are managed as separate dialog threads. The OI thread stores E-forms in chronological order, and the timeline goes downward. The user and system take alternate dialog turns in the mixed-initiative interactions. Alert messages delivered by software agents queue at the offline delegation dialog thread. The user may interrupt the OI workflow by uttering the referring expression to the alert message. This causes the current OI dialoging state to be stored at the register and the alert E-form to be moved from the OD thread to the OI focus.

OI thread. The OD thread stores a queue of alert messages received from the software agents. Each alert message uses its stock name as a default, referring expression to itself. Hence, in order to interrupt the ongoing workflow and switch to handling an alert message, the user can simply input (or utter) the stock name of the alert message. At this point, the latest E-form on the OI dialog thread that transpired immediately before the interruption is moved into the register (also illustrated in Figure 11.8), and the E-form of the selected alert message is moved from the OD thread into the focus of the OI thread. Discourse in the E-forms is not inherited across the point of interruption such that the information attributes and values corresponding to the original task and the alerted task are kept separate.

Having processed an alert message, the user may choose to handle another alert message in the queue (again by uttering its referring expression) or to revert to the original interrupted task by the command, "Please go

Table 11.6 Example dialog illustrating the interruption of an online interaction dialog by an offline delegation alert.

User1	*Show me the news of Cheung Kong.*
ISIS1	There is no news for Cheung Kong today. (Buy/Sell icon appears)
User2	(User clicks on the Buy/Sell icon)
ISIS2	There is one alert message for you regarding a previous buy request on HSBC. If you wish to handle the alert message now, please say HSBC. Otherwise, please continue.
User3	*HSBC*
ISIS3	The stock price of HSBC is at ninety-eight dollars presently. Please confirm your previous request from May 20, 2002, at 2:00 p.m. You wish to buy three lots of HSBC at the requested price of ninety-eight dollars per share. Please say yes to confirm or no to reject the action.
User4	*Yes*
ISIS4	The buy transaction is completed. Please input a new query.
User5	*Let's go back.*
ISIS5	Previously you requested to see the past news of Cheung Kong Holdings but there is no news for Cheung Kong today. How else may I help you?

back." At this point, ISIS restores the registered E-form by moving it back into focus along the OI dialog thread. The original task is thus reinstated and resumes in the OI dialog. Table 11.6 shows an example dialog that illustrates interleaving OI and OD interactions.

Previous work by Grosz and Sidner (1985) defines a general computational structure of discourse. The model is a composite of three interacting components: (1) the structure of the linguistic expressions, (2) the structure of intentions, and (3) the attentional state. Linguistic structure involves cue words, phrases, and referring expressions. Intentional structure involves the definition of ***discourse purpose*** (DP)—that is, the objective of the discourse. It also involves the definition of *discourse segments* (DS), each of which has a single intention known as *discourse segment purpose* (DSP). Attentional structure involves a ***focus stack***, which organizes focus spaces—each of which to a DP or DSP. The order of stacking reflects the relative salience of the focus spaces. The current ISIS implementation, within the confines of its domain, verifies the necessity of many of the model components of the Grosz and Sidner model. As described previously, it was necessary to provide a handler to every alert message by means of a referring expression and the stock name is used for this purpose. The user's sequence of intended tasks may be gleaned from the E-forms in the OI dialog thread, with the latest E-form being the focus of attention. The *register* in the ISIS implementation serves to remember the discourse segment and its purpose (DS or DSP) prior to an interruption. This data structure suffices, as the current cases of interruptions are short and simple. However, more complex interruptions are conceivable. For example, users may be distracted by a later interruption while they are processing an earlier one. Handling such multiple alerts will require a more sophisticated register—for example, in the form of a stack that can store multiple discourse segments. The type of interruption presented in ISIS belongs to the category of *true interruptions* in the Grosz and Sidner model. True interruptions are kept as separate discourses, just as separate E-forms are kept for the interrupted task in the OI dialog thread and the interrupting task from the OD dialog thread, and there is no information sharing between the E-forms. It should be noted that the data structures for handling multithreaded dialogs should maintain language independence in the implementation of the dialog system.

11.7 Empirical Observations on User Interaction with ISIS

We conducted a small-scale experiment involving 15 subjects. The purpose of the experiment was to observe how users interact with the end-to-end ISIS system using dialogs that involve interruptions. The subjects were technical or research staff members who were familiar with the use of computers. All of them were bilingual and speak fluent in English as well as Cantonese Chinese. They were given a brief verbal description of the capabilities of ISIS with reference to the system overview diagram (see Figure 11.1), an explanation of the meta-commands (listed in Section 11.3.3), and a three-minute demonstration video on the project's

Web site.[12] The video illustrates a typical interaction session between a user and the ISIS system by providing an example dialog. The subjects were verbally informed that ISIS can support the following user activities: checking for a stock quote and related charts, getting market trends and financial news from a Reuters data feed, placing a buy/sell order of a stock, amending/canceling a requested transaction, monitoring the movement of a stock price, and checking the user's simulated account information. The subjects were also informed that they can freely use English or Chinese for interaction as well as changes between the two from one dialog turn to another over the course of the interaction. Each subject was then asked to formulate a set of tasks in preparation for their interaction session that involved managing their simulated portfolios and/or gathering information prior to placing a buy/sell order on stocks that are of interest to them. They were advised to follow through with the completion of the transactions. The subjects were asked to interact with the system by speaking, typing, or stylus-writing. All interactions (i.e., user inputs and system responses) were automatically logged by the system. All system operations were automatic except for one Wizard of Oz activity in triggering alerts. This use of a manual trigger is due to the practical reason that subjects may set up alert notifications with target stock prices that are different from market prices. There is no guarantee that the target stock prices will be attained within the duration of the interaction session in order for the alert to be generated. Hence, the wizard takes the liberty to overwrite the real-time stock database such that the target price can be "reached" at an arbitrary instant during the interaction session. This triggers an alert message to the user. Consider the example of the user input, "Please notify me when TVB rises above four dollars per share." After an arbitrary number of dialog turns (capped below ten turns), the wizard overwrites the stock price record in the database to be greater than four dollars. The subject receives an alert immediately afterward. Subjects were informed that alerts may be artificially triggered by a wizard.

The fifteen subjects generated fifteen dialog sessions. The input included both English and Chinese queries. A typical task list prepared by the subjects is shown in Table 11.7.

[12] www.se.cuhk.edu.hk/ isis/

Table 11.7 Example task list prepared by a participating subject.

(Start by speaking English) Check the latest information for Sun Hung Kai (e.g., stock price, turnover, market trends, news) to help decide on a purchase price. Then place a purchase order for two lots of Sun Hung Kai. (Change to speaking Cantonese) Check on the portfolio to see holdings of Hang Seng or another stock. Check the latest information of the stock to decide on a selling price and the number of lots, then place an appropriate sell order for the stock.

The dialog sessions averaged over 19 dialog turns in length. One of the dialog sessions was conducted entirely in English, while the remaining ones involve between one to six language switches over the course of the dialog. The ISIS system logged a total of 291 dialog turns. After each dialog session, the system log was presented to the subjects, who were then asked to examine each system response and mark whether they considered it coherent or incoherent with regard to the original subject request. Out of the 291 dialog turns, 259 (about 89%) were deemed coherent by the subjects. The incoherent dialog turns were due to nonparsable input sentences or speech recognition errors, especially for out-of-vocabulary words in spoken input (which ISIS does not support). These incoherent dialog turns showed no direct correlation with the language used. Based on the system logs and the ISIS task goal definitions (see Section 11.3.1), we also evaluated whether the intended tasks were completed successfully. Most of the users tried to repeat and rephrase their input requests in order to complete the intended tasks. Other users dismissed the ongoing task and pursued another because they were frustrated by successive errors in the system responses. Among the 120 tasks in total, 101 were successfully completed, 17 were successfully completed after one or more repeated attempts and two were dismissed as failures by the subjects. This corresponds to a task completion rate of 98%. These performance measurements suggest that the end-to-end ISIS system is a viable prototype, but there is room for further improvement.

Among the fifteen dialog sessions, the subjects attempted to set up 29 alert messages, which in turn triggered 29 alert arrivals (indicated by visual icons). In all except two cases, the subjects chose to *immediately* interrupt the online interaction workflow to handle the alert. The two special cases correspond to the behavior of one subject who chose to complete her task in the OI dialog before switching to the alert message. Having handled the alert, the subjects needed to return to the original, interrupted task. We examined the 27 interruptions and found that only nine of them utilized the explicit spoken/textual request "go back" to resume the interrupted discourse (i.e., restoring the discourse state from the register to the OI dialog thread). For the 18 remaining cases, the user simply restarted the interrupted task afresh (three cases) or initiated another new task afresh (15 cases). User behavior, as observed from these dialog sessions, shows an inclination toward *frequent focus shifting*—that is, they tend to switch quickly from one task to another, especially when there are interruptions by alert messages. Based on the limited amount of pilot data, one may draw a preliminary conclusion that such user behavior is characteristic of the financial domain, in which information is extremely time sensitive and actions need to be taken promptly. Frequent focus shifting presents a challenge in proper handling of discourse contexts among the interrupted and interrupting tasks. The use of visual icons to indicate the presence of alert messages in ISIS (see Figure 11.7) forces the user to *explicitly* communicate an intended focus shift by clicking the icon. This signifies to the system to prepare for discourse handling when the interruption actually occurs. This option presents a simplification advantage and is therefore preferred to an alternative that requires the system to automatically infer the occurrence of a focus shift/task switch. The outcome of automatic inference is not guaranteed to be correct every time and thus presents greater ambiguity in discourse handling. The user may also *explicitly* communicate an intended focus shift *back* to a previously interrupted task via the "go back" command. This command also signifies to the system to reinstate the interrupted discourse state. However as noted previously, few subjects took advantage of this command and instead went through redundant dialog turns to set up the interrupted task from scratch. Possible remedial measures include (1) raising the user's awareness of the "go back" command or representing it as a visual icon that appears as a just-in-time reminder (Cutrell et al., 2001); and (2) automatically and transparently reinstating the interrupted discourse state immediately upon completion of the interrupting task. Should the usage context evolve such that the typical user tends to be unfocused and frequently shifts his/her focus without prior communication, a more complex discourse interpretation framework—for example, one that involves plan recognition and a focus stack (Lesh et al., 2001)—will be necessary.

11.8 Implementation of Multilingual SDS in VXML

This section addressing implementation issues for multilingual SDS takes a slight divergence by presenting an example implementation of a *simple* multilingual SDS in VXML that supports displayless voice browsing of Web content. We assume that the Web content has been transcoded and/or transformed into VXML by software platforms that support universal accessibility, such as that shown in Figure 11.9. This illustrates the AOPA (Author Once, Present Anywhere) platform, which adopts W3C standards for content specification in Extensible Markup Language (XML) and presentation specification in Extensible Stylesheet Language (XSL). Web content can be automatically transcoded from HTML to XML to be hosted in a unified content repository. The XSL automatically transforms the content to suitable layouts for different client devices with different form factors (e.g., medium to small display screens or displayless devices). More specifically, Web browsers for desktop PCs render documents in HTML, mobile mini-browsers for PDAs render documents in HTML3.2, WAP browsers render documents in WML, and (displayless) voice browsers render documents

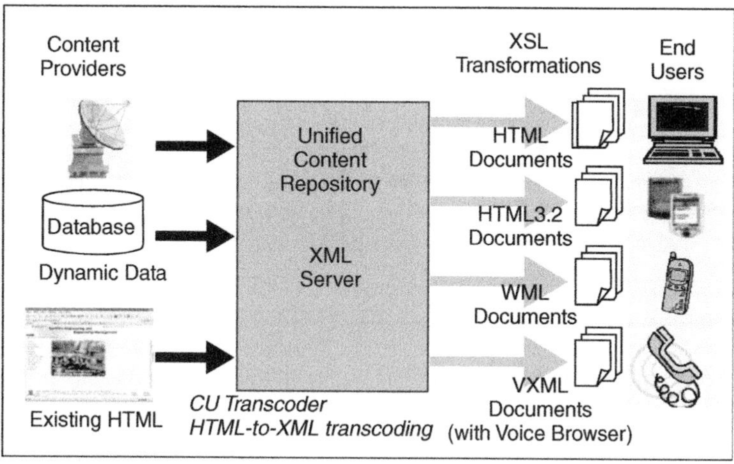

Figure 11.9: The AOPA software platform supports universal accessibility.

in Voice Extensible Markup Language (VXML).[13] In each case, the XSL stylesheets automatically repurpose the documents for layouts that suit information visualization using the various client devices. We currently focus on displayless voice browsing of Web content through the telephone channel. The content has been transcoded and transformed into VXML documents, which can be accessed by a voice gateway that connects to the Internet and to the public switch telephone network. The voice browser is software that runs on the voice gateway and can interpret VXML documents. A VXML document specifies a human-computer dialog in which human speech input is supported by speech recognition; and computer speech output is supported by speech synthesis/digitized audio. These core speech engines (for recognition and synthesis) are invoked by the voice browser. In addition, the voice browser can also handle touch-tone (telephone keypad) input. Figure 11.10 illustrates voice browsing of Web content.

Existing voice browsers generally support a single, primary language—mostly English—for voice browsing. In the following sections, we will present an example of the implementation of a monolingual SDS, followed by a bilingual voice browser, with a bilingual (English and Cantonese) SDS implemented with VXML.

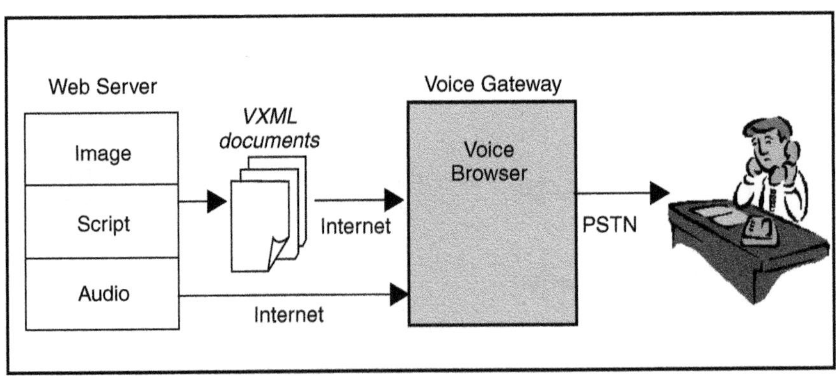

Figure 11.10: Browsing Web content by voice (see www.vxml.org).

[13] www.voicexml.org

Table 11.8 A human-computer dialog in the weather domain (C: computer, H: human).

C:	Welcome to CU Weather. Today is 22/11/2003. Please select Hong Kong weather or world weather.
H:	*World weather*
C:	Please say the city name.
H:	*Shanghai*
C:	The current weather conditions for Shanghai are ...

11.8.1 Example of a Monolingual SDS Implemented in VXML

This section presents a simple example of an English dialog in the weather domain. The dialog flow is presented in Table 11.8, and the associated VXML document is presented in Table 11.9.

11.8.2 A Bilingual Voice Browser

The Chinese University (CU) voice browser (see Figure 11.11) was developed to support bilingual human-computer dialogs in VXML. The browser is implemented based on OpenVXI 2.0,[14] a voice browser developed by SpeechWorks and hosted at CMU. OpenVXI contains a task manager (including the VXML interpreter) and five component interfaces. The interfaces are:

1. An Internet module to fetch VXML documents via HTTP
2. An XML parser to parse a VXML document into a DOM object
3. A computer telephony module for telephone speech I/O
4. A Prompt interface to pass the VXML prompt text to the corresponding text-to-speech engines (we use CU Vocal [Meng et al., 2002; Meng et al., 2000] for Cantonese and Speechify[15] for English)

[14]OpenVXI: http://fife.speech.cs.cmu.edu/openvxi/

[15]Speechify, Scansoft Inc. http://www.scansoft.com/speechify

Table 11.9 The VXML document that specifies the English dialog in Table 11.8 (explanations are boldface).

```
<?xml version="1.0"?>
<vxml version= "1.0">
<link next="#exit">
<!–This is a link at the VXML level and so the grammar remains
active throughout the document. The user can exit the dialog by saying
goodbye or byebye –>
<grammar>goodbye |byebye </grammar>
</link>
<form id="welcome" scope="document">
<block>
<!– The [date] information below is automatically filled in as XSL
transformation takes place to generate the VXML document from
XML content –>
<prompt>Welcome to CU Weather. Today is [date]
</prompt>
<goto next="#type" />
</block>
</form>

<menu id="type" scope="document">
<prompt>Please select Hong Kong weather or world
weather </prompt>
<choice next="#local">
<!–Jumps down to the form with id local below –>
<grammar>hong kong weather |hong kong<grammar>
</choice>
<choice next="#world">

<!–Jumps down to the form with id world –>
<grammar>world weather |world<grammar>
</choice>
</menu>

<form id="local" scope="document">
<!– form with id local –>
<block>
<prompt>Hong Kong's current weather condition is
...
</prompt>
```

(*continued*)

Table 11.9 Continued.

```
<goto next="#type" />
</block>
<form>

<form id="world" scope="document">
<!– form with id world –>
<field namelist="other_city">
<grammar src="eng_city.gram"/>
<!–the list of city names is obtained from the grammar file eng_city.gram
for filling in the variable other_city. –>
<prompt>Please say the city name </prompt>
<!– Based on the filled variable, the system retrieves the relevant weather
information. –>
...
</form>

<form id="exit">
<block>
<prompt>Thank you for calling.</prompt>
<exit expr="success" />
</block>
</form>
</vxml>
```

5. A Rec interface to pass the VXML grammar to the corresponding speech recognition engines (we use CURSBB[16] for Cantonese and OpenSpeech Recognizer[17] for English)

The dialogic component is a modularized computer telephony interface to handle phone calls. As indicated by the shading in Figure 11.11, major enhancements in the CU voice browser reside in the Prompt and Rec interfaces for bilingual speech recognition and speech synthesis. The CU voice browser sets the default language to Cantonese. It also interprets the value

[16]CURSBB (Chinese University Recognition Software Building Blocks) http://dsp.ee.cuhk.edu.hk/speech/cursbb/

[17]OpenSpeech Recognizer, Scansoft Inc. http://www.scansoft.com/openspeech/recognizer

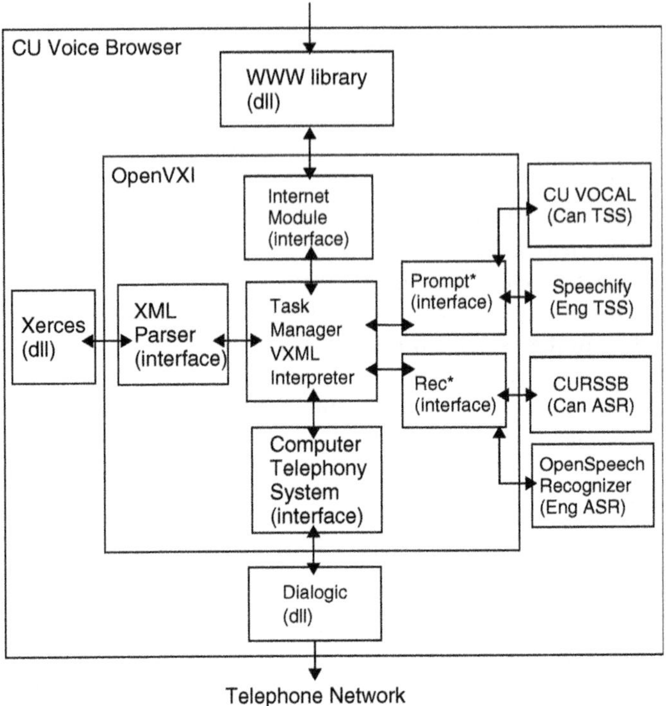

Figure 11.11: Architecture of the bilingual CU voice browser.

of the xml:lang attribute in VXML—the values "en-US" indicates English and "zh-HK" indicates Chinese. Hence the CU voice browser can invoke the appropriate (English versus Cantonese) text-to-speech engine as well as activate the appropriate grammar for speech recognition. If *both* grammars are activated, the Cantonese recognizer will, by default, be applied first to the speech input. However, if the Cantonese recognizer cannot search for any suitable recognition hypothesis in its lattice, it will reject the speech input, causing the CU voice browser to pass the speech input to the English recognizer.

We extend the example provided in Table 11.8 to illustrate the bilingual CU Weather application that is implemented with VXML. CU Weather

Table 11.10 A bilingual human-computer dialog implemented with VXML in the CU weather system.

C:	歡迎致電中文大學天氣熱線, 今日係 22/11/2003 Welcome to CU Weather. Today is 22/11/2003.
C:	請選擇語言, 請講廣東話或者英文 Please select language. Please say Cantonese or English.
H:	廣東話 *(translation: Cantonese)*
C:	請問你想查詢本地天氣, 定係世界城市天氣 *(translation: Would you like local weather or world weather information?)*
H:	世界城市 *(translation: World weather.)*
C:	請講出城市名稱 *(translation: Please say the city name.)*
H:	上海 *(translation: Shanghai)*
C:	上海既天氣而家係…… *(translation: The weather information for Shanghai is ...)*
H:	*Change language.*
C:	請選擇語言, 請講廣東話或者英文 Please select language. Please say Cantonese or English.

is a research prototype that is available for experimentation.[18] Real-time weather information is sourced from the Hong Kong Observatory Web sites.[19] The dialog flow is presented in Table 11.10, and associated VXML is presented in Table 11.11.

11.9 Summary and Conclusions

In this chapter, we have presented an introduction to spoken dialog systems, including the typical system components, the challenges and necessary extensions for supporting multiple languages, the components that remain language independent, as well as those that are language dependent. We

[18]Callers can dial +852.2603.7884 to experiment with the CU Weather system.

[19]The Hong Kong Observatory Web sites provide weather information in both English and Chinese for both local and world weather (www.hko.gov.hk).

then described in detail the implementation of an SDS known as ISIS which resembles a virtual stockbroker that can provide the user with real-time stock market information, manage simulated personal portfolios, and handle simulated financial transactions via mixed-initiative dialogs. The conversational interface is augmented with a screen display that can capture mouse-clicks, and users can also input textual queries by typing or stylus-writing. Textual input/output can be in English or Chinese, while spoken input/output can be in English, Cantonese, or Putonghua. Therefore, ISIS is a trilingual system.

We presented the CORBA architecture in ISIS and traversed each ISIS component in detail, highlighting the efforts dedicated to language-dependent development to cover trilingualism. Such efforts include speech recognition, natural language understanding, text-dependent speaker authentication, and text-to-speech synthesis. As parallel monolingual components are developed to support multilingualism, we strive to achieve the maximum amount of sharing and leverage across parallel components. We also described the language-independent components of ISIS, primarily that of dialog modeling and novel dialog features, such as adaptivity that automatically expands the knowledge base of the SDS to keep it

Table 11.11 VXML document that specifies the bilingual CU weather dialog in Table 11.10 (explanations are boldface).

<?xml version="1.0"?>
<vxml version="1.0">
<link next="exit">
<!–This is a link at the <vxml>level and so the grammar remains active throughout the document. The user can exit the dialog by saying goodbye, byebye, 再見 or 拜拜. Recognition grammars for both languages are activated since xml:lang is set to both "en-US" and "zh-HK". The target of this <link>is specified in the "next" attribute, the "#" sign means the target is located at the same document. It will trigger the <form>or <menu>element with "id" attribute equals to "exit" in this example –>
<grammar xml:lang="en-US">goodbye |byebye</grammar>
<grammar xml:lang="zh-HK">再見|拜拜</grammar>
</link>

(continued)

Table 11.11 Continued.

```
<link next=“#chlang”>
<!– Add a link to allow user to change language at any time by saying
change language or 轉語言–>
<grammar xml:lang=“en-US”>change language</grammar>
<grammar xml:lang=“zh-HK ”>轉語言</grammar>
</link>

<form id=“welcome”>
<!– The xml:lang attribute attached with the prompt tag specifies the
synthesizer to invoke for the delimited text–>
<block>
<prompt xml:lang=“zh-HK”>歡迎致電中文大學天氣熱線,
今日係 22/11/2003</prompt>
<prompt xml:lang=“en-US”>Welcome to CU Weather.
Today is 22/11/2003.</prompt>
<!– Jumps down to the menu with id chlang below –>
<goto next=“#chlang”/>
</block>
</form>

<menu id=“chlang”>
<!– menu that supports language selection, id is chlang –>
      <prompt xml:lang=“zh-HK”>
請選擇語言,請講廣東話或者英文</prompt>
<prompt xml:lang=“en-US”>Please select language.
Please say Cantonese or English.</prompt>
<!– provide a list of choices for user to select –>
      <choice next=“#chiworld”>
<!– two ways of referring to “Cantonese” –>
<grammar xml:lang=“zh-HK”>廣東話|廣府話</grammar>
      </choice>
      <choice next=“#engworld”>
<!– two ways of referring to “English” –>
<grammar xml:lang=“zh-HK”>英文|英語</grammar>
      </choice>
<choice next=“#chiworld”>
      <grammar xml:lang=“en-US”>Cantonese</grammar>
      </choice>
<choice next=“#engworld”>
```

(continued)

Table 11.11 Continued.

```
      <grammar xml:lang="en-US">English</grammar>
      </choice>
</menu>

<form id="chiworld">
<!—we try to fill in the variable "name" in this form –>
      <field name="cityname">
<!– grammar lists city names in Chinese that can be used to fill in the
variable cityname. –>
<grammar xml:lang="zh-HK">北京|長沙|洛杉磯|紐約|三藩市|蒙特利
爾|東京|札幌|倫敦|阿姆斯特丹|...
         </grammar>
         <prompt xml:lang="zh-HK">請講出城市名稱</prompt>
<!– Based on the filled variable, the system retrieves the relevant
weather information. –>
...
</form>

<form id="engworld">
      <field name="cityname">
<!– grammar lists city names in English that can be used to fill in the
variable cityname. –>
<grammar xml:lang="en-US">Beijing |Changsha |Los Angeles |New
York |San Francisco |Montreal |Tokyo |Sapporo |London |Amsterdam |...
</grammar>
         <prompt xml:lang="en-US">Please say the city name.</prompt>
...
</form>

<form id="exit">
<block>
<prompt xml:lang="zh-HK">多謝你既致電, 拜拜.</prompt>
<prompt xml:lang="en-US">Goodbye, thank you for calling.</prompt>
<!– set the exit status to success to exit the dialog –>
<exit expr="success" />
</block>
</form>
</vxml>
```

up-to-date; delegation dialogs related to asynchronous human-computer interaction; and interruptions that alert the user of late-breaking events as well as the subsequent handling of multithreaded dialogs. The dialog model, including its novel extensions, is critical for human-computer interaction but also constitutes the language-independent components in the system. Thereafter, we focused on the implementation of multilingual SDS that is centered on the VXML platform. We provided a reference implementation of a bilingual voice browser (based on one that supports a displayless voice browser) of Web content in both English and Chinese.

Promising future directions include investigation of the portability and scalability of ISIS to multidomain dialogs and multiple interruption messages. We also plan to set up a wireless communication infrastructure so that users can interface with the ISIS system in a mobile setting with smart phones and PDAs. In this context, heavier reliance will be placed on pointing and clicking to enrich our test bed for exploring multimodal and mixed-modal interactions (Oviatt et al., 1997; Oviatt and Cohen, 2000). Another possibility is to migrate the CORBA architecture toward a Web Services architecture[20] (Meng et al., 2003), which is an emerging standard based on XML that promotes modular system integration with a high degree of interoperability.

[20] http://www.w3.org/2002/ws/

Bibliography

Ackermann, U., Angelini, B., Brugnara, F., Federico, M., Giuliani, D., Gretter, R., Lazzari, G., Niemann, H. (1996). Speedata: Multilingual spoken data-entry. In: *Proceedings of the International Conference on Spoken Language Processing*. Philadelphia, pp. 2211–2214.

Ackermann, U., Angelini, B., Brugnara, F., Federico, M., Giuliani, D., Gretter, R., Niemann, H. (1997). Speedata: A prototype for multilingual spoken data-entry. In: *Proceedings of the European Conference on Speech Communication and Technology (EUROSPEECH)*. Rhodes, Greece, pp. 355–358.

Adda, G., Adda-Decker, M., Gauvain, J., Lamel, L. (1997). Text normalization and speech recognition in French. In: *Proceedings of the European Conference on Speech Communication and Technology (EUROSPEECH)*. Rhodes, Greece, pp. 2711–2714.

Adda-Decker, M. (1999). Towards multilingual interoperability in automatic speech recognition. In: *Proceedings of the ESCA-NATO Tutorial Research Workshop on Multi-lingual Interoperability in Speech Technology*. Leusden, Netherlands.

Adda-Decker, M. (2003). A corpus-based decompounding algorithm for German lexical modeling in LVCSR. In: *Proceedings of the European Conference on Speech Communication and Technology (EUROSPEECH)*. Geneva, Switzerland.

Adda-Decker, M., de Mareuil, P. B., Adda, G., Lamel, L. (2005). Investigating syllabic structures and their variation in spontaneous French. *Speech Communication* 46 (2), 119–139.

Adda-Decker, M., Lamel, L. (1999). Pronunciation variation across system configuration, language and speaking style. *Speech Communication* 29, 83–98.

AECMA (2004). The AECMA simplified English standard, psc-85-16598. http://www.simplifiedenglish-aecma.org/Simplified_English.htm.

Akiba, Y., Watanabe, T., Sumita, E. (2002). Using language and translation models to select the best among outputs from multiple MT systems. In: *Proceedings of COLING*. pp. 8–14.

Akmajian, A., Harnish, R., Demers, R., Farmer, A. (1995). *Linguistics: An Introduction to Language and Communication.* MIT Press, Cambridge, MA.

Allauzen, A., Gauvain, J. (2003). Adaptation automatique du modèle de langage d'un système de transcription de journaux parèls. *Traitement Automatique des langues* 44 (1), 11–31.

Allauzen, A., Gauvain, J. (2005b). Open vocabulary ASR for audiovisual document indexation. In: *Proceedings of the IEEE International Conference on Acoustics, Speech and Signal Processing*. Philadelphia.

Allauzen, A., Gauvain, J. (2005a). Diachronic vocabulary adaptation for broadcast news transcription. In: *Proceedings of the European Conference on Speech Communication and Technology (EUROSPEECH-INTERSPEECH)*. Lisbon, Portugal.

Allen, J., Becker, J. (Eds.) (2004). *The Unicode Standard*, Version 4.0. Addison Wesley.

Allen, J., Hunnicut, S., Klatt, D. (1987). *Text-to-Speech: The MITalk system.* Cambridge University Press, Cambridge, U.K.

Alshawi, H., Bangalore, S., Douglas, S. (2000). Learning dependency translation models as collections of finite-state head transducers. *Computational Linguistics* 26 (1), 45–60.

Amdall, I., Korkmazskiy, F., Surendran, A. C. (2000). Joint pronunciation modeling of non-native speakers using data-driven methods. In: *Proceedings of the International Conference on Spoken Language Processing*, Beijing, China.

AMSR2001, Sophia-Antipolis, France (2001). Workshop on Adaptation Methods for Speech Recognition.

Andersen, O., Dalsgaard, P. (1997). Language-identification based on cross-language acoustic models and optimised information combination. In: *Proceedings of the European Conference on Speech Communication and Technology (EUROSPEECH)*. Rhodes, Greece, pp. 67–70.

Andersen, O., Dalsgaard, P., Barry, W. (1993). Data-driven identification of poly- and mono-phonemes for four European lanugages. In: *Proceedings of the European Conference on Speech Communication and Technology (EUROSPEECH)*. Berlin, Germany, pp. 759–762.

Andersen, O., Dalsgaard, P., Barry, W. (1994). On the use of data-driven clustering techniques for language identification of poly- and mono-phonemes for Four European languages. In: *Proceedings of the IEEE International Conference on Acoustics, Speech and Signal Processing*. Adelaide, pp. 121–124.

Angkititrakul, P., Hansen, J. (2003). Use of trajectory models for automatic accent classification. In: *Proceedings of the European Conference on Speech Communication and Technology (EUROSPEECH)*. Geneva, Switzerland.

Anon., March 10, 2005. The real digital divide. *The Economist.*

Aretoulaki, M., Harbeck, S., Gallwitz, F., Nöth, E., Niemann, H., Ivanecky, J., Ipsic, I., Pavesic, N., Matousek, V. (1998). SQEL: A multilingual and multifunctional dialogue system. In: *Proceedings of the International Conference on Spoken Language Processing*. Sydney, Australia.

Armstrong, S., Kempen, M., McKelvie, D., Petitpierre, D., Rapp, R., Thompson, H. (1998). Multilingual corpora for cooperation. In: *Proceedings of the First International Conference on Language Resources and Evaluation (LREC)*. Granada, Spain, pp. 975–980.

Association, I. P. (1999). *Handbook of the International Phonetic Association: A Guide to Use of the International Phonetic Alphabet*. Cambridge University Press, Cambridge, UK.

Auberg, S., Correa, N., Locktionova, V., Molitor, R., Rothenberg, M. (1998). The accent coach: An English pronunciation training system for Japanese speakers. In: *Proceedings of the ESCA Workshop on Speech Technology in Language Learning (STiLL)*. Stockholm, Sweden.

Aust, H., Oerder, M., Seide, F., Steinbiss, V. (1995). The Phillips Automatic Train Timetable Information System. Speech Communication 17, pp. 249–262.

BABYLON (2003). Web site: http://www.darpa.mil/ipto/research/babylon/approach.html.

Bacchiani, M., Ostendorf, M. (1998). Joint acoustic unit design and lexicon generation. In: *Proceedings of the ESCA Workshop on Speech Synthesis*. Rolduc, The Netherlands, pp. 7–12.

Bahl, L., Brown, P., DeSouza, P., Mercer, R. (1989). A tree-based statistical language model for natural language speech recognition. In: *Proceedings of the IEEE International Conference on Acoustics, Speech and Signal Processing*. pp. 1001–1008.

Barnard, E., Cloete, J., Patel, H. (2003). Language and technology literacy barriers to accessing government services. Vol. 2739 of Lecture notes in Computer Science, 37–42.

Barnett, J., Corrada, A., Gao, G., Gillick, L., Ito, Y., Lowe, S., Manganaro, L., Peskin, B. (1996). Multilingual speech recognition at Dragon Systems. In: *Proceedings of the International Conference on Spoken Language Processing*, pp. 2191–2194.

Bayer, S., Doran, C., George, B. (2001). Exploring speech-enabled dialog with the GALAXY communicator infrastructure. In: *Proceedings of the Human Language Technologies Conference* (HLT). San Diego, CA. pp. 79–81.

Bazzi, J., Glass, J. (2000). Modeling out-of-vocabulary words for robust speech recognition. In: *Proceedings of the International Conference on Spoken Language Processing* (ICSLP-00). Beijing, China, pp. 401–404.

Beaugendre, F., Clase, T., van Hamme, H. (2000). Dialect adaptation for Mandarin Chinese. In: *Proceedings of the International Conference on Spoken Language Processing*. Beijing, China.

Beebe, L. M. (1987). Myths about interlanguage phonology. In: Ioup, G., Weinberger, S. H. (Eds.), *Interlanguage Phonology: The Acquisition of a Second Language Sound System. Issues in Second Language Research.* Newbury House, Cambridge, MA.

Bellegarda, J. (2004). Statistical language model adaptation: Review and perspectives. *Speech Communication* 42 (1), 93–108.

Bellegarda, J. R. (2000). Exploiting latent semantic information in statistical language modeling. *Proceedings of the IEEE* 88 (8), 1279–1296.

Berkling, K., Arai, T., Barnard, E., Cole, R. (1994). Analysis of phoneme-based features for language identification. In: *Proceedings of the IEEE International Conference on Acoustics, Speech and Signal Processing*. Adelaide, Australia, pp. 289–292.

Berry, M. W., Dumais, S. T., O'Brien, G. W. (1995). Using linear algebra for intelligent information retrieval. *SIAM Review* 37 (4), 573–595.

Besling, S. (1994). Heuristical and statistical methods for grapheme-to-phoneme conversion. In: *Konvens*. Vienna, Austria, pp. 23–31.

Billa, J., Ma, K., McDonough, J. W., Zavaliagkos, G., Miller, D. R., Ross, K. N., El-Jaroudi, A. (1997). Multilingual speech recognition: The 1996 Byblos Callhome System. In: *Proceedings of the European Conference on Speech Communication and Technology (EUROSPEECH)*. Rhodes, Greece, pp. 363–366.

Billa, J., Noamany, M., Srivastava, A., Liu, D., Stone, R., Xu, J., Makhoul, J., Kubala, F. (2002). Audio indexing of Arabic broadcast news. In: *Proceedings of the IEEE International Conference on Acoustics, Speech and Signal Processing*. Orlando, FL, pp. 5–8.

Bisani, M., Ney, H. (2003). Multigram-based grapheme-to-phoneme conversion for LVCSR. In: *Proceedings of the European Conference on Speech Communication and Technology (EUROSPEECH)*. Geneva, Switzerland, pp. 933–936.

Black, A., Font Llitjós, A. (2002). Unit selection without a phoneme set. In: *IEEE Workshop on Speech Synthesis*. Santa Monica, CA.

Black, A., Hunt, A. (1996). Generating F_0 contours from ToBI labels using linear regression. In: *Proceedings of the International Conference on Spoken Language Processing*. Philadelphia, pp. 1385–1388.

Black, A., Lenzo, K. (2000a). *Building Voices in the Festival Speech Synthesis System*, http://festvox.org/bsv/.

Black, A., Lenzo, K. (2000b). Limited domain synthesis. In: *Proceedings of the International Conference on Spoken Language Processing*. Beijing, China, pp. 411–414.

Black, A., Lenzo, K. (2001). Optimal data selection for unit selection synthesis. In: *Proceedings of the 4th ESCA Workshop on Speech Synthesis*. Scotland.

Black, A., Lenzo, K., Pagel, V. (1998a). Issues in building general letter to sound rules. In: *Proceedings of the ESCA Workshop on Speech Synthesis*. Jenolan Caves, Australia., pp. 77–80.

Black, A., Taylor, P. (1997). Automatically clustering similar units for unit selection in speech synthesis. In: *Proceedings of the European Conference on Speech Communication and Technology (EUROSPEECH)*. Rhodes, Greece, pp. 601–604.

Black, A., Taylor, P., Caley, R. (1998b). *The Festival Speech Synthesis System*, http://festvox.org/festival.

Black, A. W. (1997). Predicting the intonation of discourse segments from examples in dialogue speech. In: Sagisaka, Y., Campbell, N., Higuchi, N. (Eds.), *Computing Prosody*. Springer Verlag, pp. 117–128.

Black, A. W., Campbell, N. (1995). Optimising selection of units from speech databases for concatenative synthesis. In: *Proceedings of the European Conference on Speech Communication and Technology (EUROSPEECH)*. Madrid, Spain, pp. 581–584.

Black, A. W., Taylor, P. (1994). CHATR: A genetic speech synthesis system. In: *Proceedings of COLING*. Kyoto, Japan, pp. 983–986.

Bonaventura, P., Gallocchio, F., Micca, G. (1997). Multilingual Speech Recognition for Flexible Vocabularies. In: *Proceedings of the European Conference on Speech Communication and Technology (EUROSPEECH)*. Rhodes, Greece, pp. 355–358.

Boula de Mareil, P., Habert, B., Bnard, F., Adda-Decker, M., Baras, C., Adda, G., Paroubek, P. (2005). A quantitative study of disfluencies in French broadcast interviews. In: *ISCA DiSS '05*—Disfluency in Spontaneous Speech. Aix-en-Provence, France.

Braun, J., Levkowitz, H. (1998). Automatic language identification with perceptually guided training and recurrent neural networks. In: *Proceedings of the International Conference on Spoken Language Processing*. Sydney, Australia, pp. 3201–3205.

Breiman, L., Friedman, J., Olshen, R., Stone, C. (1984). *Classification and Regression Trees*. Wadsworth & Brooks, Pacific Grove, CA.

Brière, E. (1966). An investigation of phonological interference. *Language* 42 (4), 768–796.

Brill, E. (1995). Transformation-based error-driven learning and natural language processing: A case study in part-of-speech tagging. *Computational Linguistics* 21(4), 543–566.

Brown, P., Cocke, J., Pietra, S. A. D., Pietra, V. J. D., Jelinek, F., Lafferty, J. D., Mercer, R. L., Roossin, P. S. (1993a). A statistical approach to machine translation. *Computational Linguistics* 16, 79–85.

Brown, P., Pietra, S. D., Pietra, V. D., Mercer, R. (1993b). The mathematics of statistical machine translation: Parameter estimation. *Computational Linguistics* 19 (2), 263–311.

Bub, U., Köhler, J., Imperl, B. (1997). In-service adaptation of multilingual hidden-Markov-models. In: *Proceedings of the IEEE International*

Conference on Acoustics, Speech and Signal Processing. Munich, Germany, pp. 1451–1454.

Buckwalter, T. (2002). *Arabic morphological analyzer*. Available at http://www.ldc.upenn.edu/Catalog/CatalogEntry.jsp?catalogId=LDC2002L49 from the Linguistic Data Consortium.

Byrne, W., Doermann, D., Franz, M., Gustman, S., Hajič, J., Picheny, D. O. M., Psutka, J., Ramabhadran, B., Soergel, D., Ward, T., Zhu, W.-J. (2004). Automatic recognition of spontaneous speech for access to multilingual oral history archives. *IEEE Transactions on Speech and Audio Processing* 12 (4), 420–435.

Byrne, W., Hajič, J., Ircing, P., Jelinek, F., Khudanpur, S., Krbec, P., Psutka, J. (2001). On large vocabulary conversational speech recognition of a highly inflectional language—Czech. In: *Proceedings of the European Conference on Speech Communication and Technology (EUROSPEECH)*. pp. 487–489.

Byrne, W., Hajič, J., Ircing, P., Jelinek, F., Khudanpur, S., McDonough, J., Peterek, N., Psutka, J. (1999). Large vocabulary speech recognition for read and broadcast Czech. In: Matousek, V., Mautner, P., Ocelikova, J., Sojka (Eds.), *Text Speech and Dialog. Lecture Notes in Computer Science*. Vol. 1692, Springer Verlag, pp. 235–240.

Byrne, W., Hajič, J., Krbec, P., Ircing, P., Psutka, J. (2000). Morpheme based language models for speech recognition of Czech. In: Sojka, P., Kopecek, I., Pala, K. (Eds.), *Text Speech and Dialog. Lecture Notes in Computer Science*. Vol. 1902, Springer Verlag, pp. 211–216.

Byrne, W., Knodt, E., Khudanpur, S., Bernstein, J. (1998a). Is automatic speech recognition ready for non-native speech? A data collection effort and initial experiments in modeling conversational Hispanic speech. In: *Proceedings of the ESCA-ITR Workshop on Speech Technology in Language Learning*. Stockholm, Sweden, pp. 37–40.

Byrne, W., Knodt, E., Khudanpur, S., Bernstein, J. (1998b). Is automatic speech recognition ready for non-native speech? A data collection effort and initial experiments in modeling conversational Hispanic English. In: *Proceedings of the ESCA Workshop on Speech Technology in Language Learning (STiLL)*.

Campbell, W. (2002). Generalized linear discriminant sequence kernels for speaker recognition. In: *Proceedings of the IEEE International Conference on Acoustics, Speech and Signal Processing*. Orlando, FL.

Campbell, W., E., S., P.A., T.-C., Reynolds, D. (2004). Language recognition with support vector machines. In: *Speaker Odyssey—The Speaker and Language Recognition Workshop*. Toledo, Spain.

Canavan, A., Zipperlen, G., Graff, D. (1997). *The CALLHOME Egyptian Arabic speech corpus*. Available at http://www.ldc.upenn.edu/Catalog/CatalogEntry.jsp?catalogId=LDC97S45 from the Linguistic Data Consortium.

Candea, M., Vasilescu, I., Adda-Decker, M. (2005). Inter- and intra-language acoustic analysis of autonomous fillers. In: *Proceedings of the DiSS'05—ISCA Tutorial and Research Workshop on Disfluency in Spontaneous Speech*. Aix-en-Provence, France.

Çarki, K., Geutner, P., Schultz, T. (2000). Turkish LVCSR: Towards better speech recognition for agglutinative languages. In: *Proceedings of the IEEE International Conference on Acoustics, Speech and Signal Processing*. Istanbul, Turkey, pp. 1563–1566.

Caseiro, D., Trancoso, I. (1998). Identification of spoken European languages. In: *Proceedings of the Ninth European Signal Processing Conference (EUSIPCO)*. Rhodes, Greece.

Chan, J., Ching, P., Lee, T., Meng, H. (2004). Detection of language boundary in code-switching utterances by bi-phone probabilities. In: *Proceedings of the International Conference on Spoken Language Processing*. Jeju Island, South Korea.

Charniak, E. (2001). Immediate-head parsing for language models. In: *Proceedings of the 39th Annual Meeting of the Association for Computational Linguistics*. Toulouse, France.

Chase, L. (1998). A review of the American Switchboard and Callhome speech recognition evaluation programs. In: Rubio, A., Gallardo, N., Castro, R., Tejada, A. (Eds.), *Proceedings of the First International Conference on Language Resources and Evaluation (LREC)*. Granada, Spain, pp. 789–793.

Chelba, C., Jelinek, F. (2000). Structured language modeling. *Computer Speech and Language* 14 (4), 283–332.

Chen, L., Gauvain, J.-L., Gilles Adda, L. L. (2004). Dynamic language modeling for broadcast news. In: *Proceedings of the International Conference on Spoken Language Processing*. Jeju Island, South Korea, pp. 1281–1284.

Chen, L., Gauvain, J.-L., Lamel, L., Adda, G. (2003). Unsupervised language model adaptation for broadcast news. In: *Proceedings of the IEEE International Conference on Acoustics, Speech and Signal Processing*. Hong Kong, China, pp. I–220–223.

Chen, S., Goodman, J. (1998). *An empirical study of smoothing techniques for language modeling*. Tech. Rep. TR-10-98, Computer Science Group, Harvard University, available at http://www-2.cs.cmu.edu/~sfc/papers/h015a-techreport.ps.gz.

Choi, W. N., Wong, Y. W., Lee, T., Ching, P.C. (2000). Lexical tree decoding with class-based language models for Chinese speech recognition. In: *Proceedings of the International Conference on Spoken Language Processing*. Beijing, China. pp. 174–177.

Chotimongkol, A., Black, A. (2000). Statistically trained orthographic to sound models for Thai. In: *Proceedings of the International Conference on Spoken Language Processing*. Beijing, China., pp. 551–554.

Choukri, K., Mapelli, V. (2003). *Deliverable 5.2: Report contributing to the design of an overall coordination and strategy in the field of LR*. ENABLER project report.

Chow, Y., Schwartz, R., Roukos, S., Kimball, O., Price, P., Kubala, F., Dunham, M., Krasner, M., Makhoul, J. (1986). The role of word-dependent coarticulatory effects in a phoneme-based speech recognition system. In: *Proceedings of the IEEE International Conference on Acoustics, Speech and Signal Processing*. Tokyo, Japan, pp. 1593–1596.

Chung, G., Seneff, S., Wang, C. (2003). Automatic acquisition of names using speak and spell mode in spoken dialogue systems. In: *Proceedings of the Human Language Technologies Conference* (HLT). Edmonton, Canada, pp. 197–200.

Cimarusti, D., Ives, R. (1982). Development of an automatic identification system of spoken languages: Phase 1. In: *Proceedings of the IEEE International Conference on Acoustics, Speech and Signal Processing*. Paris.

Clarkson, P. R., Robinson, A. J. (1997). Language model adaptation using mixtures and an exponentially decaying cache. In: *Proceedings of the IEEE International Conference on Acoustics, Speech and Signal Processing*. Munich, Germany, pp. 799–802.

CMU (1998). Carnegie Mellon Pronuncing Dictionary, http://www.speech.cs.cmu.edu/cgi-bin/cmudict.

Coccaro, N., Jurafsky, D. (1998). Toward better integration of semantic predictors in statistical language modeling. In: *Proceedings of the International Conference on Spoken Language Processing*. Sydney, Australia, pp. 2403–2406.

Cohen, M. (1989). *Phonological structures for speech recognition*. Ph.D. thesis, University of California, Berkeley, Berkeley, CA.

Cohen, P., Dharanipragada, S., Gros, J., Monkowski, M., Neti, C., Roukos, S., Ward, T. (1997). Towards a universal speech recognizer for multiple languages. In: *Proceedings of the IEEE Workshop on Automatic Speech Recognition and Understanding*. Santa Barbara, CA, pp. 591–598.

Cohen, P. R., Cheyer, J., Wang, M., Baeg, S. C. (1994). An Open Agent Architecture. In: *Working Notes of the AAAT spring Symposium: Software Agents*. Menlo Park. CA, pp. 1–8.

Compernolle, D. V. (2001). Recognizing speech of goats, wolves, sheep and... non-natives. *Speech Communication* 35 (1–2), 71–79.

Comrie, B. (1989). *Language Universals and Linguistic Typology*, 2nd ed. University of Chicago Press.

Comrie, B. (1990). *The World's Major Languages*. Oxford University Press, Oxford, UK.

Constantinescu, A., Chollet, G. (1997). On cross-language experiments and data-driven units for ALISP (automatic language independent speech processing).

In: *Proceedings of the IEEE Workshop on Automatic Speech Recognition and Understanding*. Santa Barbara, CA, pp. 606–613.

Corredor-Ardoy, C., Gauvain, J., Adda-Decker, M., Lamel, L. (1997). Language identification with language-independent acoustic models. In: *Proceedings of the European Conference on Speech Communication and Technology (EUROSPEECH)*. Rhodes, Greece.

Costantini, E., Burger, S., Pianesi, F. (2002). NESPOLE!'s multi-lingual and multi-modal corpus. In: *Proceedings of Third International Conference on Language Resources and Evaluation (LREC)*. Las Palmas, Spain, pp. 165–170.

Coulmas, F. (1996). *The Blackwell Encyclopedia of Writings Systems*. Blackwell Publishing, Oxford.

Cremelie, N., ten Bosch, L. (2001). Improving the recognition of foreign names and non-native speech by combining multiple grapheme-to-phoneme converters. In: *AMSR2001*, Sophia-Antipolis, France (2001), pp. 151–154.

Cremelie, N., Martens, J. (1998). In search of pronunciation rules. In: *ESCA Workshop on Modeling Pronunciation Variation for Automatic Speech Recognition*. Rolduc, The Netherlands, pp. 23–28.

Croft, W. (1990). *Typology and Universals*. Cambridge University Press, Cambridge, UK.

Cucchiarini, C., Strik, H., Boves, L. (1998). Quantitative assessment of second language learners' fluency: An automatic approach. In: *Proceedings of the International Conference on Spoken Language Processing*. Sydney, Australia.

Culhane, C. (1996). DoD workshops on conversational and multilingual speech recognition. In: *Proceedings of the DARPA Speech Recognition Workshop*. pp. 148–153.

Cummins, F., Gers, F., Schmidhuber, J. (1999). Language identification from prosody without explicit features. In: *Proceedings of the European Conference on Speech Communication and Technology (EUROSPEECH)*. Budapest, Hungary.

Cutler, A. (1997). The comparative perspective on spoken-language processing. *Speech Communication* 21 (1-2), 3–16.

Cutrell, E., Czerwinski, M., Horvitz, E. (2001). Notification, disruption and memory: Effects of messaging on memory and performance. In: *Proceedings of the Human-Computer Interaction (Interact-01)*. Tokyo. IOS Press. pp. 263–269.

Dalsgaard, P., Andersen, O. (1992). Identification of mono- and poly-phonemes using acoustic-phonetic features derived by a self-organising neural network. In: *Proceedings of the International Conference on Spoken Language Processing*. Banff, Alberta, Canada, pp. 547–550.

Dalsgaard, P., Andersen, O. (1994). Application of inter-language phoneme similarities for language identification. In: *Proceedings of the International*

Conference on Spoken Language Processing. Yokohama, Japan, pp. 1903–1906.

Dalsgaard, P., Andersen, O., Barry, W. (1998). Cross-language merged speech units and their descriptive phonetic correlates. In: *Proceedings of the International Conference on Spoken Language Processing*. Sydney, Australia.

Dalsgaard, P., Andersen, O., Hesselager, H., Petek, B., (1996). Language-identification using language-dependent phonemes and language-independent speech units. In: *Proceedings of the International Conference on Spoken Language Processing*. Philadelphia.

Daniels, P., Bright, W. (1996). *The World's Writing Systems*. Oxford University Press, Oxford.

Dantzich, M., Robbins, D., Horvitz, E., Czerwinski, M. (2002). Scope: Providing awareness of multiple notifications at a glance. In: *Proceedings of the ACM International Working Conference on Advanced Visual Interfaces* (AVI-02).

Davel, M., Barnard, E. (2003). Bootstrapping for language resource generation. In: *Proceedings of the Symposium of the Pattern Recognition Association of South Africa*. South Africa, pp. 97–100.

Davel, M., Barnard, E. (2004a). The efficient generation of pronunciation dictionaries: Machine learning factors during bootstrapping. In: *Proceedings of the International Conference on Spoken Language Processing*. Jeju, South Korea.

Davel, M., Barnard, E., (2004b). The efficient generation of pronunciation dictionaries: Human factors during bootstrapping. In: *Proceedings of the International Conference on Spoken Language Processing*. Jeju, South Korea, pp. 2797–2800.

de Mareil, P. B., Habert, B., Banard, F., Adda-Decker, M., Baras, C., Adda, G., Paroubek, P. (2005). A quantitative study of disfluencies in French broadcast interviews. In: *Proceedings of the DISS'05—ISCA Tutorial and Research Workshop on Disfluency in Spontaneous Speech*. Aix-en-Provence, France.

Decadt, B., Duchateau, J., Daelemans, W., Wambacq, P. (2002). Transcription of out-of-vocabulary words in large vocabulary speech recognition based on phoneme-to-grapheme conversion. In: *Proceedings of the IEEE International Conference on Acoustics, Speech and Signal Processing*. Orlando, FL, pp. 861–864.

Deerwester, S., Dumais, S. T., Landauer, T. K., Furnas, G. W., Harshman, R. A. (1990). Indexing by latent semantic analysis. *Journal of the American Society for Information Science* 41 (6), 391–407.

DeFrancis, J. (1984). *The Chinese Language: Fact and Fantasy*. University of Hawaii Press, Honolulu.

Deng, L. (1997b). Integrated-multilingual speech recognition using universal phonological features in a production model. In: *Proceedings of the IEEE International Conference on Acoustics, Speech and Signal Processing*. Munich, Germany, pp. 1007–1010.

Deng, L. (1997a). A dynamic feature-based approach to speech modeling and recognition. In: *Proceedings of the IEEE Workshop on Automatic Speech Recognition and Understanding*. Santa Barbara, CA, pp. 107–114.

Deng, L., Sun, D. X. (1994). A statistical approach to automatic speech recognition using the atomic speech units constructed from overlapping articulatory features. *Journal of the Acoustical Society of America* (JASA) 95, 2702–2719.

Deng, Y., Khudanpur, S. (2003). Latent semantic information in maximum entropy language models for conversational speech recognition. In: *Proceedings of the Human Language Technology Conference of the North American Chapter of the Association for Computational Linguistics*. Edmonton, Canada, pp. 56–63.

Dharanipragada, S., Jelinek, F., Khudanpur, S. (1996). *Language model adaptation using the minimum divergence principle*. In research notes no 1, 1995 language modeling summer research workshop technical report, Center for Language and Speech Processing, Johns Hopkins University.

Dijkstra, J., van Son R. J. J. H., Pols, L. (2004). Frisian TTS, an example of bootstrapping TTS for minority languages. In: *Proceedings of the 5th ISCA Workshop on Speech Synthesis*. Pittsburgh, PA.

Doddington, G. (2001). Speaker recognition based on idiolectal differences between speakers. In: *Proceedings of the European Conference on Speech Communication and Technology (EUROSPEECH)*. Aalborg, Denmark.

Dugast, C., Aubert, X., Kneser, R. (1995). The Philips large-vocabulary recognition system for American English, French and German. In: *Proceedings of the European Conference on Speech Communication and Technology (EUROSPEECH)*. Madrid, Spain, pp. 197–200.

Dung, L. (1997). *Sprachen identifikation mittels prosodie*. Master's thesis, Technische Universität Ilmenau, 114-96 D-03.

Durand, J., Laks, B. (2002). *Phonetics, Phonology, and Cognition*. Oxford University Press.

Dutoit, T., Pagel, V., Pierret, N., van der Vreken, O., Bataille, F. (1996). The MBROLA project: Towards a set of high-quality speech synthesizers free of use for non-commercial purposes. In: *Proceedings of the International Conference on Spoken Language Processing*. Philadelphia, PA., pp. 1393–1397.

Eady, S. (1982). Differences in the F_0 patterns of speech: Tone language versus stress language. *Language and Speech* 25 (1), 29–42.

Ehsani, F., Bernstein, J., Najimi, A., Todic, O. (1997). SUBARASHII: Japanese Interactive Spoken Language Education. In: *Proceedings of the European Conference on Speech Communication and Technology (EUROSPEECH)*. Rhodes, Greece.

Eide, E. (2001). Distinctive features for use in an automatic speech recognition system. In: *Proceedings of the European Conference on Speech Communication and Technology (EUROSPEECH)*. Aalborg, Denmark.

ELDA, 2005. Evaluations and Language Resources Distribution Agency. http://www.elda.org.

Elsnet, 1994. European Corpus Initiative Multilingual Corpus I (ECI/MCI), part mul06. http://www.elsnet.org/eci.html.

E-MELD (2005). *Endangered data vs enduring practice: Creating linguistic resources that last*. E-MELD Web site. Retrieved May 25, 2005, from http://emeld.org/events/lsa_symposium.cfm.

Eskenazi, M., Hansma, S. (1998). The fluency pronunciation trainer. In: *Proceedings of the ESCA Workshop on Speech Technology in Language Learning (STiLL)*. Stockholm, Sweden.

Farinas, J., Pellegrino, F., Rouas, J.-L., Andre-Obrecht, R. (2002). Merging segmental and rhythmic features for automatic language identification. In: *Proceedings of the IEEE International Conference on Acoustics, Speech and Signal Processing*. Orlando, FL.

Federico, M., Bertoldi, N. (2004). Broadcast news LM adaptation over time. *Computer Speech and Language* 18 (4), 417–435.

Ferguson, C. A. (1989). Language teaching and theories of language. In: Alatis, J. E. (Ed.), *Georgetown University Round Table on Languages and Linguistics* 1989. Georgetown University Press.

Fernandez, F., De Cordoba, R., Ferreiros, J., Sama, V., D'Haro, L.F., Macias-Guarasa, J. (2004). Language identification techniques base on full recognition in an air traffic control task. In: *Proceedings of the International Conference on Spoken Language Processing*. Jeju Island, South Korea.

Finke, M., Waibel, A. (1997). Speaking mode dependent pronunciation modeling in large vocabulary conversational speech recognition. In: *Proceedings of the European Conference on Speech Communication and Technology (EUROSPEECH)*. Rhodes, Greece, pp. 2379–2382.

Fischer, V., Gao, Y., Janke, E. (1998). Speaker-independent upfront dialect adaptation in a large vocabulary continuous speech recognizer. In: *Proceedings of the International Conference on Spoken Language Processing.* Sydney, Australia.

Fisher, W., Doddington, G., Goudie-Marshall, K. (1986). The DARPA speech recognition research database: Specifications and status. In: *Proceedings of the DARPA Workshop on Speech Recognition*. pp. 93–99.

Fitt, S. (1995). The pronunciation of unfamiliar native and non-native town names. In: *Proceedings of the European Conference on Speech Communication and Technology (EUROSPEECH)*. Madrid, Spain, pp. 2227–2230.

Flege, J. (1995). *Speech Perception and Linguistic Experience: Theoretical and Methodological Issues in Cross-Language Speech Research*. York Press Inc.,

Timonium, MD, Ch. Second-language speech learning: Theory, findings and problems, pp. 233–272.

Foil, J. (1986). Language identification using noisy speech. In: *Proceedings of the IEEE International Conference on Acoustics, Speech and Signal Processing.* Tokyo, pp. 861–864.

Fox, R. A., Flege, J. E. (1995). The perception of English and Spanish vowels by native English and Spanish listeners. *J. Acoust. Soc. Am.* 97 (4), 2540–2551.

Franco, H., Neumeyer, L., Kim, Y., Ronen, O. (1997). Automatic Pronunciation Scoring for Language Instruction. In: *Proceedings of the IEEE International Conference on Acoustics, Speech and Signal Processing*. Munich, Germany.

Fromkin, V., Rodman, R., Hyams, N. (2003). *An Introduction to Language*, 7th ed. Heinle, Boston.

Fügen, C., Stüker, S., Soltau, H., Metze, F., Schultz, T. (2003). Efficient handling of multilingual language models. In: *Proceedings of the IEEE International Conference on Acoustics, Speech and Signal Processing*. St. Thomas, Virgin Islands, pp. 441–446.

Fukunaga, K. (1972). *Introduction to Statistical Pattern Recognition*. Academic Press, New York and London.

Funahashi, K. (1989). On the approximate realization of continuous mapping by neural networks. *Neural Networks* 2, 183–192.

Fung, P., Liu, W. K. (1999). Fast accent identification and accented speech recognition. In: *Proceedings of the IEEE International Conference on Acoustics, Speech and Signal Processing*. Phoenix, AZ.

Fung, P., Shi, B. E., Wu, D., Lam, W. B., Wong, S. K. (1997). Dealing with multilinguality in a spoken language query translator. In: *Proceedings of the ACL/EACL Workshop on Spoken Language Translation*. pp. 40–47.

Fung, T. Y., Meng, H. (2000). Concatenating syllables for response generation in domain-specific applications. In: *Proceedings of the IEEE International Conference on Acoustics, Speech and Signal Processing*. Istanbul, Turkey, pp. 933–936.

Furuse, O., Iida, H. (1994). Constituent boundary parsing for example-based machine translation. In: *Proceedings of COLING*. pp. 105–111.

Furuse, O., Iida, H. (1996). Incremental translation utilizing constituent boundary patterns. In: *Proceedings of COLING*. pp. 412–417.

García-Romero, D., Fiérrez-Aguilar, J., Gozález-Rodríguez, J., Ortega-García, J. (2003). Support vector machine fusion of idiolectal and acoustic speaker information in Spanish conversational speech. In: *Proceedings of the IEEE International Conference on Acoustics, Speech and Signal Processing*. Hong Kong, pp. 229–232.

Garrett, N. (1995). ICALL and Second Language Acquisition. In: Holland, V. M., Kaplan, J. D., Sams, M. R. (Eds.), *Intelligent Language Tutors: Theory Shaping Technology*. Lawrence Erlbaum, Mahwah, NJ.

Gauvain, J., Adda, G., Adda-Decker, M., Allauzen, A., Gendner, V., Lamel, L., Schwenk, H. (2005). Where are we in transcribing French broadcast news. In: *Proceedings of the European Conference on Speech Communication and Technology (EUROSPEECH-INTERSPEECH)*. Lisbon, Portugal.

Gauvain, J., Adda, G., Lamel, L., Adda-Decker, M. (1997). Transcribing broadcast news: The LIMSI nov96 Hub4 System. In: *Proceedings of the ARPA Spoken Language Systems Technology Workshop*. Chantilly, VA, pp. 56–63.

Gauvain, J., Lamel, L. (1996). Large vocabulary continuous speech recognition: From laboratory systems towards real-world applications. *Trans. IEICE J79-D-II, 2005–2021*.

Gauvain, J., Lamel, L. F., Adda, G., Matrouf, D. (1996). The LIMSI 1995 Hub3 System. In: *Proceedings of the ARPA Spoken Language Technology Workshop*. pp. 105–111.

Gauvain, J., Messaoudi, A., Schwenk, H. (2004). Language recognition using phone lattices. In: *Proceedings of the International Conference on Spoken Language Processing*. Jeju Island, South Korea.

Gauvain, J.-L., Lamel, L., Adda, G. (2002). The LIMSI broadcast news transcription system. *Speech Communication* 37 (1–2), 89–108.

Gavalda, M. (2000). *Growing semantic grammars*. Ph.D. dissertation, Language Technologies Institute. Carnegie Mellon University.

Gavalda, M. (2004). Soup: A parser for real-world spontaneous speech. In: H. Bunt, J. Carroll, G. S. (Eds.), *New Developments in Parsing Technology*. Kluwer.

Gendrot, C., Adda-Decker, M. (2005). Impact of duration on F1/F2 formant values of oral vowels: An automatic analysis of large broadcast news corpora in French and German. In: *Proceedings of the European Conference on Speech Communication and Technology (EUROSPEECH-INTERSPEECH)*. Lisbon, Portugal.

Germann, U., Jahr, M., Knight, K., Marcu, D., Yamada, K. (2001). Fast decoding and optimal decoding for machine translation. In: *Proceedings of the 39th Annual Meeting of the Association for Computational Linguistics*. Toulouse, France, pp. 220–227.

Geutner, P. (1995). Using morphology towards better large-vocabulary speech recognition systems. In: *Proceedings of the International Conference on Spoken Language Processing*. pp. 445–448.

Geutner, P., Finke, M., Scheytt, P. (1998). Adaptive vocabularies for transcribing multilingual broadcast news. In: *Proceedings of the IEEE International Conference on Acoustics, Speech and Signal Processing*. pp. 925–928.

Glass, J., Flammia, G., Goodine, D., Phillips, M., Polifroni, J., Sakai, S., Seneff, S., Zue, V. (1995). Multilingual spoken-language understanding in the MIT Voyager system. *Speech Communication* 17 (1–2), 1–18.

Goddeau, D., Meng, H., Polifroni, J., Seneff, S., Busayapongchai, S. (1996). A form-based dialog manager for spoken language applications.

In: *Proceedings of the International Conference on Spoken Language Processing*. Philadelphia. pp. 701–704.

Godfrey, J., Holliman, E., McDaniel, J. (1992). SWITCHBOARD: Telephone speech corpus for research and development. In: *Proceedings of the IEEE International Conference on Acoustics, Speech and Signal Processing*. pp. 517–520.

Gokcen, S., Gokcen, J. (1997). A multilingual phoneme and model set: towards a universal base for automatic speech recognition. In: *Proceedings of the IEEE Workshop on Automatic Speech Recognition and Understanding*. Santa Barbara, CA, pp. 599–603.

Gordon, R. (Ed.) (2005). *Ethnologue: Languages of the World*. SIL International, Dallas, TX.

Goronzy, S. (2002). *Robust Adaptation to Non-Native Accents in Automatic Speech Recognition. Lecture Notes on Artificial Intelligence* Vol. 2560. Springer Verlag.

Goronzy, S., Eisle, K. (2003). Automatic pronunciation modeling for multiple non-native accents. In: *Proceedings of the IEEE Workshop on Automatic Speech Recognition and Understanding*. St. Thomas, Virgin Islands.

Goronzy, S., Rapp, S., Kompe, R. (2004). Generating non-native pronunciation variants for lexicon adaptation. *Speech Communication*, 42 (1) 109–123.

Graff, D., Brid, S. (2000). Many uses, many annotations for large corpora: Switchboard and TDT as case studies. In: *Proceedings of the First International Conference on Language Resources and Evaluation (LREC)*. Athens, Greece.

Graff, D., Cieri, C., Strassel, S., Martey, N. (2000). The TDT-3 text and speech corpus. In: *1999 TDT Evaluation System Summary Papers*. National Institutes of Standards and Technology, available at http://www.itl.nist.gov/iaui/894.01/tests/tdt/tdt99/papers/LDC.ps.

Grosz, B., Sidner, C. (1985). Discourse structure and the proper treatment of interruptions. In: *Proceedings of the International Joint Conference on Artificial Intelligence* (IJCAI-85). pp. 832–839.

Hain, T. (2005). Implicit modelling of pronunciation variation in automatic speech recognition. *Speech Communication* 46 (2), 171–188.

Hajič, J. (1999). Morphological analysis of Czech word forms. Available at http://nlp.cs.jhu.edu/~hajic/morph.html from the Center for Language and Speech Processing, Johns Hopkins University.

Hajič, J., Brill, E., Collins, M., Hladká, B., Jones, D., Kuo, C., Ramshaw, L., Schwartz, O., Tillman, C., Zeman, D. (1998). *Core natural language processing technology applicable to multiple languages*. Final report of the 1998 summer workshop on language engineering, Center for Language and Speech Processing, Johns Hopkins University, available at http://www.clsp.jhu.edu/ws98/projects/nlp/report/.

Harper, M. P., Jamieson, L. H., Mitchell, C. D., G. Ying, S. P., Srinivasan, P. N., Chen, R., Zoltowski, C. B., McPheters, L. L., Pellom, B., Helzerman, R. A. (1994). Integrating language models with speech recognition. In: *Proceedings of the 1994 American Association for Artificial Intelligence Workshop on the Integration of Natural Language and Speech Processing*. Seattle, WA.

Hartmann, C., Varshney, P., Mehrotra, K., Gerberich, C. (1982). Application of information theory to the construction of efficient decision trees. *IEEE Trans. on Information Theory* IT-28 (4), 565–577.

Hazen, T. (1993). *Automatic language identification using a segment-based approach*. Master's thesis, Massachusetts Institute of Technology.

Hazen, T., Bazzi, I. (2001). A comparison and combination of methods for OOV word detection and word conference scoring. In: *Proceedings of the IEEE International Conference on Acoustics, Speech and Signal Processing*. Salt Lake City, UT. pp. 397–400.

Hazen, T., Zue, V. (1994). Recent improvements in an approach to segment-based automatic language identification. In: *Proceedings of the International Conference on Spoken Language Processing*. Yokohama, Japan, pp. 1883–1886.

Hazen, T., Zue, V. (1997). Segment-based automatic language identification. *Journal of the Acoustical Society of America* 101 (4), 2323–2331.

He, D., Oard, D., Wang, J., Luo, J., Demner-Fushman, D., Darwish, K., Resnik, P., Khudanpur, S., Nossal, M., Subotin, M., Leuski, A. (2003). Making MIRACLEs: Interactive translingual search for Cebuano and Hindi. *ACM Transactions on Asian Language Information Processing* 2 (3), 219–244.

He, X., Zhao, Y. (2001). Model complexity optimization for nonnative English speakers. In: *Proceedings of the European Conference on Speech Communication and Technology (EUROSPEECH)*. Aalborg, Denmark, pp. 1461–1463.

Hieronymus, J. (1994). *ASCII phonetic symbols for the world's languages: Worldbet*. Tech. rep., AT&T Bell Laboratories.

Hieronymus, J., Kadambe, S. (1996). Spoken language identification using large vocabulary speech recognition. In: *Proceedings of the International Conference on Spoken Language Processing*. Philadelphia.

Hieronymus, J., Kadambe, S. (1997). Robust spoken language identification using large vocabulary speech recognition. In: *Proceedings of the IEEE International Conference on Acoustics, Speech and Signal Processing*. Munich, Germany, pp. 1111–1114.

Hofstede, G. (1997). *Cultures and Organizations: Software of the Mind*. McGraw-Hill, New York.

Holter, T., Svendsen, T. (1997). Incorporating linguistic knowledge and automatic baseform generation in acoustic subword unit based speech recognition. In: *Proceedings of the European Conference on Speech Communication and Technology (EUROSPEECH)*. Rhodes, Greece, pp. 1159–1162.

Holter, T., Svendsen, T. (1999). Maximum likelihood modelling of pronunciation variation. *Speech Communication* 29, 177–191.

Honal, M., Schultz, T. (2003). Correction of disfluencies in spontaneous speech using a noisy-channel approach. In: *Proceedings of the European Conference on Speech Communication and Technology (EUROSPEECH)*. Geneva, Switzerland.

Honal, M., Schultz, T. (2005). Automatic disfluency removal on recognized spontaneous speech—rapid adaptation to speaker-dependent disfluencies. In: *Proceedings of the IEEE International Conference on Acoustics, Speech and Signal Processing*. Philadelphia.

Horvitz, E. (1999). Principles of mixed-initiative user interfaces. In: *Proceedings of the ACM Conference on Human Factors in Computing Systems* (CHI-99). Pittsburgh, PA. ACM Press. pp. 159–166.

Horvitz, E., Kadie, C., Paek, T., Hovel, D. (2003). Models of Attention in Computing and Communication: From Principles to Applications., pp. 52–59.

Huang, C., Chang, E., Zhou, J., Lee, K.-F. (2000). Accent Modeling Based on Pronunciation Dictionary Adaptation for Large Vocabulary Mandarin Speech Recognition. In: *Proceedings of the International Conference on Spoken Language Processing*. Beijing, China, pp. 818–821.

Huang, X., Hwang, M.-Y., Jiang, L., Mahajan, M. (1996). Deleted interpolation and density sharing for continuous hidden Markov models. In: *Proceedings of the IEEE International Conference on Acoustics, Speech and Signal Processing*. Atlanta, GA.

Huang, X. D., Acero, A., Hon, H.-W. (2001). *Spoken Language Processing*. Prentice Hall, NJ.

Humphries, J. J., Woodland, P. C., Pearce, D. (1996). Using Accent-Specific Pronunciation Modelling for Robust Speech Recognition. In: *Proceedings of the International Conference on Spoken Language Processing*. pp. 2324–2327.

Hunt, A., Black, A. (1996). Unit selection in a concatenative speech synthesis system using a large speech database. In: *Proceedings of the IEEE International Conference on Acoustics, Speech and Signal Processing*. Atlanta, GA, pp. 373–376.

Huot, H. (2001). *Morphologie—forme et sens des mots du francais*. Armand Colin, Paris, France.

Ikeno, A., Pellom, B., Cer, D., Thornton, A., Brenier, J. M., Jurafsky, D., Ward, W., Byrne, W. (2003). Issues in recognition of Spanish-accented spontaneous English. In: *Proceedings of the ISCA and IEEE Workshop on Spontaneous Speech Processing and Recognition*. Tokyo Institute of Technology, Tokyo, Japan.

Imamura, K. (2001). Hierarchical phrase alignment harmonized with parsing. In: *Proceedings of the Sixth Natural Language Processing Pacific Rim Symposium*. Tokyo, Japan, pp. 377–384.

Imamura, K. (2002). Application of translation knowledge acquired by hierarchical phrase alignment. In: *Proceedings of the Sixth International Conference on Theoretical and Methodological Issues in Machine Translation*. pp. 74–84.

Imamura, K. (2003). Feedback cleaning of machine translation rules using automatic evaluation. In: *Proceedings of the 41st Annual Meeting of the Association for Computational Linguistics*. Sapporo, Japan, pp. 447–454.

Imperl, B. (1999). Clustering of Context Dependent Speech Units for Multilingual Speech Recognition. In: *Proceedings of the ESCA-NATO Tutorial Research Workshop on Multi-lingual Interoperability in Speech Technology*. Leusden, Netherlands, pp. 17–22.

Imperl, B., Horvat, B. (1999). The clustering algorithm for the definition of multilingual set of context dependent speech models. In: *Proceedings of the European Conference on Speech Communication and Technology (EUROSPEECH)*. Budapest, Hungary, pp. 887–890.

International, W. (2004). http://www.wordwave.co.uk/linguistic data consortium.

IPA (1993). The International Phonetic Association—IPA chart (revised to 1993). *Journal of the International Phonetic Association* 1 (23).

Ircing, P., Psutka, J. (2001). Two-pass recognition of Czech speech using adaptive vocabulary. In: Matousek, V., Mautner, P., Moucek, R., Tauser, K. (Eds.), *Text Speech and Dialog. Lecture Notes in Computer Science*, Vol. 2166, Springer Verlag, pp. 273–277.

ISLE (1998). http://nats-www.informatik.uni-hamburg.de/isle/.

Issar, S. (1996). Estimation of language models for new spoken language applications. In: *Proceedings of the International Conference on Spoken Language Processing*. pp. 869–872.

Itahashi, S., Liang, D. (1995). Language identification based on speech fundamental frequency. In: *Proceedings of the European Conference on Speech Communication and Technology (EUROSPEECH)*. Madrid, Spain, pp. 1359–1362.

Itahashi, S., Zhou, J., Tanaka, K. (1994). Spoken language discrimination using speech fundamental frequency. In: *Proceedings of the International Conference on Spoken Language Processing*. Yokohama, Japan, pp. 1899–1902.

Iyer, R., Ostendorf, M. (1999). Modeling long distance dependence in language: Topic mixtures versus dynamic cache models. *IEEE Transactions on Speech and Audio Processing* 7 (1), 30–39.

Iyer, R., Ostendorf, M., Gish, H. (1997). Using out-of-domain data to improve in-domain language models. *IEEE Signal Processing Letters* 4 (8), 221–223.

James, C. (1980). *Contrastive Analysis*. Longman, London.

Jekat, S., Hahn, W. (2000). Multilingual Verbmobil-dialogs: Experiments, data collection and data analysis. In: W. Wahlater (Ed.) *Vermobil: Foundations of Speech-to-Speech Translations*. Springer, Berlin, pp. 575–582.

Jelinek, F. (1997). *Statistical Methods for Speech Recognition*. MIT Press, Cambridge, MA.

Jelinek, F., Lafferty, J. (1991). Computation of the probability of initial substring generation by stochastic context-free grammars. *Computational Linguistics* 17 (3), 315–323.

Jelinek, F. Mercer, R., Bahl, L., Baker, J. (1997). Perplexity—a measure of difficulty of speech recognition tasks. In: *94th Meeting of the Acoustic Society of America*. Miami Beach, FL.

Jin, Q., Navrátil, J., Reynolds, D., Campbell, J., Andrews, W., Abramson, J. (2003). Combining cross-stream and time dimensions in phonetic speaker recognition. In: *Proceedings of the IEEE International Conference on Acoustics, Speech, and Signal Processing*. Hong Kong, China.

Jitsuhiro, T., Matsui, T., Nakamura, S. (2003). Automatic generation of non-uniform context-dependent HMM topologies based on the MDL criterion. In: *Proceedings of the European Conference on Speech Communication and Technology (EUROSPEECH)*. Geneva, Switzerland, pp. 2721–2724.

Jones, R. J., Downey, S., Mason, J. J. (1997). Continuous speech recognition using syllables. In: *Proceedings of the European Conference on Speech Communication and Technology (EUROSPEECH)*. Rhodes, Greece, pp. 1215–1218.

Jurafsky, D., Ward, W., Jianping, Z., Herold, K., Xiuyang, Y., Sen, Z. (2001). What kind of pronunciation variation is hard for triphones to model? In: *Proceedings of the IEEE International Conference on Acoustics, Speech and Signal Processing*. pp. 577–580.

Kain, A. (2001). *High resolution voice transformation*. Ph.D. thesis, OGI School of Science and Engineering, Oregon Health and Science University.

Kaji, H., Kida, Y., Morimoto, Y. (1992). Learning translation templates from bilingual text. In: *Proceedings of the Conference on Computational Linguistcs*. pp. 672–678.

Kanthak, S., Ney, H. (2002). Context-dependent acoustic modeling using graphemes for large vocabulary speech recognition. In: *Proceedings of the IEEE International Conference on Acoustics, Speech and Signal Processing*. Orlando, FL, pp. 845–848.

Kanthak, S., Ney, H. (2003). Multilingual acoustic modeling using graphemes. In: *Proceedings of the European Conference on Speech Communication and Technology (EUROSPEECH)*. Geneva, Switzerland, pp. 1145–1148.

Katzner, K. (2002). *The Languages of the World*. Routledge, London/New York.

Kawai, G. (1999). *Spoken language processing applied to nonnative language pronunciation learning*. Ph.D. thesis, University of Tokyo.

Khudanpur, S., Kim, W. (2004). Contemporaneous text as side-information in statistical language modeling. *Computer Speech and Language* 18 (2), 143–162.

Kiecza, D., Schultz, T., Waibel, A. (1999). Data-driven determination of appropriate dictionary units for Korean LVCSR. In: *Proceedings of the International*

Conference on Speech Processing (ICSP '99). Acoustic Society of Korea, Seoul, South Korea, pp. 323–327.

Kikui, G., Sumita, E., Takezawa, T., Yamamoto, S. (2003). Creating corpora for speech-to-speech translation. In: *Proceedings of the European Conference on Speech Communication and Technology (EUROSPEECH)*. pp. 381–384.

Killer, M., Stüker, S., Schultz, T. (2003). Grapheme based speech recognition. In: *Proceedings of the European Conference on Speech Communication and Technology (EUROSPEECH)*. Geneva, Switzerland.

Kim, J., Flynn, S. (2004). What makes a non-native accent?: A study of Korean English. In: *Proceedings of the International Conference on Spoken Language Processing*. Jeju Island, South Korea.

Kim, W., Khudanpur, S. (2004). Lexical triggers and latent semantic analysis for cross-lingual language model adaptation. *ACM Transactions on Asian Language Information Processing* 3 (2), 94–112.

Kim, Y., Franco, H., Neumeyer, L. (1997). Automatic pronunciation scoring of specific phone segments for language instruction. In: *Proceedings of the European Conference on Speech Communication and Technology (EUROSPEECH)*, Rhodes, Greece.

Kimball, O., Kao, C., Iyer, R., Arvizo, T., Makhoul, G. (2004). Using quick transcriptions to improve conversational speech models. In: *Proceedings of the International Conference on Spoken Language Processing*. Jeju Island, South Korea.

Kipp, A., Lamel, L., Mariani, J., Schiel, F. (1993). *Translanguage English Database* (TED). LDC, ISBN 1-58563-218-X.

Kirchhoff, K. (1998). Combining articulatory and acoustic information for speech recognition in noisy and reverberant environments. In: *Proceedings of the International Conference on Spoken Language Processing*. Sydney, Australia.

Kirchhoff, K., Bilmes, J., Das, S., Duta, N., Egan, M., Ji, G., He, F., Henderson, J., Liu, D., Noamany, M., Schone, P., Schwartz, R., Vergyri, D. (2003). Novel approaches to Arabic speech recognition: Report from the 2002 Johns-Hopkins summer workshop. In: *Proceedings of the IEEE International Conference on Acoustics, Speech and Signal Processing*. pp. 344–347.

Kirchhoff, K., Bilmes, J., Das, S., Egan, M., Ji, G., He, F., Henderson, J., Noamany, M., Schone, P., Schwartz, R. (2002a). *Novel speech recognition models for Arabic. Final report of the 2001 summer workshop on language engineering*, Center for Language and Speech Processing, Johns Hopkins University, available at http://www.clsp.jhu.edu/ws2002/groups/arabic/arabic-final.pdf.

Kirchhoff, K., Parandekar, S. (2001). Multi-stream statistical N-gram modeling with application to automatic language identification. In: *Proceedings of the European Conference on Speech Communication and Technology (EUROSPEECH)*.Orlando, FL.

Kirchhoff, K., Parandekar, S., Bilmes, J. (2002b). Mixed-memory Markov models for automatic language identification. In: *Proceedings of the IEEE International Conference on Acoustics, Speech and Signal Processing*.

Knight, K. (1997). Automating knowledge acquisition for machine translation. *AI Magazine* 18 (4), 81–96.

Koehn, P., Och, F. J., Marcu, D. (2003). Statistical phrase-based translation. In: *Proceedings of the HLT/NAACL*. Edmonton, Canada.

Köhler, J. (1996). Multi-lingual phoneme recognition exploiting acoustic-phonetic similarities of sounds. In: *Proceedings of the International Conference on Spoken Language Processing*. Philadelphia, pp. 2195–2198.

Köhler, J. (1997). Multilingual phone modelling for telephone speech. In: *Proceedings of the SQEL, 2nd Workshop on Multi-Lingual Information Retrieval Dialogs*. University of West Bohemia, Plzeň, Czech Republic, pp. 16–19.

Köhler, J. (1998). Language adaptation of multilingual phone models for vocabulary independent speech recognition tasks. In: *Proceedings of the IEEE International Conference on Acoustics, Speech and Signal Processing*. Seattle, WA, pp. 417–420.

Köhler, J. (1999). Comparing three methods to create multilingual phone models for vocabulary independent speech recognition tasks. In: *Proceedings of the ESCA-NATO Tutorial Research Workshop on Multi-lingual Interoperability in Speech Technology*. Leusden, Netherlands, pp. 79–84.

Kominek, J., Bennett, C., Black, A. (2003). Evaluating and correcting phoneme segmentation for unit selection synthesis. In: *Proceedings of the European Conference on Speech Communication and Technology (EUROSPEECH)*. Geneva, Switzerland.

Kominek, J., Black, A. (2003). *The CMU ARCTIC speech databases for speech synthesis research*. Tech. Rep. CMU-LTI-03-177 http://festvox.org/cmu_arctic/, Language Technologies Institute, Carnegie Mellon University, Pittsburgh, PA.

Krauwer, S. (1998). Elsnet and ELRA: A common past and a common future. *ELRA Newsletter* 3(2).

Krauwer, S., Maegaard, B., Chouki, K. (2004). Report on basic language resource kit (BLARK) for Arabic. In: *Proceedings from NEMLAR International Conference on Arabic Language Resources and Tools*. Cairo, Egypt.

Kubala, F., Austin, S., Barry, C., Makhoul, J., Placeway, P., Schwartz, R. (1991). BYBLOS Speech Recognition Benchmark Results. In: *Proceedings from DARPA Speech and Natural Language Workshop*. Pacific Grove, CA.

Kuboň, V., Plátek, M. (1994). A grammar based approach to a grammar checking of free word order languages. In: *Proceedings of the 15th International Conference on Computational Linguistics*. pp. 906–910.

Kumpf, K., King, R. (1997). Foreign speaker accent classification using phoneme-dependent accent discrimination models and comparisons with human perception benchmarks. In: *Proceedings of the European Conference on Speech Communication and Technology (EUROSPEECH)*. Rhodes, Greece.

Küpfmüller, K. (1954). Die Entropie der deutschen Sprache. FTZ 6 6, 265–272.

Kwan, H., Hirose, K. (1997). Use of recurrent network for unknown language rejection in language identification systems. In: *Proceedings of the European Conference on Speech Communication and Technology (EUROSPEECH)*. Rhodes, Greece, pp. 63–67.

Labov, W. (1966). *The Social Stratification of English in New York City*. Ph.D. thesis, Columbia University, NY.

Labov, W. Phonological Atlas of North America, 2004. http://www.ling.upenn.edu/phono_atlas/home.html.

Lamel, L., Adda, G. (1996). On designing pronunciation lexicons for large vocabulary, continuous speech recognition. In: *Proceedings of the International Conference on Spoken Language Processing*. Philadelphia, pp. 6–9.

Lamel, L., Adda-Decker, M., Gauvain, J. L. (1995). Issues in large vocabulary multilingual speech recognition. In: *Proceedings of the European Conference on Speech Communication and Technology (EUROSPEECH)*. Madrid, Spain, pp. 185–188.

Lamel, L., Adda-Decker, M., Gauvain, J., Adda, G. (1996). Spoken language processing in a multilingual context. In: *Proceedings of the International Conference on Spoken Language Processing*. Philadelphia, pp. 2203–2206.

Lamel, L., Gauvain, J. (1994). Language identification using phone-based acoustic likelihoods. In: *Proceedings of the IEEE International Conference on Acoustics, Speech and Signal Processing*. Adelaide, Australia.

Lamel, L., Gauvain, J. (2005). Alternate phone models for conversational speech. In: *Proceedings of the IEEE International Conference on Acoustics, Speech and Signal Processing*. Philadelphia.

Landauer, T. K., Littman, M. L. (1995). Fully automatic cross-language document retrieval using latent semantic indexing. In: *Proceedings of the Sixth Annual Conference of the UW Centre for the New Oxford English Dictionary and Text Research*. pp. 31–38.

Lander, T., Cole, R., Oshika, B., Noel, M. (1995). The OGI 22 language telephone speech corpus. In: *Proceedings of the European Conference on Speech Communication and Technology (EUROSPEECH)*. Madrid, Spain.

Langlais, P., Öster, A.-M., Granström, B. (1998). Automatic detection of mispronunciation in non-native Swedish speech. In: *Proceedings of the ESCA Workshop on Speech Technology in Language Learning (STiLL)*. Stockholm, Sweden.

Langner, B., Black, A. (2005). Improving the understandability of speech synthesis by modeling speech in noise. In: *Proceedings of the IEEE International Conference on Acoustics, Speech and Signal Processing*. Philadelphia.

Latorre, J., Iwano, K., Furui, S. (2005). Polyglot synthesis using a mixture of monolingual corpora. In: *Proceedings of the IEEE International Conference on Acoustics, Speech and Signal Processing*. Philadelphia.

Lau, R., Seneff, S. (1998). A unified system for sublexical and linguistic modeling using ANGIE and TINA. In: *Proceedings of the International Conference on Spoken Language Processing*. Sydney, Australia, pp. 2443–2446.

Lavie, A., Font, A., Peterson, E., Carbonell, J., Probst, K., Levin, L., Reynolds, R., Cohen, R., Vogel, S. (2003). Experiments with a Hindi-to-English transfer-based (MT) system under a miserly data scenario. In: *ACM Transactions on Asian Language Information Processing* (TALIP).

Lavie, A., Langley, C., Waibel, A., Pianesi, F., Lazzari, G., Coletti, P., Taddei, L., Balducci, F. (2001a). Architecture and design considerations in NESPOLE!: A speech translation system for e-commerce applications. In: *Proceedings of HLT: Human Language Technology*. San Diego, CA.

Lavie, A., Levin, L., Schultz, T., Waibel, A. (2001b). Domain portability in speech-to-speech translation. In: *Proceedings of HLT: Human Language Technology*. San Diego, CA.

LDC (1997). http://www.ldc.upenn.edu/Catalog/LDC97S44.html.

LDC (2000). The Linguistic Data Consortium. Internet, http://www.ldc.upenn.edu.

LDC (2005). http://www.ldc.upenn.edu.

Lee, K.-F. (1988). *Large-vocabulary speaker-independent continuous speech recognition: The sphinx system*. Ph.D. thesis, Carnegie Mellon University, Pittsburg, PA.

Lefevre, F., Gauvain, J. L., Lamel, L. (2001). Improving genericity for task-independent speech recognition. In: *Proceedings of the European Conference on Speech Communication and Technology (EUROSPEECH)*. Aalborg, Denmark, pp. 1241–1244.

Leggetter, C. J., Woodland, P. C. (1995). MLLR for speaker adaptation of CDHMMs. *Computer Speech and Language* 9, 171–185.

Lesh, N., Rich, C., Sidner, C. (2001). Collaborating with focused and unfocused users under imperfect communication. In: *Proceedings of the International Conference on User Modeling*. Sonthofen. Springer. pp. 63–74.

Levin, E., Narayanan, S., Pieraccini, R., Biatov, K., Bocchieri, E., Fabbrizio, G., D., Eckert, W., Lee, S., Pokrovsky, A., Rahim, N., Ruscitti, P., Walker, M. (2000). The AT&T DARPA Communicator Mixed-Initiative Spoken Dialog System. In: *Proceedings of the International Conference on Spoken Language Processing*. Beijing, pp. 122–125.

Levin, L., Gates, D., Lavie, A., Waibel, A. (1998). An interlingua based on domain actions for machine translation of task-oriented dialogues. In: *Proceedings of the International Conference on Spoken Language Processing*. Sydney, Australia.

Li, J., Zheng, F., Wu, W. (2000). Context-independent Chinese initial-final acoustic modeling. In: *Proceedings of the International Symposium on Chinese Spoken Language Processing*. pp. 23–26.

Li, K. (1994). Automatic language identification using syllabic spectral features. In: *Proceedings of the IEEE International Conference on Acoustics, Speech and Signal Processing*. Adelaide, Australia, pp. 297–300.

Linguistic Data Consortium (1996). Celex2. http://www.ldc.upenn.edu/Catalog/LDC96L14.html.

Linguistic Data Consortium (2005). http://www.ldc.upenn.edu/Catalog/docs/LDC97T19/ar-trans.txt.

Liu, F. H., Gao, Y., Gu, L., Picheny, M. (2003). *Noise robustness in speech to speech translation*. IBM Technical Report RC22874.

Liu, W. K., Fung, P. (2000). MLLR-based accent model adaptation without accented data. In: *Proceedings of the International Conference on Spoken Language Processing*. Beijing, China.

Livescu, K. (1999). *Analysis and modeling of non-native speech for automatic speech recognition*. MIT Department of Electrical Engineering and Computer Science.

Livescu, K., Glass, J. (2000). Lexical modeling of non-native speech for automatic speech recognition. In: *Proceedings of the IEEE International Conference on Acoustics, Speech and Signal Processing*. Istanbul, Turkey.

Lo, W. K., Meng, H., Ching, P. C. (2000). Subsyllabic acoustic modeling across Chinese dialects. In: *Proceedings of the International Symposium on Chinese Spoken Language Processing*.

Lyons, J. (1990). *Introduction to Theoretical Linguistics*. Oxford University Press, Oxford.

Ma, K., Zavaliagkos, G., Iyer, R. (1998). BBN pronunciation modeling. In: *9th Hub-5 Conversational Speech Recognition Workshop*. Linthicum Heights, MD.

Macon, M., Kain, A., Cronk, A., Meyer, H., Müeller, K., Säeuberlich, B., Black, A. (1998). *Rapid prototyping of a German TTS system*. Unpublished report Oregon Graduate Institute, http://www.cslu.ogi.edu/tts/research/multiling/de-report.html.

Maddieson, I., Vasilescu, I. (2002). Factors in human language identification. In: *Proceedings of the International Conference on Spoken Language Processing*. Denver, CO.

Maegaard, B. (2004). NEMLAR—an Arabic language resources project. In: *Proceedings of the IV International Conference on Language Resources and Evaluation*. Lisbon, Portugal.

Malfrere, F., Dutoit, T. (1997). High quality speech synthesis for phonetic speech segmentation. In: *Proceedings of the European Conference on Speech Communication and Technology (EUROSPEECH)*. Rhodes, Greece, pp. 2631–2634.

Manning, C. D., Schütze, H. (1999). *Foundations of Statistical Natural Language Processing*. The MIT Press, Cambridge, Massachusetts.

Manos, A., Zue, V. (1997). A segment-based spotter using phonetic filler models. In: *Proceedings of the IEEE International Conference on Acoustics, Speech and Signal Processing*. Munich. pp. 899–902.

Mapelli, V., Choukri, K. (2003). *Deliverable 5.1 report on a (minimal) set of LRS to be made available for as many languages as possible, and map of the actual gaps*. Internal ENABLER project report.

Marcu, D., Wong, W. (2002). A phrase-based, joint probability model for statistical machine translation. In: *Proceedings of the Conference on Empirical Methods in Natural Language Processing*. Philadelphia, PA, pp. 6–7.

Marcus, A., Gould, E. (2000). Crosscurrents: Cultural dimensions and global Web user-interface design. *ACM Interactions* 7 (4), 32–48.

Mariani, J. (2003). Proposal for an ERA-net in the field of human language technologies. In: *LangNet*.

Mariani, J., Lamel, L. (1998). An overview of EU programs related to conversational/interactive systems. In: *Proceedings of the DARPA Broadcast News Transcription and Understanding Workshop*, Landsdowne, VA. pp. 247–253.

Mariani, J., Paroubek, P. (1998). Human language technologies evaluation in the European framework. In: *Proceedings of the DARPA Broadcast News Workshop*. Herndon, VA, pp. 237–242.

Marino, J., Nogueiras, A., Bonafonte, A. (1997). The demiphone: An efficient subword unit for continuous speech recognition. In: *Proceedings of the European Conference on Speech Communication and Technology (EUROSPEECH)*. Rhodes, Greece, pp. 1171–1174.

Markov, A. (1913). An example of statistical investigation in the text of "Eugene Onegin" illustrating coupling "tests" in chains. In: *Proceedings of the Academy of Science of St. Petersburg*. pp. 153–162.

Martin, A., Doddington, G., Kamm, T., Ordowski, M., Przybocki, M. (1997). The DET curve in assessment of detection task performance. In: *Proceedings of the European Conference on Speech Communication and Technology (EUROSPEECH)*. Rhodes, Greece, pp. 1895–1898.

Maskey, S., Black, A., Mayfield Tomokiyo, L. (2004). Optimally constructing phonetic lexicons in new languages. In: *Proceedings of the International Conference on Spoken Language Processing*. Jeju Island, South Korea.

Matrouf, D., Adda-Decker, M., Lamel, L., Gauvain, J. (1998). Language identification incorporating lexical information. In: *Proceedings of the International Conference on Spoken Language Processing*. Sydney, Australia, pp. 181–184.

Matsumoto, Y., Ishimoto, H., Usturo, T. (1993). Structural matching of parallel texts. In: *Proceedings of the 31st Annual Meeting of the Association for Computational Linguistics*. pp. 23–30.

Mayfield Tomokiyo, L. (2000). Linguistic properties of non-native speech. In: *Proceedings of the IEEE International Conference on Acoustics, Speech and Signal Processing*. Istanbul, Turkey.

Mayfield Tomokiyo, L. (2001). *Recognizing Non-native Speech: Characterizing and Adapting to Non-native Usage in Speech Recognition*. Ph.D. thesis, Carnegie Mellon University.

Mayfield Tomokiyo, L., Burger, S. (1999). Eliciting natural speech from non-native users: Collecting speech data for LVCSR. In: *Proceedings of the ACL-IALL Joint Workshop in Computer-Mediated Language Assessment and Evaluation in Natural Language Processing*.

Mayfield Tomokiyo, L., Jones, R. (2001). You're not from 'round here, are you? Naive Bayes detection of non-native utterance text. In: *Proceedings of the NAACL Conference*. Seattle, WA.

Mayfield Tomokiyo, L., Waibel, A. (2001). Adaptation methods for Non-native speech. In: *Proceedings of the ISCA Workshop on Multilinguality in Spoken Language Processing*. Aalborg, Denmark.

McTear, M. (2002). Spoken dialog technology: enabling the conversational user interface. ACM Computing Survey 34 (1), 90–169.

Meng, H., Busayapongchai, S., Glass, J., Goddeau, D., Hetherington, L., Hurley, E., Pao, C., Polifroni, J., Seneff, S., Zue, V. (1996). WHEELS: A Conversational System on Electronic Automobile Classifieds. In: *Proceedings of the International Conference on Spoken Language Processing*. Philadelphia. IEEE, pp. 542–545.

Meng, H., Chan, S. F., Wong, Y. F, Chan, C. C., Wong, Y. W., Fung, T. Y., Tsui, W.C., Chen, K., Wang, L., Wu, T. Y., Li, X. L., Lee, T., Choi, W. N., Ching, P. C., Chi, H. S. (2001). ISIS: A learning system with combined interaction and delegation dialogs. In: *Proceedings of the European Conference on Speech Communication and Technology (EUROSPEECH)*. Scandinavia. pp. 1551–1554.

Meng, H., Ching, P.C., Wong, Y. F., Chan, C. C. (2002). A multi-modal, trilingual, distributed spoken dialog system developed with CORBA, JAVA, XML and KQML. In: *Proceedings of the International Conference on Spoken Language Processing*. Denver, CO. pp. 2561–2564.

Meng, H., et al. (2002). CU VOCAL: Corpus-based syllable concatenation for Chinese speech synthesis across domains and dialects. In: *Proceedings of the International Conference on Spoken Language Processing.* Denver, CO.

Meng, H., Lam, W., Wai, C. (1999). To believe is to understand. In: *Proceedings of the European Conference on Speech Communication and Technology (EUROSPEECH)*. Budapest, Hungary, pp. 2015–2018.

Meng, H., Lee, S., Wai, C. (2000). CU FOREX: A bilingual spoken dialog system for foreign exchange enquires. In: *Proceedings of the IEEE International Conference on Acoustics, Speech and Signal Processing*. Istanbul, Turkey, pp. 229–232.

Meng, H., Lo, T. H., Keung, C. K., Ho, M. C., Lo, W. K. (2003). CU VOCAL Web service: A text-to-speech synthesis Web service for voice-enabled Web-mediated application. In: *Proceedings of the 12th International World Wide Web Conference* (WWW-03), Budapest, Hungary. www2003.org/cdrom/papers/poster/p056/p56-meng.htmlkerning2.

Messaoudi, A., Lamel, L., Gauvain, J. (2004). Transcription of Arabic broadcast news. In: *Proceedings of the International Conference on Spoken Language Processing*. Jeju Island, South Korea. pp. 521–524.

Metze, F., Langley, C., Lavie, A., McDonough, J., Soltau, H., Waibel, A., Burger, S., Laskowski, K., Levin, L., Schultz, T., Pianesi, F., Cattoni, R., Lazzari, G., Mana, N., Pianta, E., Besacier, L., Blanchon, H., Vaufreydaz, D., Taddei, L. (2003). The NESPOLE! speech-to-speech translation system. In: *Proceedings of the HLT: Human Language Technology*. Vancouver, Canada.

Metze, F., Waibel, A. (2002). A flexible stream architecture for ASR using articulatory features. In: *Proceedings of the International Conference on Spoken Language Processing*. Denver, CO.

Micca, G., Palme, E., Frasca, A. (1999). Multilingual vocabularies in automatic speech recognition. In: *Proceedings of the ESCA-NATO Tutorial Research Workshop on Multi-lingual Interoperability in Speech Technology*. Leusden, Netherlands, pp. 65–68.

Miller, G. A., Fellbaum, C., Tengi, R., Wolff, S., Wakefield, P., Langone, H., Haskell, B. (2005). Wordnet a lexical database for the English language. http://wordnet.princeton.edu.

Mimer, B., Stüker, S., Schultz, T. (2004). Flexible tree clustering for grapheme-based speech recognition. In: *Elektronische Sprachverarbeitung* (ESSV). Cottbus, Germany.

Mitten, R. (1992). *Computer-Usable Version of Oxford Advanced Learner's Dictionary of Current English*, Oxford Text Archive.

Mohri, M., Pereira, F., Riley, M. (2000). Weighted finite state transducers in speech recognition. In: *ISCA ITRW Workshop on Automatic Speech Recognition: Challenges for the Millenium*. pp. 97–106.

Mohri, M., Riley, M. (1997). Weighted determinization and minimization for large vocabulary speech recognition. In: *Proceedings of the European Conference on Speech Communication and Technology (EUROSPEECH)*. Rhodes, Greece, pp. 131–134.

Morimoto, T., Uratani, N., Takezawa, T., Furuse, O., Sobashima, Y., Iida, H., Nakamura, A., Sagisaka, Y., Higuchi, N., Yamazaki, Y. (1994). A speech and language database for speech translation research. In: *Proceedings of the International Conference on Spoken Language Processing*. pp. S30–1.1 – S30–1.4.

Muthusamy, Y. (1993). *A segmental approach to automatic language identification*. Ph.D. thesis, Oregon Graduate Identification of Science and Technology, Portland, Oregon.

Muthusamy, Y., Barnard, E., Cole, R. (1994a). Reviewing automatic language recognition. *IEEE Signal Processing Magazine* 11 (4), 33–41.

Muthusamy, Y., Cole, R., Oshika, B. (1992). The OGI multi-language telephone speech corpus. In: *Proceedings of the International Conference on Spoken Language Processing*. Banff, Alberta, Canada.

Muthusamy, Y., Jain, N., Cole, R. (1994b). Perceptual benchmarks for automatic language identification. In: *Proceedings of the IEEE International Conference on Acoustics, Speech and Signal Processing*. Adelaide, Australia, pp. 333–336.

Nagao, M. (1984). A framework of a mechanical translation between Japanese and English by analogy principle. In: Elithorn, Banerji, R. (Eds.), *Artificial and Human Intelligence*. Amsterdam: North-Holland, pp. 173–180.

Nagata, M. (1994). A stochastic Japanese morphological analyzer using a forward-DP backwarad-A* N-best search algorithm. In: *Proceedings of COLING*. pp. 201–207.

Nakamura, A., Matsunaga, S., Shimizu, T., Tonomura, M., Sagisaka, Y. (1996). Japanese speech databases for robust speech recognition. In: *Proceedings of the International Conference on Spoken Language Processing*. pp. 2199–2202.

Nass, C., Brave, S. (2005). *Wired for Speech: How Voice Activates and Advances the Human-Computer Relationship*. MIT Press, Cambridge, MA.

Nass, C., Gong, L. (2000). Speech interfaces from an evolutionary perspective: Social psychological research and design implications. *Communications of the ACM* 43 (9), 36–43.

National Institute of Standards and Technology (2004). DARPA EARS Rich-Transcription workshops. Described at http://www.nist.gov/speech/tests/rt/index.htm.

NATO Non-Native Speech Corpus (2004). http://www.homeworks.be/docs/tr-ist-011-all.pdf.

Navrátil, J. (1996). Improved phonotactic analysis in automatic language identification. In: *Proceedings of the European Signal Processing Conference* (EUSIPCO-96). Trieste, Italy, pp. 1031–1034.

Navrátil, J. (1998). *Untersuchungen zur automatischen sprachen-identifikation auf der basis der phonotaktik, akustik und prosodie*. Ph.D. thesis, Technical University of Ilmenau, Germany.

Navrátil, J. (2001). Spoken language recognition—a step towards multilinguality in speech processing. *IEEE Transactions on Speech and Audio Processing*. 9 (6), 678–685.

Navrátil, J., Zühlke, W. (1997). Phonetic-context mapping in language identification. In: *Proceedings of the European Conference on Speech Communication and Technology (EUROSPEECH)*. Rhodes, Greece, pp. 71–74.

Navrátil, J., Zühlke, W. (1998). An efficient phonotactic-acoustic system for language identification. In: *Proceedings of the IEEE International Conference on Acoustics, Speech and Signal Processing*. Seattle, WA, pp. 781–784.

Ney, H. (1999). Speech translation: Coupling of recognition and translation. In: *Proceedings of the IEEE International Conference on Acoustics, Speech and Signal Processing*. Phoenix, AZ, pp. 517–520.

Ney, H. (2001). Stochastic modeling: From pattern classification to language translation. In: *Proceedings of the ACL 2001 Workshop on DDMT*. pp. 33–37.

Nießen, S., Vogel, S., Ney, H., Tillmann, C. (1998). A DP based search algorithm for statistical machine translation. In: *Proceedings of the 36th Annual Meeting of the Association for Computational Linguistics and the 17th International Conference on Computational Linguistics*. Montréal, P.Q., Canada, pp. 960–967.

NIST (1997). The 1997 Hub-5NE evaluation plan for recognition of conversational speech over the telephone, in non-English languges, http://www.nist.gov/speech/tests/ctr/hub5me_97/current-plan.htm.

NIST (2002). *Automatic evaluation of machine translation quality using N-gram co-occurence statistics*. http://www.nist.gov/speech/tests/mt/doc/ngram-study.pdf.

NIST (2003). The 2003 NIST language evaluation plan, http://www.nist.gov/speech/tests/lang/doc/LangRec_EvalPlan.v1.pdf.

Oard, D. W. (2003). The surprise language exercises. *ACM Transactions on Asian Language Information Processing* 2 (2), 79–84.

Och, F., Tillmann, C., Ney, H. (1999). Improved alignment models for statistical machine translation. In: *Proceedings of the EMNLP/WVLC*.

Och, F. J., Ney, H. (2000). Improved statistical alignment models. In: *Proceedings of the 38th Annual Meeting of the Association for Computational Linguistics*. Hong Kong, China, pp. 440–447.

Oflazer, K. (1994). Two-level description of Turkish morphology. *Literary and Linguistic Computing* 9 (2), 137–148.

Olive, J., Greenwood, A., Coleman, J., Greenwood, A. (1993). *Acoustics of American English Speech: A Dynamic Approach*. Springer Verlag.

Ostendorf, M. (1999). Moving beyond the beads-on-a-string model of speech. In: *Proceedings of the IEEE Workshop on Automatic Speech Recognition and Understanding*. Keystone, CO.

Osterholtz, L., Augustine, C., McNair, A., Rogina, I., Saito, H., Sloboda, T., Tebelskis, J., Waibel, A., Woszczyna, M. (1992). Testing generality in JANUS: A multi-lingual speech tranlslation system. In: *Proceedings of the IEEE International Conference on Acoustics, Speech and Signal Processing*.

Oviatt, S., Cohen, P. (2000). Multimodal interfaces that process what comes naturally. Communications of the ACM 43 (3), 45–53.

Oviatt, S., DeAngeli, A., Kuhn, K. (1997). Integration and synchronization of input modes during multimodal human-computer interaction. In: *Proceedings of the ACM Conference on Human Factors in Computing Systems* (CHI-97). Atlanta. ACM Press, pp. 415–422.

Pallett, D., Fiscus, J., Martin, A., Przybocki, M. (1998). 1997 broadcast news benchmark test results: English and non-English. In: *Proceedings of the DARPA Broadcast News Transcription & Understanding Workshop*, Landsdowne, VA. pp. 5–11.

Papineni, K. A., Roukos, S., Ward, R. T. (1998). Free-flow dialog management using forms. In: *Proceedings of the International Conference on Spoken Language Processing*. Sydney, Australia. pp. 1411–1414.

Papineni, K., Roukos, S., Ward, T., Zhu, W. (2002). BLEU: A method for automatic evaluation of machine translation. In: *Proceedings of the 40th Annual Meeting of the Association for Computational Linguistics*. pp. 311–318.

Parandekar, S., Kirchhoff, K. (2003). Multi-stream language identification using data-driven dependency selection. In: *Proceedings of the IEEE International Conference on Acoustics, Speech and Signal Processing*. Hong Kong, China.

PASHTO (2005). http://www.speech.sri.com/projects/translation/full.html.

Pastor, M., Sanchis, A., Casacuberta, F., Vidal, E. (2001). EuTrans: A speech-to-speech translator prototyp. In: *Proceedings of the European Conference on Speech Communication and Technology (EUROSPEECH)*. pp. 2385–2389.

Paul, D., Baker, J. (1992). The design for the *Wall Street Journal*-based CSR corpus. In: *Proceedings of the International Conference on Spoken Language Processing*. Banff, Alberta, Canada, pp. 899–902.

Pellegrino, F., Farinas, J., Obrecht, R., 1999. Comparison of two phonetic approaches to language identification. In: *Proceedings of the European Conference on Speech Communication and Technology (EUROSPEECH)*.

Pérennou, G. (1988). Le projet bdlex de base de données lexicales et phonologiques. In: *1ères journées du GRECO-PRC CHM*. Paris, France.

Pfau, T., Beham, M., Reichl, W., Ruske, G. (1997). Creating large subword units for speech recogntion. In: *Proceedings of the European Conference on Speech Communication and Technology (EUROSPEECH)*. Rhodes, Greece, pp. 1191–1194.

Pfister, B., Romsdorfer, H. (2003). Mixed-lingual text analysis for Polyglot TTS synthesis. In: *Proceedings of the European Conference on Speech Communication and Technology (EUROSPEECH)*. Geneva, Switzerland.

Rabiner, L., Juang, B. (Eds.) (1993). *Fundamentals of Speech Recognition*. Prentice Hall, Englewood Cliffs, NJ.

Ramsey, S. R. (1987). *The Languages of China*. Princeton University Press.

Ramus, F., Mehler, J. (1999). Language identification with suprasegmental cues: A study based on speech resynthesis. *Journal of the Acoustical Society of America* 105 (1), 512–521.

Rao, A. S., Georgeff, M. P. (1995). *BDI Agents: From Theory to Practice*. Tech. rep. Australian Artificial Intelligence Institute, Melbourne, Australia.

Rao, P. S., Monkowski, M. D., Roukos, S. (1995). Language model adaptation via minimum discrimination information. In: *Proceedings of the IEEE International Conference on Acoustics, Speech and Signal Processing*. pp. 161–164.

Raux, A., Black, A. (2003). A unit selection approach to F_0 modeling and its application to emphasis. In: *Proceedings of the IEEE Workshop on Automatic Speech Recognition and Understanding*. St Thomas, Virgin Islands.

Reynolds, D. A. (1992). A Gaussian mixture modeling approach to text-independent speaker identification. Ph.D. thesis, Georgia Institute of Technology, Atlanta, GA.

Reynolds, D., et al. (2003). Exploiting high-level information for high-performance speaker recognition, SuperSID project final report. Summer workshop 2002, Center for Language and Speech Processing, The Johns Hopkins University. http://www.clsp.jhu.edu/ws2002/groups/supersid/SuperSID_Final_Report_ CLSP_WS02_2003_10_06.pdf.

Riccardi, G., Gorin, A. L., Ljolje, A., Riley, M. (1997). A spoken language system for automated call routing. In: *Proceedings of the IEEE International Conference on Acoustics, Speech and Signal Processing*. Munich, Germany. pp. 1143–1146.

Riley, M., Byrne, W., Finke, M., Khudanpur, S., Ljolie, A., McDonough, J., Nock, H., Saraclar, M., Wooters, C., Zavaliagkos, G. (1999). Stochastic pronunciation modelling from hand-labelled phonetic Corpora. *Speech Communication* 29, 209–224.

Riley, M., Ljojle, A. (1996). *Automatic Speech and Speaker Recognition*. Kluwer, Ch. Automatic Generation of Detailed Pronunciation Lexicons, pp. 285–301.

Ringger, E. K. (1995). *Robust loose coupling for speech recognition and natural language understanding*. Technical Report 592, University of Rochester Computer Science Department.

Roark, B. (2001). Probabilistic top-down parsing and language modeling. *Computational Linguistics* 27 (2), 249–276.

Roark, B., Saraclar, M., Collins, M., Johnson, M. (2004). Discriminative language modeling with conditional random fields and the perceptron algorithm. In: *Proceedings of the 42nd Annual Meeting of the Association for Computational Linguistics*.

Ronen, O., Neumeyer, L., Franco, H. (1997). Automatic detection of mispronunciation for Language Instruction. In: *Proceedings of the European Conference on Speech Communication and Technology (EUROSPEECH)*. Rhodes, Greece.

Rosenfeld, R. (1994). *Adaptive statistical language modeling: A maximum entropy approach*. Ph.D. thesis, Computer Science Department, Carnegie Mellon University.

Rosenfeld, R. (2000). Two decades of statistical language modeling; where do we go from here? *Proceedings of the IEEE* 88 (8), 1270–1278.

Rosset, S., Bennacef, S., Lamel, L. (1999). Design strategies for spoken language dialog systems. In: *Proceedings of the European Conference on Speech Communication and Technology (EUROSPEECH)*, Budapest, Hungary, pp. 1535–1538.

Rouas, J.-L., Farinas, J., Pellegrino, F., Andre-Obrecht, R. (2003). Modeling prosody for language identification on read and spontaneous speech. In: *Proceedings of the IEEE International Conference on Acoustics, Speech and Signal Processing*. Hong Kong, China.

Roux, J. (1998). Xhosa: A tone or pitch-accent language? *South African Journal of Linguistics* Supp. 36, 33–50.

Rudnicky, A., Thayer, E., Constantinides, P., Tchou, C., Shern, R., Lenzo, K., Xu, W., Oh, A. (1999). Creating natural dialogs in the Carnegie Mellon Communicator System. In: *Proceedings of the European Conference on Speech Communication and Technology (EUROSPEECH)*. Budapest, Hungary, pp. 1531–1534.

Sadek, M. D., Mori, R. (1997). Dialog systems. In: de Mori, R. (Ed.) Spoken Dialog with Computers. Acadamic Press, 523–561.

Sagisaka, Y., Kaiki, N., Iwahashi, N., Mimura, K. (1992). ATR ν-talk speech synthesis system. In: *Proceedings of the International Conference on Spoken Language Processing*. Banff, Canada, pp. 483–486.

Sahakyan, M. (2001). *Variantenlexikon italienischer und deutscher Lerner des Englischen für die automatische Spracherkennung*. Master's thesis, IMS, University of Stuttgart.

Sampson, G. (1985). *Writing Systems: A Linguistic Introduction*. Stanford University Press, Stanford, CA.

Schaden, S. (2002). A database for the analysis of cross-lingual pronunciation variants of European city names. In: *Proceedings of the Third International Conference on Language Resources and Evaluation* (LREC). Las Palmas de Gran Canaria, Spain, pp. 1277–1283.

Schaden, S. (2003a). Rule-based lexical modelling of foreign-accented pronunciation variants. In: *Proceedings of the 10th Conference of the European Chapter of the Association for Computational Linguistics* (EACL). Budapest, Hungary, pp. 159–162.

Schaden, S. (2003b). Generating non-native pronunciation lexicons by phonological rules. In: *Proceedings of the 15th International Conference of Phonetic Sciences* (ICPhS). Barcelona, Spain, pp. 2545–2548.

Schillo, C., Fink, G. A., Kummert, F. (2000). Grapheme-based speech recognition for large vocabularies. In: *Proceedings of the International Conference on Spoken Language Processing*. Beijing, China, pp. 584–587.

Schukat-Talamazzini, E., Hendrych, R., Kompe, R. H. N. (1995). Permugram language models. In: *Proceedings of the European Conference on Speech Communication and Technology (EUROSPEECH)*. Madrid, Spain, pp. 1773–1776.

Schultz, T. (2002). Globalphone: A multilingual text and speech database developed at Karlsruhe University. In: *Proceedings of the International Conference on Spoken Language Processing*. Denver, CO.

Schultz, T. (2004). Towards rapid language portability of speech processing systems. In: *Conference on Speech and Language Systems for Human Communication* (SPLASH). Delhi, India.

Schultz, T., Black, A., Vogel, S., Woszczyna, M. (2005). Flexible speech translation systems. In: *IEEE Transactions on Speech and Audio Processing*, to appear.

Schultz, T., Jin, Q., Laskowski, K., Tribble, A., Waibel, A. (2002). Improvements in non-verbal cue identification using multilingual phone strings. In: *Proceedings of the 40th Annual Meeting of the Association for Computational Linguistics*. Philadelphia.

Schultz, T., Rogina, I., Waibel, A. (1996a). LVCSR-based language identification. In: *Proceedings of the IEEE International Conference on Acoustics, Speech and Signal Processing*. Atlanta, GA.

Schultz, T., Waibel, A. (1997a). Fast bootstrapping of LVCSR systems with multilingual phoneme sets. In: *Proceedings of the European Conference on Speech Communication and Technology (EUROSPEECH)*. Rhodes, Greece, pp. 371–373.

Schultz, T., Waibel, A. (1997b). The Globalphone project: Multilingual LVCSR with Janus-3. In: *Multilingual Information Retrieval Dialogs: 2nd SQEL Workshop*. Plzen, Czech Republic, pp. 20–27.

Schultz, T., Waibel, A. (1998c). Multilingual and crosslingual speech recognition. In: *Proceedings of the DARPA Workshop on Broadcast News Transcription and Understanding*. Lansdowne, VA, pp. 259–262.

Schultz, T., Waibel, A. (1998a). Adaptation of pronunciation dictionaries for recognition of unseen languages. In: *Proceedings of the SPIIRAS International Workshop on Speech and Computer*. SPIIRAS, St. Petersburg, Russia, pp. 207–210.

Schultz, T., Waibel, A. (1998b). Language independent and language adaptive large vocabulary speech recognition. In: *Proceedings of the International Conference on Spoken Language Processing*. Sydney, Australia, pp. 1819–1822.

Schultz, T., Waibel, A. (2000). Polyphone decision tree specialization for language adaptation. In: *Proceedings of the IEEE International Conference on Acoustics, Speech and Signal Processing*. Istanbul, Turkey.

Schultz, T., Waibel, A. (2001). Language independent and language adaptive acoustic modeling for speech recognition. *Speech Communication* 35 (1–2), 31–51.

Schwartz, R., Chow, Y., Roucos, S. (1984). Improved hidden Markov modelling of phonemes for continuous speech recognition. In: *Proceedings of the*

IEEE International Conference on Acoustics, Speech and Signal Processing. San Diego, CA, pp. 35.6.1–35.6.4.

Schwartz, R., Jin, H., Kubala, F., Matsoukas, S. (1997). Modeling those F-conditions—or not. In: *Proceedings of the 1997 DARPA Speech Recognition Workshop*.

Seneff, S., Chuu, C., Cyphers, D. S. (2000). Orion: From on-line interaction to off-line delegation. In: *Proceedings of the International Conference on Spoken Language Processing*. Beijing, China, pp. 142–145.

Seneff, S., Lau, R., Meng, H. (1996). ANGIE: A new framework for speech analysis based on morpho-phonological modeling. In: *Proceedings of the International Conference on Spoken Language Processing*. Philadelphia, pp. 110–113.

Seneff, S., Lau, R., Polifroni, J. (1999). Organization, communication and control in the Galaxy-II Conversational System. In: *Proceedings of the European Conference on Speech Communication and Technology (EUROSPEECH)*. Budapest, Hungary, pp. 1271–1274.

Seneff, S., Wang, C., Peabody, M., Zue, V. (2004). Second language acquisition through human-computer dialogue. In: *Proceedings of the International Symposium on Chinese Spoken Language Processing*. Hong Kong, China.

Seymore, K., Rosenfeld, R. (1997). Using story topics for language model adaptation. In: *Proceedings of the European Conference on Speech Communication and Technology (EUROSPEECH)*. pp. 1987–1990.

Shafran, I., Ostendorf, M. (2003). Acoustic model clustering based on syllable structure. *Computer Speech and Language* 17 (4), 311–328.

Shimohata, M., Sumita, E., Matsumoto, Y. (2004). Building a paraphrase corpus for speech translation. In: *Proceedings of the First International Conference on Language Resources and Evaluation (LREC)*. pp. 453–457.

Shoup, J. (1980). Phonological aspects of speech recognition. In: Lea, W. (Ed.), *Trends in Speech Recognition*. Prentice-Hall, Englewood Cliffs, NJ, pp. 125–138.

Shozakai, M. (1999). Speech interface VLSI for car applications. In: *Proceedings of the IEEE International Conference on Acoustics, Speech and Signal Processing*. Phoenix, AZ, pp. I–220–223.

Shriberg, E., Stolcke, A. (1996). Word predictability after hesitations. In: *Proceedings of the International Conference on Spoken Language Processing*. Philadelphia.

Silberztein, M. (1993). *Dictionnaires électroniques et analyse automatique de textes: le système INTEX*. Masson.

SimplifiedEnglish (2004). http://www.userlab.com/se.html.

Singer, E., Torres-Carrasquillo, P., Gleason, T., Campbell, W., Reynolds, D. (2003). Acoustic, phonetic, and discriminative approaches to automatic language identification. In: *Proceedings of the European Conference on Speech Communication and Technology (EUROSPEECH)*. Geneva, Switzerland.

Singh, R., Raj, B., Stern, R. (2002). Automatic generation of subword units for speech recognition systems. In: *IEEE Transactions on Speech and Audio Processing*. 10. pp. 98–99.

Somers, H. (1999). Review article: Example-based machine translation. *Journal of Machine Translation*, 14 (2), 113–157.

SPICE (2005). http://www.cmuspice.org.

Spiegel, M. (1993). Using the Orator synthesizer for a public reverse-directory service: Design, lessons, and recommendations. In: *Proceedings of the European Conference on Speech Communication and Technology (EUROSPEECH)*. Berlin, Germany, pp. 1897–1900.

Spiegel, M., Macchi, M. (1990). Development of the Orator synthesizer for network applications: Name pronunciation accuracy, morphological analysis, custimization for business listing, and acronym pronunciation. In: *Proceedings of the AVOIS*. Bethesda, MD.

Sproat, R. Emerson, T. (2003). The first international Chinese word segmentation bakeoff. In: *Proceedings of the Second SIGHAN Workshop on Chinese Language Processing*. pp. 1–11.

Sproat, R., Zheng, F., Gu, L., Li, J., Zheng, Y., Su, Y., Zhou, H., Bramsen, P., Kirsch, D., Shafran, I., Tsakalidis, S., Starr, R., Jurafsky, D. (2004). *Dialectal Chinese speech recognition: Final report*. Final report of the 2004 summer workshop on language engineering, Center for Language and Speech Processing, Johns Hopkins University, available at http://www.clsp.jhu.edu/ws2004/groups/ws04casr/report.pdf.

Stolcke, A., Bratt, H., Butzberger, J., Franco, H., Gadde, Rao, V. R., Plauche, M., Richey, C., Shriberg, E., Sonmez, K., Weng, F., Zheng, J. (2000). The SRI March 2000 Hub-5 conversational speech transcription system. In: *Proceedings of the NIST Speech Transcription Workshop*. College Park, MD.

Strange Corpus (1995). http://www.phonetik.uni-muenchen.de/bas/bassc1deu.html.

Strassel, S., Glenn, M. (2003). Creating the annotated TDT-4 Y2003 evaluation corpus. In: *2003 TDT Evaluation System Summary Papers*. National Institutes of Standards and Technology, available at http://www.nist.gov/speech/tests/tdt/tdt2003/papers/ldc.ppt.

Strassel, S., Maxwell, M., Cieri, C. (2003). Linguistic resource creation for research and technology development: A recent experiment. *ACM Transactions on Asian Language Information Processing* 2 (2), 101–117.

Stüker, S., Metze, F., Schultz, T., Waibel, A. (2001). Integrating multilingual articulatory features into speech recognition. In: *Proceedings of the European Conference on Speech Communication and Technology (EUROSPEECH)*. Geneva, Switzerland.

Stüker, S., Schultz, T. (2004). A grapheme based speech recognition system for Russian. In: *Proceedings of the Specom*. St. Petersburg, Russia.

Stüker, S., Schultz, T., Metze, F., Waibel, A. (2003). Multilingual articulatory features. In: *Proceedings of the IEEE International Conference on Acoustics, Speech and Signal Processing*. Hong Kong, China.

Stylianou, Y., Cappé, O., Moulines, E. (1995). Statistical methods for voice quality transformation. In: *Proceedings of the European Conference on Speech Communication and Technology (EUROSPEECH)*. Madrid, Spain, pp. 447–450.

Sugaya, F., Takezawa, T., Yokoo, A., Sagisaka, Y., Yamamoto, S. (2000). Evaluation of the ATR-MATRIX speech translation system with a pair comparison method between the system and humans. In: *Proceedings of the International Conference on Spoken Language Processing*. pp. 1105–1108.

Sugiyama, M. (1991). Automatic language identification using acoustic features. In: *Proceedings of the IEEE International Conference on Acoustics, Speech and Signal Processing*. pp. 813–816.

Sumita, E. (2001). Example-based machine translation using DP-matching between word sequences. In: *Proceedings of the Association for Computational Linguistics Workshop on DDMT*. pp. 1–8.

Sumita, E., Iida, H. (1991a). Experiments and prospects of example-based machine translation. In: *Proceedings of the 29th Annual Meeting of the Association for Computational Linguistics*. pp. 185–192.

Sumita, E., Yamada, S., Yamamoto, K., Paul, M., Kashioka, H., Ishikawa, K., Shirai, S. (1999). Solutions to problems inherent in spoken-language translation: The ATR-matrix approach. In: *Proceedings of the MT Summit VII*. pp. 229–235.

Takami, J., Sagayama, S. (1992). A successive state splitting algorithm for efficient allophone modeling. In: *Proceedings of the IEEE International Conference on Acoustics, Speech and Signal Processing*. pp. 573–576.

Takezawa, T., Kikui, G. (2003). Collecting machine-translation-aided bilingual dialogues for corpus-based speech translation. In: *Proceedings of the European Conference on Speech Communication and Technology (EUROSPEECH)*. pp. 2757–2760.

Takezawa, T., Morimoto, T., Sagisaka, Y. (1998b). Speech and language databases for speech translation research in ATR. In: *Proceedings of the 1st International Workshop on East-Asian Language Resources and Evaluation* (EALREW).

Takezawa, T., Morimoto, T., Sagisaka, Y. (1998c). Speech and language database for speech translation research in ATR. In: *Proceedings of the Oriental COCOSDA Workshop'98*. pp. 148–155.

Takezawa, T., Morimoto, T., Sagisaka, Y., Campbell, N., Iida, H., Sugay, F., Yokoo, A., Yamamoto, S. (1998a). A Japanese-to-English speech translation system: ATR-MATRIX. In: *Proceedings of the International Conference on Spoken Language Processing*.

Takezawa, T., Sumita, E., Sugaya, F., Yamamoto, H., Yamamoto, S. (2002). Toward a broad-coverage bi-lingual corpus for speech translation of travel conversations in the real world. In: *Proceedings of the Third International Conference on Language Resources and Evaluation (LREC)*. pp. 147–152.

Tan, G., Takechi, M., Brave, S., Nass, C. (2003). Effects of voice vs. remote on U.S. and Japanese user satisfaction with interactive HDTV systems. In: *Extended Abstracts of the Computer-Human Interaction (CHI) Conference*. pp. 714–715.

Tarone, E., Cohen, A. D., Dumas, G. (1983). A closer look at some interlanguage terminology: A framework for communication strategies. In: Færch, C., Kasper, G. (Eds.), *Strategies in Interlanguage Communication*. Longman.

Taylor, P., Black, A. (1998). Assigning phrase breaks from part-of-speech sequences. *Computer Speech and Language* 12, 99–117.

Taylor, P., Black, A., Caley, R. (1998). The architecture of the festival speech synthesis system. In: *Proceedings of the 3rd ESCA/COCOSDA Workshop on Speech Synthesis*. Jenolan Caves, Australia. pp. 147–151.

Teixeira, C., Trancoso, I., Serralheiro, A. (1996). Accent identification. In: *Proceedings of the International Conference on Spoken Language Processing*. Philadelphia.

Thyme-Gobbel, A., Hutchins, S. (1996). On using prosodic cues in automatic language identification. In: *Proceedings of the International Conference on Spoken Language Processing*. Philadelphia.

Tian, J., Kiss, I., Viikki, O. (2001). Pronunciation and acoustic model adaptation for improving multilingual speech recognition. In: *Proceedings of the IEEE Workshop on Automatic Speech Recognition and Understanding*. Madonna di Campiglo, Italy.

Toda, T., Kawai, H., Tsuzaki, M. (2004). Optimizing sub-cost functions for segment selection based on perceptual evaluations in concatenative speech synthesis. In: *Proceedings of the IEEE International Conference on Acoustics, Speech and Signal Processing*. Montreal, Canada, pp. 657–660.

Toda, T., Kawai, H., Tsuzaki, M., Shikano, K. (2003). Segment selection considering local degradation of naturalness in concatenative speech synthesis. In: *Proceedings of the IEEE International Conference on Acoustics, Speech and Signal Processing*. Hong Kong, China, pp. 696–699.

TOEIC (2005). Web site: http://www.ets.org/toeic/.

Tokuda, K., Kobayashi, T., Imai, S. (1995). Speech parameter generation from HMM using dynamic features. In: *Proceedings of the IEEE International Conference on Acoustics, Speech and Signal Processing*. Detroit, MI. pp. 660–663.

Tokuda, K., Yoshimura, T., Masuko, T., Kobayashi, T., Kitamura, T. (2000). Speech parameter generation algorithms for HMM-based speech synthesis. In: *Proceedings of the IEEE International Conference on Acoustics, Speech and Signal Processing*. Istanbul, Turkey. pp. 1315–1318.

Tomokiyo, L., Badran, A. (2003). *Egyptian Arabic romanization conventions*, internal report, Carnegie Mellon, 2003.

Tomokiyo, L., Black, A., Lenzo, K. (2005). Foreign accents in synthesis: Development and evaluation. In: *Proceedings of the European Conference on Speech Communication and Technology (EUROSPEECH)*. Lisbon, Portugal.

Torres-Carrasquillo, P., Reynolds, D., Deller, J. Jr. (2002a). Language identification using Gaussian mixture model tokenization. In: *Proceedings of the IEEE International Conference on Acoustics, Speech and Signal Processing*. Orlando, FL.

Torres-Carrasquillo, P., Singer, E., Kohler, M. Greene, R., Reynolds, D., Deller, J., Jr. (2002b). Approaches to language identification using Gaussian mixture models and shifted delta cepstral features. In: *Proceedings of the International Conference on Spoken Language Processing*. Denver, CO.

Tsopanoglou, A., Fakotakis, N. (1997). Selection of the most effective set of subword units for an HMM-based speech recognition system. In: *Proceedings of the European Conference on Speech Communication and Technology (EUROSPEECH)*. Rhodes, Greece, pp. 1231–1234.

Tsuzuki, R., Zen, H., Tokuda, K., Kitamura, T., Bulut, M., Narayanan, S. (2004). Constructing emotional speech synthesizers with limited speech database. In: *Proceedings of the International Conference on Spoken Language Processing*. Jeju Island, South Korea.

Übler, U., Schüßler, M., Niemann, H. (1998). Bilingual and dialectal adaptation and retraining. In: *Proceedings of the International Conference on Spoken Language Processing*. Sydney, Australia.

Uebler, U., Boros, M. (1999). Recognition of non-native German speech with multilingual recognizers. In: *Proceedings of the European Conference on Speech Communication and Technology (EUROSPEECH)*. Budapest, Hungary. pp. 907–910.

Ueda, Y., Nakagava, S. (1990). Prediction for phoneme/syllable/word-category and identification of language using HMM. In: *Proceedings of the European Conference on Speech Communication and Technology (EUROSPEECH)*. Kobe, Japan. pp. 1209–1212.

Vapnik, V. (1999). *The Nature of Statistical Learning Theory*, 2nd Ed. Springer.

Venugopal, A., Vogel, S., Waibel, A. (2003). Effective phrase translation extraction from alignment models. In: *Proceedings of the 41st Annual Meeting of the Association for Computational Linguistics*. Sapporo, Japan. pp. 319–326.

Vergyri, D., Kirchhoff, K. (2004). Automatic diacritization of Arabic for acoustic modeling in speech recognition. In: *Proceedings of the COLING Workshop on Arabic Script Based Languages*. Geneva, Switzerland.

Vergyri, D., Kirchhoff, K., Duh, K., Stolcke, A. (2004). Morphology-based language modeling for Arabic speech recognition. In: *Proceedings of the International Conference on Spoken Language Processing*. Jeju Island, South Korea.

Vogel, S. (2003). SMT decoder dissected: Word reordering. In: *Proceedings of the International Conference on Natural Language Processing and Knowledge Engineering* (NLP-KE). Beijing, China.

Vogel, S., Hewavitharana, S., Kolss, M., Waibel, A. (2004). The ISL statistical translation system for spoken language translation. In: *International Workshop on Spoken Language Translation*. Kyoto, Japan, pp. 65–72.

Vogel, S., Ney, H., Tillmann, C. (1996). HMM-based word alignment in statistical translation. In: *Proceedings of the 16th International Conference on Computational Linguistics*. Copenhagen, Denmark, pp. 836–841.

Vogel, S., Zhang, Y., Huang, F., Tribble, A., Venogupal, A., Zhao, B., Waibel, A. (2003). The CMU statistical translation system. In: *Proceedings of the MT-Summit IX*. New Orleans, LA.

Waibel, A., Badran, A., Black, A. W., Frederking, R., Gates, D., Lavie, A., Levin, L., Lenzo, K., Mayfield Tomokiyo, L. M., Reichert, J., Schultz, T., Wallace, D., Woszczyna, M., Zhang, J. (2004). Speechalator: Two-way speech-to-speech translation on a consumer PDA. In: *Proceedings of the European Conference on Speech Communication and Technology (EUROSPEECH)*. Geneva, Switzerland.

Waibel, A., Geutner, P., Mayfield Tomokiyo, L., Schultz, T., Woszczyna, M. (2000). Multilinguality in speech and spoken language systems. *Proceedings of the IEEE* 88 (8), 1297–1313.

Walter, H. (1997). *L'aventure des mots venus d'ailleurs*. Robert Laffont, Paris, France.

Wang, Y., Waibel, A. (1998). Fast decoding for statistical machine translation. In: *Proceedings of the International Conference on Spoken Language Processing*. Sydney, Australia.

Wang, Y.-Y., Waibel, A. (1997). Decoding algorithm in statistical translation. In: *Proceedings of the ACL/EACL '97*. Madrid, Spain, pp. 366–372.

Wang, Z., Schultz, T. (2003). Non-native spontaneous speech recognition through polyphone decision tree specialization. In: *Proceedings of the International Conference on Spoken Language Processing*. Geneva, Switzerland. pp. 1449–1452.

Wang, Z., Topkara, U., Schultz, T., Waibel, A. (2002). Towards universal speech recognition. In: *4th IEEE International Conference on Multimodal Interfaces*. pp. 247–252.

Ward, T., Roukos, S., Neti, C., Epstein, M., Dharanipragada, S. (1998). Towards speech understanding across multiple languages. In: *Proceedings of the International Conference on Spoken Language Processing*. Sydney, Australia.

Watanabe, T., Sumita, E. (2002). Bidirectional decoding for statistical machine translation. In: *Proceedings of COLING*. Taipei, Taiwan, pp. 1079–1085.

Watanae, T., Sumita, E. (2002). Statistical machine translation based on hierarchical phrase alignment. In: *Proceedings of the TMI-2002*. pp. 188–198.

Watanae, T., Sumita, E. (2003). Example-based decoding for statistical machine translation. In: *Proceedings of the IX-th MT Summit.*

Weingarten, R. (2003). http://www.ruediger-weingarten.de/texte/latinisierung.pdf.

Wells, J. (1997). SAMPA computer readable phonetic alphabet. In: Gibbon, D., Moore, R., Winski, R. (Eds.) *Handbook of Standards and Resources for Spoken Language Systems*. Mouton de Gruyter, Berlin and New York.

Weng, F., Bratt, H., Neumeyer, L., Stolke, A. (1997). A study of multilingual speech recognition. In: *Proceedings of the European Conference on Speech Communication and Technology (EUROSPEECH)*. Rhodes, Greece, pp. 359–362.

Whaley, L. (1997). *Introduction to Typology: The Unity and Diversity of Language*. Sage, Thousand Oaks, CA.

Wheatley, B., Kondo, K., Anderson, W., Muthusamy, Y. (1994). An evaluation of cross-language adaptation for rapid HMM development in a new language. In: *Proceedings of the IEEE International Conference on Acoustics, Speech and Signal Processing*. Adelaide, Australia, pp. 237–240.

Whittaker, E. W. D. (2000). *Statistical language modelling for automatic speech recognition of Russian and English.* Ph.D. thesis, Cambridge University Engineering Department, Cambridge, UK.

Williams, G., Terry, M., Kaye, J. (1998). Phonological elements as a basis for language-independent ASR. In: *Proceedings of the International Conference on Spoken Language Processing*. Sydney, Australia.

Witt, S., Young, S. (1999). Offline acoustic modeling of non-native accents. In: *Proceedings of the European Conference on Speech Communication and Technology (EUROSPEECH)*. Budapest, Hungary.

Wong, E., Sridharan, S. (2002). Methods to improve Gaussian mixture model based language identification system. In: *Proceedings of the International Conference on Spoken Language Processing*. Denver, CO.

Woodland, P. C. (1999). Speaker adaptation: Techniques and challenges. In: *Proceedings of the IEEE Workshop on Automatic Speech Recognition and Understanding*. Keystone, CO, pp. 85–90.

Woodland, P. C. (2001). Speaker adaptation for condtinuous density HMMs: A review. In: *AMSR2001*, Sophia-Antipolis, France (2001).

Workshop (1994). *Proceedings of the ARPA Spoken Language Technology Workshop*. Plainsboro, NJ.

Wu, D. (1995). Stochastic inversion transduction grammars, with application to segmentation, bracketing, and alignment of parallel corpora. In: *Proceedings of the 14th International Joint Conference on Artificial Intelligence* (IJCAI-95). Montreal, Canada.

Xu, P., Jelinek, F. (2004). Using random forests in the structure language model. In: *Advances in Neural Information Processing Systems* (NIPS). Vol. 17. Vancouver, Canada.

Yamamoto, H., Isogai, S., Sagisaka, Y. (2003). Multi-class composite N-gram language model. *Speech Communication* 41, 369–379.

Yan, Y., Barnard, E. (1995). An approach to automatic language identification based on language-dependent phone recognition. In: *Proceedings of the IEEE International Conference on Acoustics, Speech and Signal Processing*. Detroit, MI.

Yan, Y., Barnard, E., Cole, R. (1996). Development of an approach to automatic language identification based on phone recognition. *Computer Speech and Language* 10 (1), 37–54.

Yoshimura, T., Tokuda, K., Masuku, T., Kobayashi, T., Kitamura, T. (1999). Simultaneous modeling of spectrum, pitch and duration in HMM-based speech synthesis. In: *Proceedings of the European Conference on Speech Communication and Technology (EUROSPEECH)*. Budapest, Hungary, pp. 2347–2350.

Young, S., Adda-Decker, M., Aubert, X., Dugast, C., Gauvain, J.-L., Kershaw, D., Lamel, L., Leeuwen, D., Pye, D., Robinson, A., Steeneken, H., Woodland, P. (1997). Multilingual large vocabulary speech recognition: The European SQALE project. *Computer Speech and Language* 11, 73–89.

Young, S., Odell, J., Woodland, P. (1994). Tree-based state tying for high accuracy acoustic modeling. In: *Proceedings of the ARPA Spoken Language Technology Workshop*. Princeton, NJ, pp. 307–312.

Yu, H., Schultz, T. (2003). Enhanced tree clustering with single pronunciation dictionary for conversational speech recognition. In: *Proceedings of the European Conference on Speech Communication and Technology (EUROSPEECH)*. Geneva, Switzerland.

Zellig, H. (1955). From phoneme to morpheme. *Language* 31 (2), 190–222.

Zhang, J. S., Markov, K., Matsui, T., Nakamura, S. (2003a). A study on acoustic modeling of pauses for recognizing noisy conversational speech. *IEICE Trans. on Inf. & Syst.* 86-D (3).

Zhang, J. S., Mizumachi, M., Soong, F., Nakamura, S. (2003). An introduction to ATRPTH: A phonetically rich sentence set based Chinese Putonghua speech database developed by ATR. In: *Proceedings of the ASJ Meeting*. pp. 167–168.

Zhang, J. S., Zhang, S. W., Sagisaka, Y., Nakamura, S. (2001). A hybrid approach to enhance task portability of acoustic models in Chinese speech recognition. In: *Proceedings of the European Conference on Speech Communication and Technology (EUROSPEECH)*. pp. 1661–1663.

Zhang, Y., Vogel, S., Waibel, A. (2003b). Integrated phrase segmentation and alignment model for statistical machine translation. In: *Proceedings of the International Conference on Natural Language Processing and Knowledge Engineering* (NLP-KE). Beijing, China.

Zissman, M. (1993). Automatic language identification using Gaussian mixture and hidden Markov models. In: *Proceedings of the International Conference on Acoustics, Speech and Signal Processing*. pp. 399–402.

Zissman, M. (1996). Comparison of four approaches to automatic language identification. *IEEE Transactions on Speech and Audio Processing*. 4 (1), 31–44.

Zissman, M. (1997). Predicting, diagnosing and improving automatic language identification performance. In: *Proceedings of the European Conference on Speech Communication and Technology (EUROSPEECH)*. Rhodes, Greece.

Zissman, M., Singer, E. (1994). Automatic language identification of telephone speech messages using phoneme recognition and N-gram modeling. In: *Proceedings of the IEEE International Conference on Acoustics, Speech and Signal Processing*. Adelaide, Australia, pp. 305–308.

Zissman, M., Singer, E. (1995). Language identification using phoneme recognition and phonotactic language modeling. In: *Proceedings of the IEEE International Conference on Acoustics, Speech and Signal Processing*. Detroit, MI, pp. 3503–3506.

Zue, V., Seneff, S., Glass, J., Polifroni, J., Pao, C., Hazen, T. J., Hetherington, L. (2000). JUPITER: A telephone-based conversational interface for weather information. *IEEE Transactions on Speech and Audio Processing*. 8 (1), 147–151.

Index

A

B

C

D

E

F

G

H

I

J

K

L

N

O

T

U

V

W

X

Printed and bound by CPI Group (UK) Ltd, Croydon, CR0 4YY

03/10/2024

01040410-0016